CG-263
US Coast Guard
Paint and Color Manual
December 1952

The CG-263 Paint and Color manual was prepared by the US Coast Guard to document the basic instructions for color schemes used by the service. It includes color diagrams of the color schemes, but is largely in black and white. It includes the technical details of how to mix, store, prepare, and apply various paints that were in use at the time. It is being reprinted as a convenience to operators, enthusiasts and private owners.

Should you have suggestions or feedback on ways to improve this book please send email to Books@OcotilloPress.com

---

Edited 2021 Ocotillo Press
ISBN 978-1-954285-61-3

Ocotillo Press
Houston, TX 77017
Books@OcotilloPress.com

COAST GUARD SEAL

The U. S. Coast Guard Seal as shown on this cover sheet is that which the President adopted by Executive Order 10707 of 6 May 1957.

Reproduction of the official seal of the United States Coast Guard is authorized for use on decalcomanias, letterhead stationery, covers of manuals and booklets and otherwise as approved by the Commandant. Reproduction for the above purposes is authorized in the five color combinations or in any two contrasting color combinations as appropriate.

At present all the colors for the seal are described in fabric cable numbers. Federal Standard Color Numbers will be specified as obtainable. Cable number colors are described as follows:

Description	*Cable No.*	*Use*
White	65005	Stripes, disc, border
Scarlet	65006	Red Stripes, motto
Ultramarine	65010	Blue letters, chief of shield
Grotto Blue	70022	Blue background
Gold	-----	Anchors, border

COAST GUARD EMBLEM

Coast Guard General Order No. 7 of 3 April 1957 prescribes the emblem of the Coast Guard which replaced the emblem adopted in 1927.

The emblem in full color is as shown by the pressure-sensitive labels furnished with amendment No. 5 for placement on the manual's binder cover.

The emblem, in the color combinations illustrated above, is authorized for use on the Coast Guard Ensign and on official flags and pennants in accordance with specifications for each particular flag or pennant. Reproduction of the emblem in full color or two color combinations of blue and white or black and white is authorized for use on decalcomanias, printed recruiting material, or as approved by the Commandant.

469384 O—58

COVER DESIGN

The U. S. Coast Guard Emblem on the cover of this Manual is shown in one of the approved color combinations to be used for the official configuration.

The emblem may be displayed in one of the following color combinations as may be suitable for the purpose:

(a) Four-Color Scheme: Gold, No. 34 Emblem Blue, No. 13 Fire Red, and white as shown on the cover.

(b) Three-Color Scheme: No. 34 Emblem Blue, No. 13 Fire Red, and white. Same as the Four-Color Scheme except that the anchors shall be No. 34 Emblem Blue, the background of the shield shall be white, and a No. 34 Emblem Blue ring shall encircle the shield and the legend SEMPER PARATUS. Illustrated above.

(c) Two-Color Scheme: No. 29 Bright Blue and white. Same as the Three-Color Scheme except that No. 29 Bright Blue is used and the stripes on the shield shall also be No. 29 Bright Blue. Illustrated above.

(d) Open Type: For use on stationery and forms when a small size is desired and where an open type emblem will lend itself best to the reproductive process used. The use of any suitable color for the emblem is authorized in this case. Illustrated above.

The difference in the shades of blue specified above is to permit the use of the No. 29 Bright Blue stocked for painting the Two-Color emblem on signs, etc. The No. 34 Emblem Blue corresponds to the blue color most used in the reproduction of Three and Four-Color Scheme emblems on printed posters and decals. This color is shown for shade purposes only and is not stocked as a paint in the supply system. When it is desired to paint a Three or Four-Color emblem, a small amount of a blue paint matching No. 34 Emblem Blue may be purchased on the open market.

The color for the emblem when used in flags shall be in accordance with the specification for each particular flag.

The use of stars in the chief of the shield shall be optional. In cases where it is desirable, and the emblem will be large enough that it is suitable to do so, stars may be used. In such cases they shall be arranged in two horizontal rows, 7 in the upper and 6 in the lower, as shown on the cover.

UNITED STATES COAST GUARD

ADDRESS REPLY TO:
COMMANDANT
U. S. COAST GUARD
HEADQUARTERS
WASHINGTON 25, D. C.

ETD
--- 1 December 1952

LETTER OF PROMULGATION

1. The U. S. Coast Guard Paint and Color Manual, CG 263, contains the basic Coast Guard instructions in the field of color, paint materials, and painting methods. The publication is effective upon receipt.

2. The publication "Instructions for Painting United States Coast Guard Vessels, Boats, and Stations, 1935", along with all amendments and revisions thereof, is hereby cancelled. The instructions previously promulgated in the Painting Instructions Memorandum series have been included in the Manual and all Painting Instructions Memorandums issued will be cancelled. Changes to the Manual will be issued as serially numbered amendments.

3. The Manual has been prepared to apply to Coast Guard Units the scientific use of color in painting to reduce accidents and to improve working and living environments. Effort has been made in each case to select the best available material for each use. A substantial reduction in the total costs to paint Coast Guard Units should result due to the use of more durable materials, standardization of colors and materials, and stocking a reduced variety of materials.

4. Throughout the Manual the standardization has been accomplished with a realization of the need of flexibility to meet the problems individual to the various districts. The principle adopted is one of district standardization within a framework of service-wide standardization.

ETD
--- 1 December 1952

5. The promulgation of the Manual does not constitute authority for wide scale painting solely for the purpose of changing to standard color schemes contained therein. Conversion to color schemes contained in the Manual shall be accomplished in accordance with the existing need for repainting, present schedules for repainting, and the availability of regularly allotted funds for the purpose. However, all painting accomplished after receipt of the Manual shall comply with the instructions contained therein.

6. The Safety Color Code, Section 3-16, applicable to shore units only, shall be placed in effect at the earliest practicable date and shall be completed by all shore units prior to 1 January 1954.

7. It is expected that all units will be in compliance with the Manual by 1 January 1957.

Merlin O'Neill

MERLIN O'NEILL
Vice Admiral, U. S. Coast Guard
Commandant

Dist. (SDL No. 50)
A: a aa b c d dd(3); e f i (2); remainder (1)
B: f (10); c (7); e i j (5); d g h k l (3); b m (1)
C: a (3); d g (2); b e f h i j jj k l n p q r (1)
s u v (1)
D: a b d e f (1)

GENERAL CONTENTS

CHAPTER 1 — GENERAL INFORMATION

General information concerning the purpose of painting, composition and storage of paint materials, preparation of surfaces for painting, and instructions for proper application of paint materials.

CHAPTER 2 — COATING SYSTEMS

Coating systems (complete paint films from subsurfaces to topcoat) are prescribed for practically all types of surfaces and exposure conditions found in the Coast Guard.

CHAPTER 3 — COLOR PRACTICE

Color specifications, marking and safety color codes are described and illustrated for shore establishments, vehicles, vessels, aircraft and miscellaneous items.

CHAPTER 4 — MATERIALS

Information on individual materials concerning their use, instructions for application and means of procurement.

GENERAL INDEX

For quick and convenient location of the contents of this Manual, the Index should be freely consulted for references and cross-references.

HOW TO USE THIS MANUAL

(A) Chapter 1 should be read by all supervisory personnel and used as a reference for problems concerning surface preparation and paint application.

(B) Selection of color is made by reference to Chapter 3, referring to section 3-1 and to section covering unit or equipment concerned.

(C) Selection of coating systems (types of materials) for different items is next made by reference to pertinent sections of Chapter 2.

(D) Selection of individual paint items is accomplished by reference to Chapter 4 where stock numbers are given for colors and types of materials. Sections 1–11–3 and 1–11–7 contain information to assist in arriving at quantities of each item to be ordered.

In the preparation of this Manual the Office of Engineering, U. S. Coast Guard Headquarters, received the assistance and advice of—
Faber Birren and Company
Industrial Color Consultants,
under Coast Guard Contract Tcg–38526.

CHAPTER 1, GENERAL INFORMATION

TABLE OF CONTENTS

SECTION 1–1, INTRODUCTION

Chapter 1 of the Manual is informatory in character and is designed to acquaint personnel with the different types of paints and their composition, how to prepare surfaces for painting and how to apply paint properly. In addition, miscellaneous general background information on painting materials and painting is included.

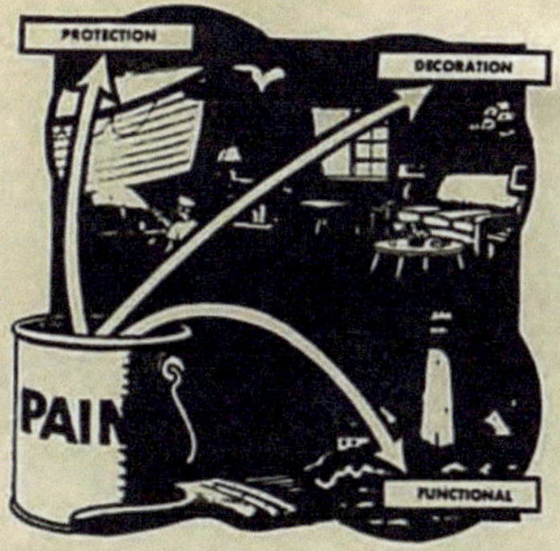

SECTION 1–2

THE PURPOSE OF PAINTING

Painting is an expensive procedure. Not only are paints, paint brushes and other paint materials expensive but the many man-hours consumed each year in painting Coast Guard vessels and stations could well be used for other maintenance work. In order to employ paint materials and painting man-hours effectively and economically the fundamental purposes of painting must be borne in mind. The importance of each of these purposes depends, of course, on the particular surface which is to be painted. Following is a brief discussion of each of these fundamental purposes of painting.

1–2–1, PROTECTION

Resistance to moisture from rain, snow, ice, spray and condensation constitutes perhaps the greatest single protective characteristic of paint. All things made of metal corrode. Moisture causes wood to swell, warp and rot. Interior wall finishes of buildings are ruined by neglect of exterior surfaces. Porous masonry is attacked and destroyed by moisture. Paint films must therefore be as impervious to moisture as possible in order to provide a protective waterproof film over the surface to which applied. Paint can also act as a protective film against attack by acids, alkalies, or marine organisms.

1–2–2, DECORATION

The word "paint" comes from Sanskrit, the oldest known written language, and means "to adorn." Painted color as a decorative medium, long recognized as an art, has now moved into the realm of science. Scientific tests prove that, correctly used, painted color on interior surfaces has a soothing effect on the nervous system. A compartment or room painted in pastel tints is easy to live in, but a room painted a brilliant red or orange would offer far too much stimulation. It can be readily seen that the decorative function of paint involves more than mere color.

1–2–3, FUNCTIONAL USES

Painting is used as a sanitary measure. A smooth, washable, painted surface, which can be cleaned easily, helps produce a clean and healthful atmosphere. Therefore, a painted compartment is a healthier place to live in than one that is unfinished and unpainted.

Another purpose of paint is to reflect light. Used in the interior of a ship, light-colored paints reflect and distribute both natural and artificial light better, and thus help secure maximum efficiency from the ship's lighting system. Correct illumination helps you do your job better and easier. The reflecting quality of paint is also used in camouflage to confuse an observer and make it difficult for him to locate the painted object.

Still another function of colored paint is the identification of objects. Red is used to identify fire fighting equipment. Yellow means caution, green means safety. Colors are used to identify compressed gas cylinders, highway traffic lanes, signs of all types and for many other applications. One of the most important uses of color as an identifying medium in the Coast Guard is in Aids to Navigation. Without buoy colors, the effectiveness of our buoy system would be reduced.

SECTION 1-3

THE COMPOSITION OF PAINT

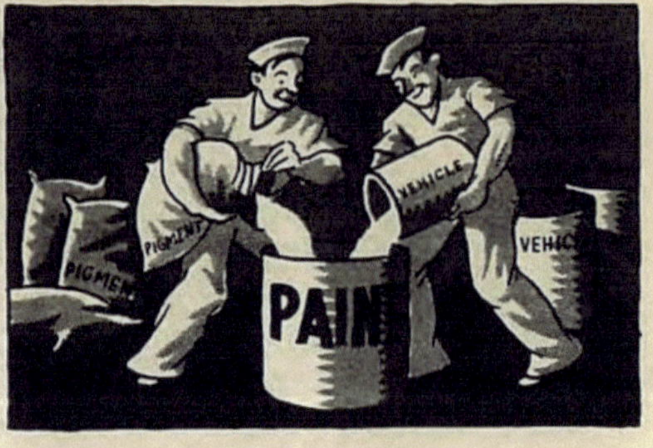

While all paints used in the Coast Guard are ready-mixed, a discussion of paint composition is presented for general background information and to provide personnel with a better understanding of the potentialities and limitations of paint coatings. Technical paint terminology must be used in this and following discussions for purposes of accuracy and brevity.

Paint is made essentially of two components—the pigment and the vehicle.

The *pigment* is a substance in fine powder form which imparts color and opacity to a paint and to a limited degree determines its consistency and general characteristics such as adhesion, durability, ease of application, and in special cases, fire retardancy and corrosion resistance. The pigment component of a paint may be a single pigment but generally is a combination of two or more pigments.

The *vehicle* is the liquid component of a paint. The pigment is suspended in or carried by the vehicle. The vehicle is made up of a non-volatile and a volatile portion. The function of the non-volatile is to bind the pigment particles together, leaving, after evaporation of the volatiles, a continuous solid film having the required characteristics of adhesion, durability, corrosion resistance (in metal primers), gloss and gloss retention. The function of the volatile portion is to thin the paint to the proper consistency for application as received. However, further thinning may be necessary in some cases. The volatile portion of the vehicle evaporates shortly after the paint has been applied. Small amounts of driers are incorporated in the vehicle to accelerate the drying process which converts the non-volatile vehicle to a solid. They are usually compounds of lead, manganese, or cobalt naphthanates and act as catalysts in the oxidation of the vehicle.

A factor which plays an important part in the characteristics of the dried paint film is the ratio by volume of the pigment to the non-volatile portion of the vehicle. In the dried paint film this determines the proportion of the pigment powder to the binder. Paints with a relatively large proportion of pigment exhibit certain characteristics that make them especially suitable for interior use. These paints have a flat or semigloss finish. Paints with a relatively large proportion of properly selected vehicle are generally tough, durable, weather resistant, and have a gloss finish.

The proper selection of the raw materials and the quantities used of each depends upon the end use of the product and is a job for the paint technologist.

1–3–1, COLOR PIGMENTS

(A) TITANIUM DIOXIDE

Titanium dioxide is the most opaque and whitest of all white pigments. Practically all white and light-colored paints contain titanium dioxide pigment in varying amounts.

(B) ZINC OXIDE

Zinc oxide is a white pigment of relatively low opacity, but does add important properties such as hardness and color retention to a paint film. There are many grades of zinc oxide, each having properties suitable for use in various types of paints.

(C) IRON OXIDE

Iron oxide pigments range in color from yellow to reds, browns and blacks. They possess very good opacity and durability and are comparatively inexpensive. The ores that occur naturally are commonly called earth colors. The manufactured types are termed synthetics.

(D) IRON BLUES

Iron blues are chemical compositions of sodium and potassium ferriferrocyanide, deep blue in color. This group has fairly good durability and color retention when used as the major portion of the pigment in a paint. When used in combination with other pigments, particularly white pigments, iron blues have a tendency to reduce in the package and give a lighter color. However, they will oxidize when exposed to air and assume their original color.

(E) ULTRAMARINE BLUE

Ultramarine blue is made by calcining mixtures of clays and silicas with sodium salts, sulphur, and carbonaceous material. It yields a brilliant, clear blue tone, is resistant to alkali and heat but very sensitive to acids. It has low hiding power and is not suitable for outside paints.

(F) CHROME YELLOW

Chrome yellow pigments range in color from light yellow to deep orange and may vary in composition from normal lead chromate ($PbCrO_4$) to basic lead chromate ($PbCrO_4$-PbO). They have good opacity, good durability, and reasonably good color retention. Basic lead chromate is used in the formulation of primers for steel.

(G) CHROME GREEN

Chrome greens are a combination of lead chromate and iron blues and range in color from light shades of green to very deep greens. Due to their lack of chemical resistance, these pigments are not used in paints intended for use in corrosive locations.

(H) CHROMIUM OXIDE AND CHROMIUM HYDROXIDE

Chromium oxide and chromium hydroxide pigments are green in color and are exceptionally durable and permanent. Chromium oxide has a somewhat dirty hue but has good opacity. Chromium hydroxide is a very clean light green with low opacity. These pigments are employed in the Spruce Green color used so widely at shore establishments.

(I) TOLUIDINE RED TONERS

Toluidine red toners are a group of organic red pigments covering a fairly wide range of red. They are quite durable and permanent when not diluted with white pigment. They are cleaner and brighter in color than the iron oxides and are used extensively in paints where color is an important factor. The most widespread single use of toluidine red toner in the Coast Guard is in the red paint for buoys.

(J) LAMPBLACK

Lampblack is one of the most widely used pigments. It is made by burning oils or tars under carefully controlled conditions. Lampblack is used for darkening the color of a paint and is used extensively in the various gray paints.

1–3–2, EXTENDER PIGMENTS

Extender pigments are filler materials with little or no opacity. They are used primarily to increase the pigment portion of paints in order to control gloss, diffuse or dilute other colored pigments and to retard settling.

(A) MAGNESIUM SILICATE

Magnesium silicate is a fibrous type talc in which the particle shape is a long needle-shaped crystal. When used in paint it tends to reinforce and strengthen the film. It also retards settling of the pigment. It is widely used in primers for metal surfaces.

(B) CALCIUM CARBONATE

Calcium carbonate is an extender used mainly to control the amount of pigment required in a coating in excess of that needed for hiding power and color. Commonly called whiting.

1–3–3, SPECIAL PURPOSE PIGMENTS

The purpose served by the preceding pigments

is primarily that of adding color and solids to the paint. The following pigments serve various special purposes as discussed below.

(A) RED LEAD

Red lead has for many years been used to protect iron and steel against corrosion. The ability of red lead to inhibit corrosion is considered to be partly due to its slight alkaline nature. Steel and iron corrode less rapidly in an alkaline environment and it is believed that corrosion under a red lead film is suppressed because any moisture penetrating a red lead film is rendered alkaline by the red lead.

When used in conjunction with linseed oil or other paint vehicle, red lead has the property of penetrating into pits and crevices forming tough, elastic adhesive paint film. Thus, red lead paints afford physical and chemical protection to iron and steel surfaces.

Red lead being low in the electromotive series may have some slight reaction when in direct contact with aluminum, zinc and magnesium. While the extent of this reaction is open to question it is considered better practice to use a zinc chromate primer for priming these metals.

(B) ZINC CHROMATE

The electrolytic theory of the corrosion of steel created a special interest in chromium compounds. Because of the soluble chromate and mildly basic nature of zinc chromate, attention was focused on it as a corrosion-inhibiting pigment, and it became the subject of extensive investigations in primers for steel. Its use in conjunction with linseed oil has met with only moderate success because of the sensitivity of the binder to the water soluble portion of the pigment. Zinc chromate in an alkyd vehicle has given good performance when applied to very well prepared surfaces. Its general use as a primer on steel, however, has declined due to poor performance experienced when it is applied to the average metal surface.

(C) ZINC DUST

A possible explanation for the favorable results obtained in zinc dust pigmented primers for steel is the generally accepted theory that the finely divided zinc dust presents enough zinc surface to react with ingredients of the vehicle that would otherwise be free to attack and corrode the surface to be painted and disturb the bond between the metal and paint film. Another factor to consider is that the inhibiting qualities of metallic zinc and the mild neutralizing qualities of zinc oxide, which tends to prevent an acidic condition, are both present in zinc dust. Zinc dust paint films are non-toxic. This combination of adherence, corrosion inhibiting of steel, and non-toxic qualities is unique and has made zinc dust the foremost pigment for use in coatings for the interior of metal drinking water tanks.

(D) ALUMINUM POWDER

One of the most important properties of aluminum pigments is their ability to "leaf" when suspended in a vehicle of oil or varnish. When aluminum paint is applied some of the flakes are carried to the surface of the paint film by convection currents within the wet paint film and are held there by surface tension forming an almost continuous metal leaf at the surface. The remainder of the flakes are distributed throughout the film in more or less parallel layers. Because of this arrangement it is difficult for air, light and moisture to penetrate the film.

Aluminum paint is more impermeable to moisture and oils than most interior or exterior enamels. Its hiding power is also extremely high. The combination of these two properties makes it an excellent sealer for wood treated with creosote or other petroleum preservatives and for sealing knots and pitch pockets. It may also be used as a combined primer-sealer for overall application on wood surfaces although products formulated for this particular function are generally preferred.

Its thermal emissivity is relatively low. This property makes it particularly valuable as a coating for boiler fronts and other heated surfaces where it is desired to reduce the amount of heat radiated.

Aluminum paint is sometimes used to reflect light although its use for this purpose is declining. While the initial reflectance may approach that of ivory colored paints, discoloration of the vehicle with time reduces its reflectance. Also since its reflectance is high in specular reflectance, producing highlights, and low in diffuse reflectance, producing dark areas which contrast with the highlights, it is unsuitable for interior use as a high reflectance paint. White paint will reflect more light and heat than aluminum paint.

(E) COPPER AND MERCURY COMPOUNDS

These compounds have proved to be the most effective toxic ingredients for use in antifouling paints for both steel and wooden hulls.

Cuprous oxide is the least expensive of the compounds and is the principal toxic constituent in most antifouling paints.

Mercuric oxide or mercuric chloride is frequently combined with cuprous oxide since the mercury compounds are more toxic to grass fouling. Antifouling paints used on wooden boats should not contain the mercury compounds since they may react with and destroy boat fastenings made of brass or bronze.

1-3-4, OIL VEHICLES

Oil vehicles consist of a drying oil (an oil which dries to a hard non-tacky film), volatile thinner and dryer. A combination of drying oils is sometimes used to obtain certain properties. All or a portion of the oil may be specially treated to improve gloss and leveling or to control penetration. Oil vehicles are used mainly in exterior paints for wood surfaces. Some of the more important oils used in paint formulations follow:

(A) LINSEED OIL

This oil is obtained from flax seeds and has the ability to absorb oxygen from the air and to eventually dry, forming a tough hard film.

Linseed oil is of two basic types, i.e., raw and boiled. The raw or untreated linseed oil dries very slowly and has the ability to penetrate into the surface upon which it is applied. This material is used as the principal ingredient in the vehicle of "old-fashioned" slow drying red lead primers. These primers may take several weeks to dry thoroughly.

Boiled linseed oil is made by heating the raw linseed oil in the presence of air and driers to carry out a partial oxidation and polymerization of the oil thus accelerating its drying.

Linseed oil is also frequently heat bodied by treating with heat to carry out a partial polymerization and produce a more viscous liquid. These treated oils (boiled and heat bodied) are often used as all or part of the vehicle in oil type paints.

(B) TUNG (CHINA WOOD) OIL

This oil is obtained from the nut of the tung tree and must be heat treated before it is suitable for use in paints. It is often used in combination with resins to form an oleoresinous vehicle. This oil is less flexible than linseed oil but is more waterproof and chemical resistant. This oil is used in conjunction with linseed oil and phenolic resins in the formulation of many spar varnishes and ready-mixed aluminum paints.

(C) SOYBEAN OIL

This oil is obtained by crushing soybeans. It is relatively inexpensive and has very good retention of color and elasticity. It is used extensively in modifying alkyd resin vehicles to impart non-yellowing properties to the finished paint. Its slow-drying properties are used to advantage in retaining the plasticity of calking compounds.

(D) FISH OIL

An oil extracted from fish which has the property of withstanding high temperatures and is used in heat-resisting paints. It has rather poor drying properties and poor water resistance. This oil is usually used in combination with other oils and resins in the production of oleoresinous coatings.

(E) CASTOR OIL

This is a non-oxidizing oil obtained from the castor bean and is used as a plasticizer in lacquer and other coatings. It can be converted into a drying oil by dehydration. Dehydrated castor oil has very good water and chemical resistance and is widely used in combination with resins to form oleoresinous vehicles. It is also used to some extent in synthetic alkyd finishes.

1-3-5, OLEORESINOUS VEHICLES

Oleoresinous vehicles consist of oils and natural resins cooked and blended together in various proportions. Paints can be formulated for either interior or exterior use depending on the kind of resin and oil used and their proportions.

Oleoresinous vehicles are usually classified as short, medium or long oil vehicles according to the proportions of oil and resin. A short vehicle contains less oil in proportion to resin than a medium or long oil vehicle. Short oil vehicles have high abrasion resistance but low resistance to moisture and hence are used primarily in interior paints. Conversely, long oil vehicles have low abrasion resistance but relatively high water resistance and are, therefore, used primarily in exterior paints.

1-3-6, ALKYD VEHICLES

Alkyd vehicles are made by the reaction of a polybasic acid, such as phthalic, maleic or succinic with a polyhydric alcohol, such as glycerine or penterythritol, and vegetable oils or fatty acids from such vegetable oils as linseed oil, soybean oil, dehydrated castor oil and to a lesser extent, tung oil. The properties of this type of vehicle can be varied through a wide range depending on the type of polybasic acid, polyhydric alcohol and vegetable oil or fatty acid used.

The reaction of these components must be carried out in a closed retort under controlled conditions in order to produce a uniform vehicle. When the reaction has been completed, the volatile portion is added to give the viscosity and brushing and spraying properties required in the finished vehicle.

Pigments, driers, and thinners as used in straight oil paints, are usually used in alkyd resin paints. Straight alkyd resin vehicles should not be confused with vinyl-alkyd resin vehicles which require special thinners. Alkyd resin paints may be thinned with mineral spirits.

The choice of alkyd vehicle to use in a given paint is controlled by the film properties and performance expected of the product. Many alkyd vehicles are compatible with oils and oleoresinous vehicles and are often used in this manner to give improved durability and service characteristics to the finished coating. Alkyd paints, properly formulated, have excellent durability, water and chemical resistance, and generally give better performance than the older drying oil vehicles.

1-3-7, PHENOLIC RESIN VEHICLES

Phenolic resins are made by condensing phenol with formaldehyde or similar aldehydes. These resins are commonly used in combination with one of the oils to produce a vehicle which has excellent water and chemical resistance. Phenolic resins have a marked effect on the polymerization of oils which is evidenced by a more rapid drying of the film. They also have an inhibiting effect upon the oxidation of the oil after the initial drying cycle. This tends to prolong the life of a film, since all paint films invariably fail due to a long, slow oxidation period. An example of phenolic resin vehicle is Spar Varnish which is formulated of phenolic resin, linseed oil, tung oil, petroleum spirits, xylene, and driers. This formulation shows excellent durability on exposure, good abrasion resistance and gloss retention.

1-3-8, VINYL RESIN VEHICLES

The term "vinyl resin" is usually employed to define a class of synthetic resins derived from a vinyl ester, such as vinyl acetate, vinyl chloride, vinylidene chloride, etc. by polymerization of the monomeric compounds under conditions which are controlled to yield products of the desired characteristics. In contrast to the preparation of phenolic, urea and alkyd resins, which are formed by a condensation process, the vinyls are formed by polymerization. As a result, each polyvinyl resin molecule consists of a linear chain in which the monomers have reacted with one another to form high molecular weight polymers.

The characteristics of the vinyl resins are closely associated with their molecular weights. Certain properties of the vinyl resins vary with the molecular weight, while others seem to be completely independent of it. For example, tensile and impact strength, abrasion resistance and solution viscosity increase with the molecular weight of the polymer, whereas water absorption, refractive index, hardness, and electrical properties are practically independent of molecular size. The solubility of the resins in organic solvents varies inversely with the molecular weight.

There are six types of vinyl resins which find applications in the plastics and coatings fields: (1) polyvinyl acetate, (2) polyvinyl alcohol, (3) polyvinyl chloride, (4) copolymers of vinyl chloride and vinyl acetate, (5) polyvinyl butyral, and (6) polyvinylidene chloride and copolymers with vinyl chloride, vinyl cyanide, etc.

Pretreatment-Wash Primer is formulated from the polyvinyl butyral resin which is alcohol soluble. Cleaning brush and spray equipment and thinning when necessary require denatured ethyl alcohol.

Vinyl anticorrosive paint and vinyl alkyd topcoat paints are formulated from the vinyl chloride-vinyl acetate copolymer. Cleaning brush and spray equipment, and thinning when necessary, requires methyl isobutyl ketone and toluene. These solvents are conveniently combined in one container listed in the catalog as Vinyl Paint Thinner.

Vinyl paints form an exceedingly tough and durable film. They are more resistant to alkalies than most paints and may be scrubbed with high alkaline cleaners. Vinyl paints must have a chemically clean surface for proper adhesion.

1-3-9, CHLORINATED RUBBER VEHICLES

Chlorinated rubber resin is a rubber solution vehicle obtained by dissolving milled rubber in toluol and introducing chlorine gas. The chlorinated product is then precipitated by discharging the solution into water which produces a white granular solid. The precipitate is then dissolved in coal tar or chlorinated solvents and necessary plasticizers added. Coatings made with chlorinated rubber vehicles are quick drying and resistant to water and most chemicals. They have a limited use in maintenance work to fulfill certain conditions such as coating rubber sound-projector domes. Chlorinated rubber base paint may be thinned when necessary with aromatic petroleum naphtha.

1-3-10, RUBBER SOLUTION VEHICLES OTHER THAN CHLORINATED RUBBER

Rubber solution vehicles may consist of certain resins, such as polystyrene butadiene, properly plasticized and dissolved in combinations of coal tar and petroleum solvents. Protective coatings formulated with these vehicles are quick drying, have excellent water and chemical resistance and are used most widely where corrosive conditions such as acid, alkali and brine are to be overcome. They also show good resistance to abrasion and scrubbing.

Rubber solution paints may be thinned with Synthetic Enamel Thinner. These paints are especially recommended for such application as the interior of concrete swimming pools, concrete showers, etc.

1-3-11, SILICONE VEHICLES

Silicone vehicles are synthetically produced from siloxane intermediates and a wide range of silicone vehicles can be produced by varying the molecular properties of the various intermediates. The chief value in silicone paints lies in their unusual resistance to high and low temperatures. Due to its high cost, the use of silicone paints is limited to application where severe conditions of corrosion and high temperature are present. Protective coatings formulated of this vehicle require the use of coal tar solvents, such as toluol, if thinning is necessary for proper application consistency. Properly formulated silicone vehicle insulating varnishes exhibit remarkable electrical insulating properties at high temperatures.

1-3-12, EMULSION VEHICLES

Emulsion vehicles are pigmented vehicles consisting of oils, resins, nitrocellulose or other organic binders emulsified in water. The pigmented vehicle is dispersed throughout the emulsion in fine droplets and is called the dispersed phase—the water being called the continuous phase.

Emulsion paints have the following components each of which is briefly discussed in the following paragraphs:

Hiding Pigment

- Extender Pigment
- Drying Oils
- Oleoresinous Material
- Emulsifying Agent
- Preservative
- Stabilizers
- Antifoam Agent
- Driers
- Solvent
- Water

Pigments used depend upon the end use of the paint. In general, most pigments used in oil paints are suitable for use in emulsion paints.

The choice of oil, resin or varnish for use in emulsion paint is governed only by cost, availability and performance requirements. Oils and resins are generally used without organic solvents to avoid odor and fire hazard.

The purpose of the emulsifying agent is to prevent the coalescence of the droplets of the dispersed phase. They are principally soaps and may be added as such or formed during the emulsification. A film of the agent is formed around each droplet of the dispersed phase only a few molecules thick. The emulsifying agent must, therefore, be resistant to physical and chemical changes.

The stabilizer is used to reduce consistency change or actual breaking of the emulsion caused by variations in temperature, mechanical working or other abnormal conditions. Stabilizers used include casein, soya protein, methyl cellulose, starches and others.

Preservatives are included to prevent decomposition of proteins used and to eliminate fungus growth on emulsion paint in partly filled containers or on the dried film. More than one type of preservative is frequently used to take care of all types of decomposition in either the water soluble materials or the water insoluble portion of the paint.

Antifoaming agents are required to reduce the large quantity of foaming produced by agitation in the manufacturing process. A similar condition exists when the painter brushes the paint on a wall. The brushing action produces considerable air bubbles which do not break until the film is partly set. This produces a pitted surface in the dry film and is undesirable.

Metallic driers are necessary in emulsion paints which contain oxidizable oils and resins for the same reasons that they are necessary in oil paints.

The volatile material in emulsion paints is principally water; although volatile emulsifying agents such as ammonia are frequently used and there may be small amounts of organic solvents derived from driers or antifoaming agents present.

A properly formulated emulsion paint is stable over the normal temperature ranges but the emulsion may be broken by freezing or by temperatures in excess of 160° F. Obviously if the continuous phase (water) is frozen, the condition at the interface changes from a liquid to a solid which upsets the balanced film of emulsifying agent concentrated there. Sometimes the emulsion may be re-formed by sufficient agitation but this is not always possible. When the temperature is raised above 160° F., the viscosity and interfacial tension changes, and again the balanced film of emulsifying agent is destroyed. A proper balance of stabilizers improves heat stability but there are definite limitations in this direction.

Emulsion paints produce excellent flat finishes with very low sheen. They are easy brushing, fast drying, have good dry hiding power, lack objectionable paint odor and present no fire hazard during storage or application.

The fact that any oil, resin, copolymer or higher polymer may be emulsified and then pigmented indicates a wide field for expansion in emulsion paints.

1-3-13, SOLVENTS AND THINNERS

In the manufacture of paints and varnishes, solvents which have the ability to dissolve the resin and oil combinations and to achieve the proper consistency for application must be used. Since the solvents evaporate when the paint is applied, the proportion of solvent to non-solvent used in a paint determines the thickness of the dry film per coat. In order to obtain satisfactory dry film thickness solvents should be used only in such quantities as are required to render the proper consistency for application. Excess thinning of paint lessens its dry film thickness per coat, resulting in lower ultimate life, especially so if the finish is exposed to the elements.

Oil, oleoresinous, alkyd and phenolic vehicle paints are compatible with petroleum distilled solvents such as naphtha and mineral spirits. A solvent of this type is stocked under the name of Paint Thinner. Turpentine has been considered a general all-purpose thinner for many years. However, due to its high cost, its use is not justified since mineral spirits will satisfactorily thin most paints used in the Coast Guard and the cost is considerably less. In addition, turpentine is a vegetable oil as contrasted to petroleum solvents which are mineral oils. Turpentine is therefore more of a fire hazard due to the danger of spontaneous combustion as discussed in Section 1–4, Storage of Paint Materials.

Many coatings formulated from synthetic vehicles are not compatible with petroleum distillates and therefore require thinners of higher solvent power such as the coal tar distillates—toluol, zylol, and solvent naphtha or the hydrogenated naphthas which are made from petroleum distillates by chemical treatment. These distillates are sometimes used in combinations to obtain varying power of solvency. An example of this type is Synthetic Enamel Thinner which has intermediate solvent powers, and Vinyl Paint Thinner which has high solvent powers.

1–3–14, JOB MIXED PAINTS PROHIBITED

As previously stated this discussion of raw materials used in the manufacture of paints is purely for informational purposes and shall not be construed as authority for mixing paints "on the job" from raw materials.

The correct formulation of protective coatings is a job for a paint technologist and requires a great deal of experience and a thorough knowledge of the properties of the basic materials used in the coatings. There are a large number of pigments, oils, resins, solvents, and driers and an infinite number of combinations of these constituents. Only experienced formulators can produce coatings with the properties required to give best results. Furthermore, mixing is only one step in the process of paint manufacturing. Proper dispersion of the pigment is necessary requiring the use of paint mills and other equipment designed especially for the purpose.

For these reasons *the use of "job-mixed" paints in the Coast Guard is prohibited.* Only ready-mixed paints available through Navy and Coast Guard supply channels as specified in the materials section of this Manual are authorized for use. These paints are top quality paints and are the result of many years of research by both industrial and governmental paint laboratories. Occasionally inferior paint products find their way into supply channels. Oftener, however, paint failures result from improper application of good material.

In the event an inferior paint is received all pertinent information including the name, stock number, manufacturer's name, contract number, and lot number as stenciled on the exterior of the container together with details of the method of application and nature of the paint failure should be furnished the Commandant (ETD).

In the event an appropriate type of paint for a particular application is not available through Navy and Coast Guard supply channels, the Commandant (ETD) should be informed of this situation also giving the details of the purpose for which the paint is to be used.

1–3–15, THE MANUFACTURE OF PAINT

Navy and Coast Guard paints are manufactured to specifications. These specifications are written around formulations made by experienced paint technologists. Since paint formulation is not an exact science many trial formulations are prepared and subjected to laboratory tests. The most promising of these are then service tested for several years before a final selection is made of the best formulation for a particular job.

Specifications are then written which set forth the formulation, the requirements for the paint materials and test methods for determining whether the materials meet these requirements. These specifications are furnished to the paint manufacturer who then commences the manufacture of the paint in accordance with the specifications.

The first step in the manufacturing process consists of the blending of the pigments, oils, resins, driers and solvents in a mixing tank. The liquid ingredients are pumped from supply tanks at ground level up to the mixing tank which is generally located on the highest floor of the plant. There they are introduced into the mixing tank. Not all of the liquid ingredients are put in the mix at this time. Generally speaking the more volatile liquids such as some of the solvents are added at a later stage to prevent their loss by evaporation during the manufacturing process. The pigments are then weighed out to get the proper amounts of each and added to the liquids in the mixing tank. The mixture is then mechanically agitated to form a homogenous mix about the consistency of syrup.

This mix is then fed down to the next level in the plant where it is placed in the "mills." Mills of two types, i.e., pebble (ball) mills and roller mills, are widely used. Pebble mills are large cylinders about ⅓ filled with pebbles or steel balls approximately the size of marbles. The mix is fed into the pebble mill where it is "ground" by revolving the mill for a period of from 8 to 24 hours. Grinding is necessary to insure thorough wetting and complete dispersal of the pigment.

Roller mills achieve the same grinding result by running the mix between steel rollers. The lower rotating cylinder picks up the mix on its surface and carries it around to where it is squeezed or ground between the upper and lower cylinder. The mix emerging from between the two rollers is scraped off the cylinder by a knife edge in contact with the surface of the cylinder. This mix is run through 3 to 5 such sets of rollers. Each pair of rollers is set successively closer together. Roller mills are used where very fine grinds are desired.

After the mix is ground it is fed to the next lower level where the remaining solvents are added to the mix to achieve the desired consistency of the finished paint. This is done in a mechanically agitated mixing tank. When all the ingredients have been mixed together the paint is tinted to the proper color by adding small portions of liquid tinting material.

The finished paint is then led to the ground level of the plant where it is packaged for delivery. Either prior to or immediately after delivery random samples of the paint are selected by Navy or Coast Guard inspectors and are subjected to a series of rigorous chemical and physical tests to determine if they have been formulated in accordance with the specification. If the paint passes the inspection it is accepted and placed in stock for issue.

SECTION 1–4

STORAGE OF PAINT MATERIAL

When you store paint—whether aboard ship or at a shore station—always place it in the paint locker. The cans should be marked with the name of the paint, the formula or specification number, and the date of manufacture—and kept tightly sealed. To prevent settling, all cans should be turned "bottoms-up" at least once every three months. When new stocks of paint are received, they should be stored so that the oldest stock is used first. Overage or obsolete paints must be surveyed by the supply officer who will determine whether or not any further use can be made of them. Any can of paint that is over a year old should be inspected and, if unfit for use, surveyed. When this occurs to large quantities of paint, instructions for disposition should be requested from the Commandant (FS). In certain cases it may be possible to return the paint to the manufacturer for reworking.

Particular caution must be exercised in the storage and mixing of paint to reduce the fire hazard involved. Paint stored in a tightly closed container is not extremely hazardous. The hazard occurs when the top is left off the can. The vapors then escape into the paint locker and the paint in the can has access to the air it needs to support combustion. The principal hazard exists therefore in the area in which the paint is being mixed. It is desirable to limit the amount of paint in this area and to store the bulk of the paint in an area separated from the paint mixing area. This gives rise therefore to two separate paint lockers, i.e., the paint storage room and paint mixing room.

The paint storage room shall contain only unopened paint materials. There shall be no rags, mattresses, rope or similar material subject to spontaneous combustion stored in the compartment. Paint storage rooms should be located in an area where the ambient temperature will not rise above 90° F. The ambient temperature of paint storage rooms at shore stations should be kept as low as possible during hot summer weather. Locating the storage room on the north side of buildings, in a shady place, or in a basement along with insulation and ventilation will do much to reduce storage room temperature.

The paint mixing room should be located apart from the paint storage room. At large shore stations, separate buildings for storage and mixing are highly desirable. Where this is not possible, as on ships and at small shore stations which have only one paint locker, paint mixing should be done out in the open air whenever practicable. The paint mixing room at shore stations should never be used for storage of paint. Only that paint which is to be mixed during the day should be present. Adequate ventilation must be provided in paint mixing rooms to prevent the concentration of explosive volatile vapors. Paint left over from the day's work should be tightly closed and stored in the paint storage room. Rags, rope and other combustible material shall be kept clear of the paint mixing room. Wet brush stowage tanks shall be provided with sheet metal covers to prevent evaporation of volatile solvents. Used rags and empty containers shall be disposed of, since they constitute a fire hazard. Empty containers which are to be used for other purposes shall be thoroughly cleaned and steamed out or filled with water to expel all volatile vapors.

Spontaneous combustion must always be guarded against in paint lockers. Oils of vegetable oil origin dry by oxidation. The oxidation reaction gives off heat. If the heat cannot escape, such as where the oxidation takes place in the center of a pile of rags, it builds up to the flash point of the oil and a fire results. Linseed oil and turpentine are both of vegetable oil origin. Rags soaked with these oils are fire hazards. This is one of the reasons leading to the replacement of turpentine by mineral spirits as a solvent. Mineral spirits evaporate rather than oxidize. This action does not create heat and hence cannot cause spontaneous combustion. Since linseed oil is used extensively in paint products, oil and paint soaked rags should be disposed of and not left lying about in paint lockers.

Electrical devices inside the paint mixing room should be carefully inspected daily. Switchboxes and motors and other electrical equipment in paint mixing rooms shall be of explosion-proof design and care shall be exercised to keep all doors and cover plates closed on all such equipment. Make sure there are no sparks produced that could cause a fire. SMOKING IN THE PAINT MIXING ROOM IS ABSOLUTELY PROHIBITED.

Paint lockers are often equipped with remote controlled CO_2 systems and hand portable CO_2 type fire extinguishers. Paint locker personnel must be thoroughly familiar with the operation of these devices. Fifteen-lb. extinguishers must be weighed monthly and when the weight drops to 13½ lbs. (10% loss) they must be recharged. Cylinders of 35 and 50 lbs. must be weighed semi-annually and replaced with a full cylinder if the weight has dropped to 31½ lbs. or 45 lbs., respectively.

Remote control devices shall be tested whenever CO_2 cylinders are disconnected from the system for weighing. Upon being discharged, an extinguisher shall be immediately recharged. Never return an empty or partially empty extinguisher to its rack.

SECTION 1–5

SURFACE PREPARATION

The proper preparation of the surface to be painted is the most important single factor in securing good paint performance. Time spent in careful surface preparation will be more than repaid by the additional life of the paint film.

1–5–1, PREPARATION OF METALLIC SURFACES (Except Aluminum and Galvanized Steel)

Metallic surfaces other than aluminum and galvanized steel should be cleaned by dry sandblasting wherever possible. This removes all mill scale, rust, oil and foreign matter and leaves a chemically clean surface which will insure best paint performance. Immediately upon the completion of the sandblasting 1 liberal coat (at least 0.5 mils film thickness) of Pretreatment-Wash Primer shall be applied. If the bare sandblasted surface is allowed to stand overnight a thin film of rust, often invisible, will form on the metal nullifying to a large extent the advantages of the sandblasting. If the surface cannot be coated with Pretreatment-Wash Primer the same day it is sandblasted, it shall be washed with a rust inhibitor solution consisting of a mixture of $\frac{4}{5}$ qt. of dry diammonium phosphate, $\frac{1}{5}$ qt. of dry sodium *nitrite* and 40 gals. of water. The Pretreatment-Wash Primer shall be applied as soon thereafter as possible.

Wet sandblasting shall be used where dry sandblasting is impracticable, and where the volume of metal cleaning makes manual methods uneconomical. It is particularly well suited to use at bases and depots not having an area remote from docks or shops available for dry sandblasting. A rust inhibitor solution shall be used in both the sandblasting slurry and the final water wash. The rust inhibited slurry shall consist of $\frac{1}{5}$ qt. sodium *nitrite*, $\frac{4}{5}$ qt. diammonium phosphate, 15 gals. of water and 300 lbs. of sand. In the final operation of washing down the sand from the blasted area the water wash shall consist of $\frac{1}{5}$ qt. sodium *nitrite*, $\frac{4}{5}$ qt. diammonium phosphate and 40 gals. of water. One liberal coat (at least 0.5 mils film thickness) Pretreatment-Wash Primer should be applied as soon as the water-washed surface has begun to dry. It may be satisfactorily applied to damp but not thoroughly wet surfaces.

Where sandblasting methods are not practicable, manual methods may be used. The surface should first be roughed by the use of roughing tools such as chipping hammers and scrapers. In the use of these tools care must be taken to prevent nicking, denting or scratching of the surface. Nicks, dents and scratches provide ideal starting points for early failure of paint films. The low portion of such surface irregularities are difficult to clean and become a source of corrosion. On the sharp edged high points only a thin film of paint will adhere causing early paint failure. Chipping hammers should never have a chisel sharp edge. Thin plate (under $\frac{3}{16}$ in.) should never be chipped but should be scraped and wirebrushed.

When the bulk of rust and old paint has been removed by roughing tools, finishing tools shall be used to complete the job. The principal finishing tools in use are hand wirebrush, electric and pneumatic wirebrushes, power driven abrasive wheels, discs, or belts, and sandpaper. The use of finishing tools without prior use of roughing tools usually results in low productivity in terms of area finished per unit time, or in the case of power wirebrushing, in poor surface preparation due to "glazing" rather than removing particles of rust and old paint. The hand scraper is used on small jobs where power tools are impractical and on flat surfaces, crevices and corners. The hand wirebrush is a useful tool for light rust and for brushing around welds and in places not accessible to the power wirebrush. Sandpaper and sanding discs are used where a particularly smooth, clean surface is desired. Also where old paint is removed only in spots, the edge of the remaining paint film should be sanded so that it tapers down to the bare metal. This will give a uniform appearance to the new paint film. Steel wool should not be used as a substitute for sandpaper since small steel particles become imbedded in the paint and form a source of corrosion. After the finishing operation is completed brush the surface to remove paint chips, dirt and dust and immediately apply 1 liberal coat (at least 0.5 mils film thickness) of Pretreatment-Wash Primer.

No rust inhibiting treatment other than the above shall be used. Specifically, phosphoric acid inhibitors are no longer authorized for Coast Guard use because of the following factors: (a) the erratic performance of water solutions of phosphoric acid, (b) the necessity for careful control of more complex phosphoric acid solutions to obtain consistently favorable results, (c) phosphoric acid inhibitors reduce effectiveness of Pretreatment-Wash Primer, and (d) use of phosphoric acid seldom produces detectable improvement in the paint performance of all paint systems.

All ungalvanized steel surfaces to be painted shall be treated as above with the exception of potable water and feed water tanks. In the case of these tanks omit the use of the Pretreatment-Wash Primer. For information pertaining to interior painting of potable water tanks, see Article 2–3–7.

CAUTION. When applying Pretreatment-Wash Primer on interior surfaces, care must be taken to assure adequate ventilation and fire precautions must be strictly enforced.

1–5–2, PREPARATION OF ALUMINUM SURFACES

Sandblasting and wirebrushing aluminum surfaces are not recommended except in cases of severe corrosion. Solvent cleaning is usually all that is necessary and may be accomplished by washing the surface with Paint Thinner after brushing off dirt and dust. Follow the Paint Thinner wash by washing with Paint Cleaner and water. Rinse with fresh water.

On exterior surfaces, old defective paint coatings may be removed with Paint and Varnish Remover. Since Paint Remover contains wax, the surface, after removing the old paint, must be washed with Paint Thinner, followed by Paint Cleaner and water. Paint Remover is not recommended for use on interior surfaces.

On interior surfaces, old firmly adhering paint coatings should not be removed. Flaking, scaling or peeling patches may be removed with hand scrapers, but great care must be taken not to damage the aluminum surface. Dulling the corners of scrapers will help prevent nicking the aluminum. The edges of firmly adhering paint around cleaned patches may be faired with fine sandpaper. Do not sandpaper the aluminum surface. After scraping and sanding, wash the entire surface with Paint Thinner followed by washing with Paint Cleaner and water and a fresh water rinse.

1–5–3, PREPARATION OF GALVANIZED STEEL SURFACES

Sandblasting of galvanized steel should be avoided if possible. When it is found necessary to sandblast these surfaces, the nozzle of the sandblast hose should be held at a greater distance from the surface, air pressure should be somewhat lower, and the stream of abrasive should be played more rapidly over the surface than is done in sandblasting steel.

Chipping hammers shall not be used on galvanized surfaces. Old paint shall be removed by hand scrapers followed by hand wirebrushing.

Bare galvanized steel which has not been sandblasted or wirebrushed shall be wiped with Paint Thinner and allowed to dry for 30 minutes. This shall be followed by scrubbing with soap and water. Rinse with fresh water. As soon as the surface is dry, apply 1 liberal coat of Pretreatment-Wash Primer.

1–5–4, PREPARATION OF WOOD SURFACES

(A) BARE WOOD

Bare wood surfaces shall be planed or sanded to a smooth surface. The degree of smoothness is determined by the use to which the surface will be placed.

Plastic wood shall be used to fill cracks and holes. If the wood is to be subsequently varnished or lacquered the plastic wood can be stained to match the color of the wood. The plastic wood will harden sufficiently within four hours to withstand ordinary handling and the use to which it will normally be subjected. When dry it can be cut, sawed, bored, reamed, filed and will withstand, without cracking, the driving of nails into it.

Putty may be used for filling nail holes, worm holes, dents and cracks where no strength is required. Putty should never be applied to bare wood because the oil in the putty will be absorbed by the wood and cause the putty to dry out, crack and eventually fall out. Holes to be puttied must first be primed with either a regular primer or with paint.

Wood which will be subjected to moist conditions, placed in or close to the ground and all exterior wood shall be treated with wood preservative. Pressure treated wood shall be used where severe conditions are encountered. Where a moderate amount of protection will suffice treat the wood with 2 coats of Copper Naphthenate Wood Preservative. The preservative shall be swabbed or brushed on, applying as much of the preservative as the wood will absorb. Joints and end grain surfaces shall be thoroughly soaked with the preservative. The preservative shall be allowed to dry 72 hours before application of paint. In cases where the structure is of a temporary nature or where periodic treatment is necessary due to causes other than decay, preservative treatment may be omitted.

On interior surfaces where appearance is not important, such as wooden boat bilges, the wood shall not be painted. Painting tends to prevent the normal ventilation from reaching the wood thereby increasing the opportunity for the wood to rot. It also makes it impossible to renew wood preservative until the paint has peeled off or been removed. By omitting the paint, the surfaces may be inspected to ascertain whether or not the wood should be retreated. As long as the green color of preservative is clearly visible, the material is effective. When the green color is no longer visible, danger of rot exists. The wood shall be retreated at such intervals as to maintain a clearly visible green coloration.

Where the wood is to be subsequently varnished, it shall be treated with Wood Preservative, Chlorinated Phenols, Type B, a colorless preservative.

Wood which is to be placed in soil or submerged in water shall be pressure treated with creosote in accordance with the current edition of Federal Specification TT–W–571. Creosoted wood which is to be subsequently painted shall be weathered for ninety days and given 1 coat of Ready-Mixed Aluminum Paint.

Subsequent to preservative treatment, if used, knots and pitch pockets shall be sealed to prevent the pitch from bleeding through the paint. Remove all excess pitch by scraping and coat the effected areas with Ready-Mixed Aluminum Paint.

(B) PREVIOUSLY PAINTED WOOD SURFACES

Previously painted wood surfaces shall be thoroughly washed and cleaned of all foreign matter. The use of alkaline solution such as lye or trisodium phosphate is prohibited. Paint Cleaner, a synthetic detergent formulated especially for washing painted surfaces, shall be used. Oil and grease not removed by the washing operation may be removed by wiping with Paint Thinner. The Paint Thinner should then be washed off and the surface given a fresh water rinse. Allow the surface to dry thoroughly before applying any paint.

(C) BURNING OFF

Should the removal of old paint from wood surfaces become necessary due to scaling, blistering, peeling, checking, cracking or excessively built-up thickness, it may be removed by burning. The technique is to blister the paint with a portable gasoline blow torch, without scorching the wood, and then to scrape it off with a hand scraper while it is hot and soft. Start at the bottom of the job and work toward the top. A good job of burning off will result in a surface reasonably free from scorches and the paint so completely removed that only a slight sandpaper rub will be necessary before painting.

When burning paint off an old building, care must be taken to prevent flames entering between cracks in the siding and igniting cobwebs, litter within the wall, or the framing members themselves.

Burning off should never be done in a closed compartment because the gases and fumes are toxic. The burning gasoline from the torch and the heated paint film give off toxic gases. Wear a respirator, have a bucket of water available to extinguish small burning particles of paint and have a fire extinguisher handy in case of a fire.

(D) USE OF PAINT REMOVER

Paint remover may be used to advantage on furniture and on inside jobs where burning off cannot be accomplished. Apply the remover liberally with a full brush without brushing out any more than is necessary. Do not break the wax film which forms over the remover after it has been applied because this would allow the gases to escape without acting on the paint film. Due to the toxic nature of the volatiles in paint remover, adequate ventilation is necessary.

Let the remover stand until the old paint or varnish begins to blister or wrinkle. This takes from ten to twenty minutes depending upon the type and thickness of the old finish. Sometimes a second coat of remover will be necessary. After the old finish begins to wrinkle or blister, lift it off with a broad putty knife or hand scraper. Apply another coat of remover and rub in the direction of the grain with steel wool to remove any last traces of the old finish. Wash the surface with Paint Thinner or alcohol to get rid of any wax or acids left by the remover. Soap and water may be used but water tends to darken the wood and causes the grain to raise.

Paint remover will work faster and better on horizontal surfaces. Doors and other pieces which can be moved should be placed in a horizontal position on a bench or sawhorses to facilitate the operation.

Wood surfaces to be varnished or lacquered will probably require some sanding to obtain a fine smooth surface and expose the grain of the wood to best advantage. The surface must be dry before the application of paint, varnish or lacquer.

(E) CALKING WOODEN VESSELS

Prior to painting, hulls of wooden vessels shall be thoroughly inspected and any defective calking shall be renewed. Before applying seam compound but after calking with oakum, the seam and oakum shall be saturated with Copper Naphthenate Wood Preservative. After the preservative has soaked into the wood and oakum, pay hull seams with Oil Type Calking Compound, applied with a putty knife. Finish seams with a hollow or concave surface to prevent forcing the compound out of the seam when the planking swells. Deck planking seams are sealed by paying with Marine Glue. In those cases when proprietary seam sealers are authorized, manufacturers' instructions shall be closely followed.

(F) CALKING CRACKS ON BUILDINGS

Calking material may be used to fill holes and cracks which are too large to be filled with putty or plastic wood. Cracks between ends of siding boards, which occur due to shrinkage of the lumber, may be filled with Plastic Calking Compound prior to painting. Some cracks such as those under blind stops on windows or doors, which exist when window casing is nailed over siding boards or shingles, may be filled by calking with oakum up to about ½ in. of the outer edge of the blind stop. The remainder of the crack should be filled with calking compound which can be smoothed off to present a neat appearance. Split or loose boards should be replaced or renailed securely. Calking compound should never be used as a substitute for adequate repairs.

1–5–5, PREPARATION OF MASONRY SURFACES

Masonry surfaces are prepared for painting by cleaning, repairing and sealing the surface. In some cases a pretreatment must be given the surface prior to paint application. These steps are discussed in greater detail in the following paragraphs.

(A) CLEANING THE SURFACE

The procedure for cleaning masonry surfaces for painting will depend upon the type of paint previously applied and the type of paint which is to be used. Specifically, oil paints and emulsion paints may be applied over old, firmly adhering coatings of oil paint or cement-water paint. Rubber solution paints require the complete removal of old coatings of oil paint since the solvents in rubber solution paint will lift undercoats of oil paint. Cement-water paint requires the complete removal of all old paint unless the old coating is cement-water paint in good condition. Old whitewash must be removed completely prior to the application of any paint coating. Old coatings of cement-water paint may be covered with any type paint.

In cases where new paint is not compatible with old coatings or where old coating is whitewash or flaking and scaling paint, complete removal of the old coating will be necessary. This can be accomplished by scraping and wirebrushing. Sandblasting is the most effective method if equipment is available.

Where new and old paints are compatible and the old coating is in good condition, surface preparation will consist of removing dust, dirt, efflorescence, oil and grease. Dust and dirt can be removed by wirebrushing. All old paint should be lightly wirebrushed to make the surface uniform. Oil and grease can be effectively removed by washing with Paint Thinner followed by soap and water and a fresh water rinse.

(B) REPAIR OF MASONRY SURFACES

After the surface has been thoroughly cleaned, loose mortar between masonry courses shall be chipped and picked out, and the joint brushed thoroughly to remove all dust and loose particles. The cleaned surface shall be dampened before new mortar is applied to prevent absorption of water from the mixture.

To make repairs, a mixture of 1 part cement to 2½ parts sand, or 1 cement to 3 sand is recommended in ordinary cases. However, a 1:2 mixture is recommended for damp basements or masonry exposed to very moist conditions. A 1:2 mixture represents 1 part Portland cement to 2 parts sand, measured by volume. The proportion to be used will depend entirely upon conditions, a larger proportion of cement being necessary in cases where excessive moisture prevails.

Enough water should be used to make a fairly dry mortar about the consistency of putty. It should be thoroughly mixed and worked to insure best results. In filling cracks, the mortar should be used like calking material, that is, it should be well tamped to form a complete bond. When the crack has been tightly packed the surface should be smoothed with a trowel. In pointing up joints in masonry, the mortar may be applied with a trowel and the surface finished to conform to the old mortar.

After the material has hardened, the new work should be kept wet for several days to increase the strength of the mortar. If work has been done on outside walls, they should be covered by tarpaulins to protect them from direct exposure to the sun and drying winds.

(C) REPAIR OF PLASTER SURFACES

Cracks, holes and other defects in plaster surfaces must be repaired prior to repainting if a good finished job is to be achieved.

Small cracks, holes and indentations can easily be repaired with spackling putty. This material is applied with a putty knife or, in the case of hairline cracks, merely rubbed in with the fingers. Spackling putty is marketed under several brand names and the manufacturer's directions for use should be followed. Open market purchase of this material is authorized.

Large holes in plaster surfaces or areas where large sections of plaster have fallen out must be replastered. Remove all loose plaster around the edges of holes. Edges of firmly adhering plaster should be dovetailed so that the hole is wider next to the lath than on the surface. This will provide an anchor for the new patch material. Old, firmly adhering plaster adjacent to the hole should be wetted to prevent the absorption of water from the new plaster. Apply new plaster listed in Class 59 Catalog of Navy Material as Calcined Gypsum (Plaster of Paris) Retarded. This material sets in 40 minutes. Therefore, the job must be done as quickly as is consistent with good workmanship. Follow directions for mixing and applying which are printed on containers. The surface should be kept damp for about 24 hours after the new plaster has set.

After the new plaster has dried about 7 days, sandpaper to smoothness. The work should be scheduled to allow the new plaster to age as long as possible, allowing 2 weeks at least prior to application of Synthetic Rubber Emulsion Paint, and 1 month before application of Interior Gloss Enamel. Longer aging periods are recommended if at all practicable.

(D) SEALING AND WATERPROOFING THE SURFACE

New plaster and porous masonry must be sealed to prevent suction spots and saponification (the action of alkalies in the masonry on oils in the paint). Modern sealers are a combination of sealer and primer and are listed in Chapter 4. Cement-water paints do not require sealers or primers and may be used as sealers on porous masonry prior to the application of other types of paints, or they may be used as finish coats in themselves. However, in the case of oil paints, where cement-water paint is used as a sealer or where mortar-filled joints are new, a minimum of 1 month drying time should elapse before the application of the oil paint.

On old surfaces where minor repairs are made prior to oil painting and 1 month drying time is impractical, calking compound rather than cement mortar should be used as a crack filler.

(E) PRETREATMENT OF MASONRY SURFACES

Cement-water paints require thorough wetting of the surface prior to painting. Usually wetting the walls not more than an hour before painting with a garden hose adjusted to a fine spray is sufficient. The wall should be damp but not dripping wet when the paint is applied. If the surface dries rapidly, as it may in hot weather, it should be redampened slightly just before painting.

Emulsion paints and rubber solution paints may be applied on dry or damp masonry. In very hot weather it is advisable to dampen the surface prior to the application of emulsion paints.

Oil paints require dry surfaces. The application of oil paint to new masonry should be deferred until the walls have had time to dry. This may require 1 month to a year. Leaks around flashing of doors and windows should be repaired to exclude moisture before the application of oil paint. At least 1 week of clear, dry weather should precede the application of the first coat of oil paint to masonry. Since masonry surfaces tend to chill and collect condensed moisture, the application of oil paint early in the morning and late afternoon should be avoided except in dry climates.

1–5–6, CLEANING PAINTED SURFACES

Painted surfaces to be cleaned shall be washed with Paint Cleaner, a synthetic detergent, which is listed in Chapter 4. Where Paint Cleaner is not readily available, household detergents such as Soilax, Fab, Tide or equal may be used.

The use of free alkalies is extremely harmful to most paints. These alkalies attack the paint film, break it down and cause early paint failure. The use of lye, trisodium phosphate or other strong alkali solutions for the purpose of washing painted surfaces is therefore prohibited.

It will be found that a weak mixture of Paint Cleaner or equal synthetic detergent is entirely adequate for washing paint work and only such cleaners are authorized.

Vinyl Paints are an exception to the above, due to their exceptional alkali and abrasion resistance and may be scrubbed with the more alkaline cleaners.

1–5–7, ABRASIVE PAPER GRADING

Abrasive paper is graded by two systems, i.e., grit number symbols and commercial grit size.

The grit number symbol system (1, 1/0, 6/0, etc.) is based on the size of grit that will pass through a screen of a given mesh. The grit symbol number corresponds to the gauge of the screen. Thus, a number 3/0 abrasive paper contains grit which will pass through a number 3 gauge screen. This was the earliest system developed but is rapidly being replaced by the commercial grit size system.

The commercial grit size system is based on the number of pieces of grit of uniform size that can be placed in a single layer in an area of one square inch. The below comparative grading chart will provide information for relative comparisons. Products of individual manufacturers may vary slightly.

Grade	*Com'l Grit Size*	*Grit Number Symbol* Silicon Carbide	Aluminum Oxide	Garnet	Flint*	Emery*
	600	12/0	—	—	—	—
	500	11/0	11/0	—	—	—
	400	10/0	10/0	—	—	—
Very	360	—	—	—	—	—
Fine	320	9/0	9/0	—	7/0	—
	280	8/0	8/0	8/0	6/0	—
	240	7/0	7/0	7/0	5/0	—
	220	6/0	6/0	6/0	4/0	—
	—	—	—	—	3/0	—
	180	5/0	5/0	5/0	—	3/0
	150	4/0	4/0	4/0	2/0	2/0
Fine	120	3/0	3/0	3/0	—	1/0
	—	—	—	—	1/0	—
	100	2/0	2/0	2/0	—	—
	—	—	—	—	½	½
	80	1/0	1/0	1/0	—	1
Medium	—	—	—	—	1	—
	60	½	½	½	—	1½
	50	1	1	1	1½	2
	—	—	—	—	2	2½
	40	1½	1½	1½	—	—
Coarse	—	—	—	—	2½	—
	36	2	2	2	—	3
	30	2½	2½	2½	3	—
	24	3	3	3	3½	—
Very	20	3½	3½	3½	—	—
Coarse	16	4	4	—	—	—
	12	4½	4½	—	—	—

*Commercial Grit Size not applicable.

SECTION 1–6

PAINT MIXING

Ready-mixed paint requires some vigorous stirring and mixing before use. The first step in the preparation of paint is to open the can in such a way that the lid is not deformed and can be used to reclose the can. This is done by using a flat tool such as a screw-driver and prying the lid up only enough to loosen it. Do this all around the lid as in Figure 1 and no damage will be done to either the can or lid.

Fig. 1. Correct way to open a can.

No matter how high the quality of a paint may be, it will give poor service unless it is thoroughly mixed. When paint is settled in a can the upper portion contains too much oil and too little pigment. Paint mixing is best achieved by power shakers or power mixers. Paint shakers are a relatively expensive item and are warranted only at those establishments where large volumes of paints are used. A satisfactory paint mixer can be made by attaching a Paint Mixing Attachment, Item No. 3383, Class 40, General Stores Section of the Catalog of Navy Material to a portable electric or pneumatic drill. Vessels and stations which are allowed portable drills should procure this attachment.

When power shakers or mixers are not available, it is necessary to mix the paint by hand. To do this, pour off most of the clear liquid at the top of the container as in Figure 2 and mix the remainder thoroughly with a broad paddle.

Fig. 2. Pouring off the top liquid.

Add a small amount of the clear liquid and stir well, using a figure-8 motion with the paddle. Repeat the process, adding a small amount of liquid at a time, until all of the liquid has been added and the paint is uniform. Next, "box" the paint. Boxing is performed by pouring the paint back and forth between two containers as in Figure 3. This insures a homogenous mixture.

After the paint has been thoroughly mixed, all particles of pigment, dirt and skin should

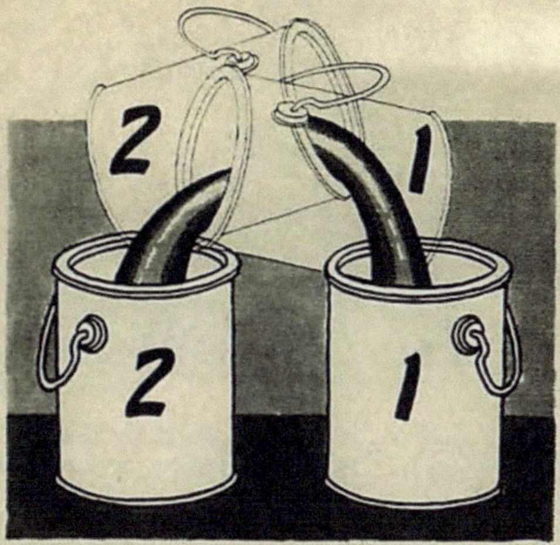

be removed by straining through a wire screen or cheesecloth. Any paint left over from a job should be strained before storing.

Turning paint containers, which are in storage, "bottoms up" periodically (at least once every 90 days) will greatly reduce the labor involved in mixing paint.

Fig. 3. The steps in "boxing."

SECTION 1–7

BRUSH PAINTING

Many types of brushes are carried in stock listed under Class 38, General Stores Section of the Catalog of Navy Material. Each brush is made for a particular purpose and, for best results from brush painting, the right brush for the job must be used. Quality brushes are hand made and are costly. Properly used, they will give good results and have a long life. In general the care of brushes is not accorded the attention it is due. The following paragraphs describe brush constructions, brush types, use of brushes and care of brushes.

1–7–1, BRUSH CONSTRUCTION

The quality of a paint brush is dependent upon five factors:

(a) The texture of the bristle which determines the paint spreading properties of the brush. Textures refer to the blending of stiff and soft bristles in the right proportion.

(b) The bristles weight content which determines the degree of paint retention. The longer the bristle the more the weight in the brush and the more costly it becomes.

(c) The solidity of the bristle which determines the degree of paint absorption. Proper solidity is the result of the right proportion of long and short bristles.

(d) The setting of the bristle which determines the degree of shedding of the bristle.

(e) The construction of the brush which determines the smoothness of the paint application.

From examination of the above factors it can be seen that the quality of the bristle determines, to a large extent, the quality of the brush. Both natural and synthetic bristles are used for paint brushes. While the quality of synthetic bristles has advanced rapidly in recent years and may some day be superior to the natural bristle, the latter is still considered the best. All quality paint brushes use Chinese hog bristles. These bristles which are black in color are gathered by natives in various regions of the world, cleaned and shipped to the United States.

Natural bristle is thick at the butt end and tapers to the flag or split end. The cross-section of the bristle is oval and it has a slight curvature in the lengthwise direction. As a comparison horse hair does not taper, has no curvature, is round in cross section and has no flag ends. See Figure 4.

In the manufacturing operation the bristle is first sorted in bundles. In the sorting operation all the butt ends are placed in the same direction. Each bundle receives its proper proportion of soft and hard bristles and long and short bristles and the curvature of the bristles is made to lie in the same direction. These bundles are hand formed by master craftsmen. The bundles are then imbedded in hard rubber or suitable solvent-resistant bed-

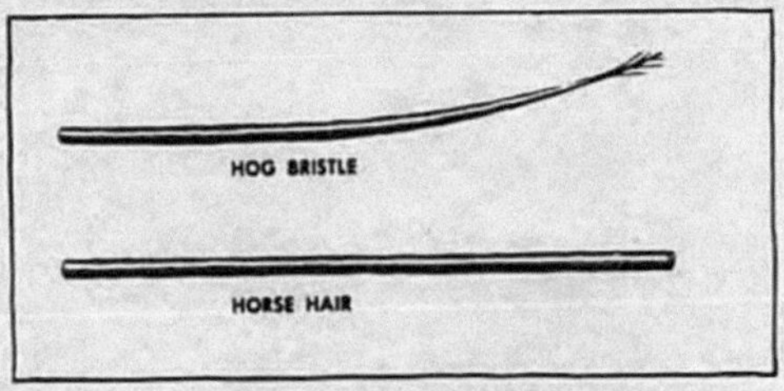

Fig. 4. Comparison of hog bristle and horse hair.

ding compound. In setting the bundles in the bedding material they are arranged in rows —the number and length of the rows forming a pattern to give the brush its desired shape, i.e., flat, oval or round. In addition, the bundles are oriented so that their curvature inclines the flag ends toward the center of the brush.

With the bristles set in the bedding compound, a metal strip, the ferrule, is wrapped around the bedding material. One edge of the ferrule is then nailed or riveted through the bedding material. A handle is slipped into the other end of the ferrule and similarly nailed or riveted.

All brushes have their bristles formed so that they taper toward the center at the flag end. This is done by orienting the natural curvature of the bristle and also by the proper blending of short and long bristles. See Figure 5.

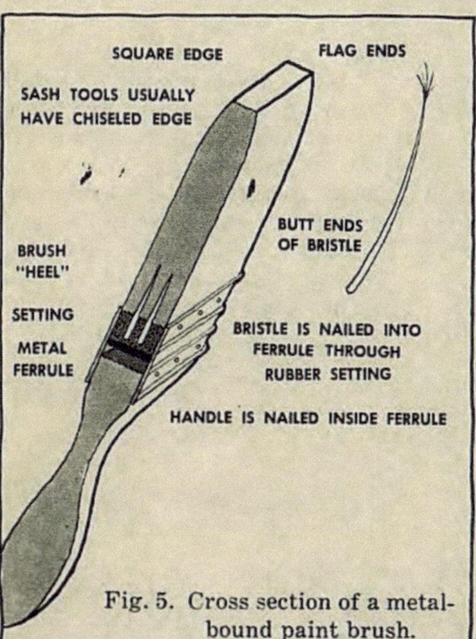

Fig. 5. Cross section of a metal-bound paint brush.

1–7–2, TYPES OF BRUSHES

(A) FLAT BRUSHES

Flat brushes have a narrow rectangular cross section. They are available in widths from 2 in. to 6 in. and in high, medium or utility grades. The length of the bristle varies with the width of the brush. The 4 in. width is the most universally used width. These brushes are recommended for paint application on very large to moderately small surfaces. They are not recommended for varnishing.

(B) OVAL SASH BRUSHES

Oval sash brushes have chisel edges with flag ends carefully preserved. They are available in 2 in. and 3 in. sizes. These brushes are for paint application on moderately small surfaces, in corners, on pipes, in pockets and other irregular surfaces that would damage a flat brush. They are not recommended for varnishing.

(C) FLAT FITCH BRUSHES

Flat fitch brushes have chisel edges and their flag ends have been carefully preserved. They come in widths from 7/16 in. to 1¼ in. These brushes are used for painting very small surfaces and for large lettering. They are not recommended for varnishing.

(D) FLAT VARNISH BRUSHES

Flat varnish brushes are available in double or triple thicknesses of bristle and in sizes from 1 in. to 2½ in. They have chisel edges with flag ends carefully preserved. These brushes provide smooth application of varnish and are used for the majority of varnish work. They may also be used for enameling.

(E) OVAL VARNISH BRUSHES

Oval varnish brushes are available in 1¼ in. and 1½ in. widths. These brushes pick up and retain more varnish than the flat brushes but do not give as smooth application. They are used for large surfaces or rough work. These may be used for either paint or varnish.

(F) FLOWING VARNISH BRUSHES

Flowing varnish brushes are made of skunk hair or 50 per cent skunk hair and 50 per cent Chinese hog bristle. They are well cupped, solid (no hollow space in the center of the brush) and are chisel edged. They are available in sizes from 1 in. to 2 in. These brushes are used for small jobs and fine work.

(G) MARKING BRUSHES

Marking brushes are made of Chinese hog bristle with ends cupped to a dome. They are available in 7/64 in., 11/64 in. and 15/64 in. sizes. These brushes are used for fine work.

(H) LETTERING BRUSHES

Lettering brushes are made of ox-ear hair and are available from 1/16 in. to 5/32 in. These brushes are used for small lettering.

(I) FLAT DUSTING BRUSHES

Flat dusting brushes contain any type of bristle suitable for dust brushes. They have no ferrule. These brushes are used for brushing the dirt and dust off surfaces to be painted. See Figure 6 for different types of brushes.

1–7–3, HOW TO BREAK IN A NEW BRUSH

Before a new brush is used, it should be well shaken to remove loose hairs and dust, and then carefully combed straight. A steel comb is recommended for the purpose. The new brush should then be soaked in boiled linseed oil for about 48 hours. This will make the brush more flexible, easier to clean, and causes the bristles to swell and be gripped more firmly by the ferrule of the brush thus preventing the bristles from working loose during painting work. When soaking the brush in linseed oil, avoid resting the brush on the bristles. This can be done by suspending the brush in a brush keeper (described later). After the required soaking of the brush, the surplus linseed oil should be pressed out. Then rinse the bristle in Paint Thinner until all the surplus oil is removed. The brush is now ready for use. Whenever a used clean brush has been stored in a dry condition for some time, it is advisable to soak it in linseed oil as described for new brushes, prior to use.

1–7–4, BRUSHING TECHNIQUE

The life of any paint brush can be greatly extended and a better paint job will result from proper brush usage. Brush painting is an art which must be mastered if good results are to be expected. Hold the brush as illustrated in Figure 7. Hold it firmly but lightly. Hold it so you can use a free and easy wrist motion plus the action of your arm.

When you pick up the brush, don't put your fingers on the bristles below the ferrule or metal band. Fingers smeared with paint promote sloppy jobs and perhaps blood poisoning should you happen to have open cuts or scratches on your hands. Besides, exerting too much pressure on the center of the brush will wear it into a fish tail shape.

When using a flat brush, don't try to paint with the narrow edge. This practice will

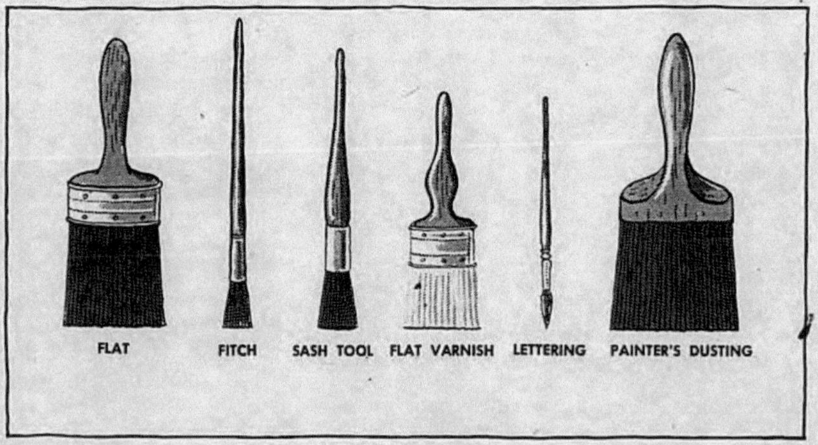

Fig. 6. Types of brushes.

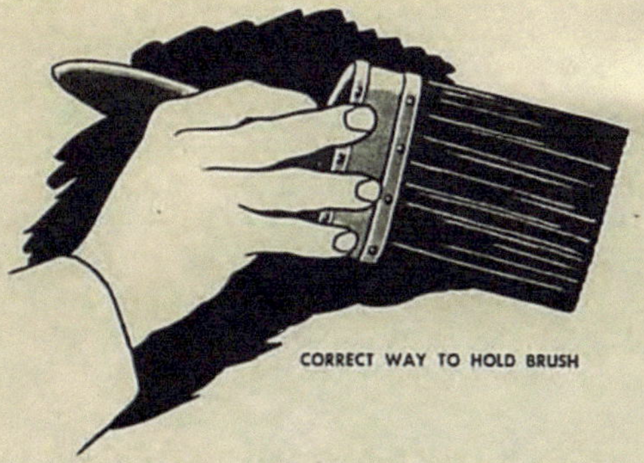

Fig. 7. Brush is held firmly, as shown.

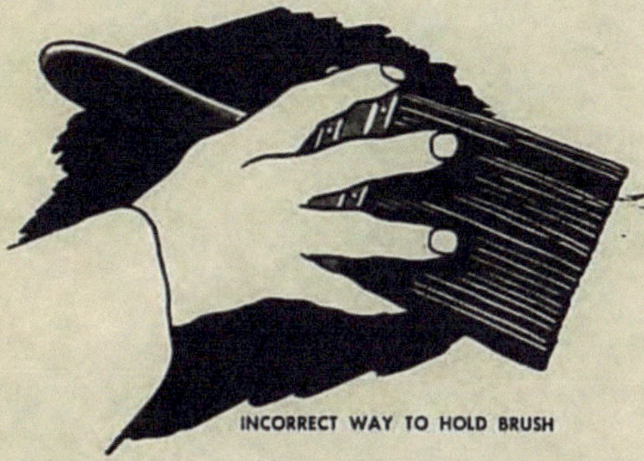

Fig. 8. Fingers are kept from bristles.

cause the bristles to spread, wear the corners down and spoil the shape and efficiency of the brush. When using an oval brush, don't let it turn in your hand. Continuous rotation of the brush will cause the bristles to wear into a pointed shape and become useless. Don't poke oversized brushes into corners and around moldings and pipes. Use a smaller brush such as an oval sash brush that will fit into such odd places.

Before you start to apply paint to the surface, work the paint well into the brush. This is done by holding the mixing paddle across the top of the paint container, dipping the brush into the paint, then wiping the flat sides of the brush clean across the edge of the paddle. Do this several times so that you can be sure that the brush is filled with paint. Do not wipe the brush across the top edge of the paint container which is a curved surface and will tend to distort the shape of the brush.

When applying paint to a surface, dip only about ½ the bristles into the paint. Tap it lightly against the inside of the can to remove excess paint and then apply it to the surface. Any brush has a certain capacity. If you overcharge it, paint will drop all around the work particularly when working overhead.

Hold the brush at right angles to the surface being painted with the end of the brush just touching the surface. Lift it clear of the surface before starting the return stroke. If the brush is held obliquely and is not lifted, the finished job will be uneven, showing lap spots and giving a "daubed" appearance. Also a brush that is held at too great an angle will wear down at the ends. A brush that is used properly will wear down evenly all around.

Since paint films tend to draw thin on corners and edges, coat the edges and sharp corners as a preliminary to applying the overall coat. Apply the overall coat immediately after this preliminary coat and without waiting for it to dry. The preliminary coating of edges and corners is not counted as a paint coat but is considered a part of each regular coat.

When applying the overall coat first "lay on," then "lay off." "Laying on" means applying the paint first in long horizontal strokes. "Laying off" means crossing the first strokes by working up and down. This method, shown in Figure 9, insures even distribution and complete coverage with a minimum amount of paint. A good rule is to "lay on" the paint the shortest distance across the area and "lay off" the longest distance. When painting bulkheads, or any vertical surface,

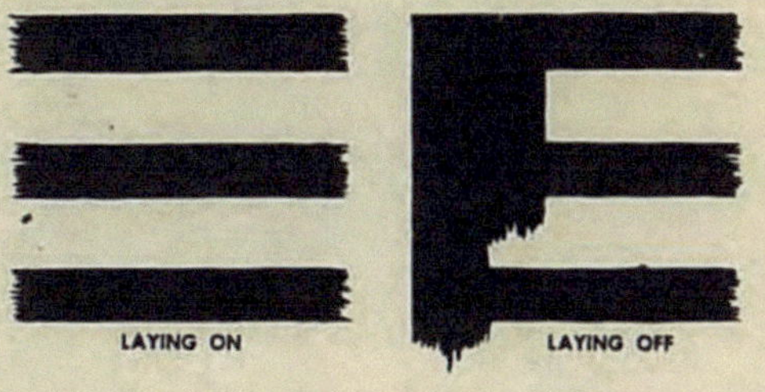

Fig. 9. Proper method of applying paint.

"lay on" in horizontal strokes and "lay off" vertically.

When two men work together on a bulkhead, the top stretch man should start at the bottom and "lay off" upward. This method results in the meeting of the wet edges which will blend into a perfect finish without showing where each man worked.

To avoid brush marks when finishing up a square, use strokes directed toward the last square finished, gradually lifting the brush near the end of the stroke while it is still in motion. Every time the brush touches the wet surface at the beginning of a stroke it leaves a mark. Therefore, always end up brushing backwards toward the area already painted.

Always paint overheads first, then bulkheads, then decks. All drippings should be wiped up as you go along. Drops of paint are too thick to dry out properly and, when painted over, become soft spots in the finished coat in addition to being unsightly.

Wait until the first coat is thoroughly dry before applying a second. Paint dries only when it is exposed to the air. It cannot dry properly if covered over too soon.

The paint in the can must be stirred frequently while painting. If the paint is left undisturbed the pigment will settle to the bottom and you will be painting with oil.

1–7–5, CARE OF BRUSHES AFTER USE

Always clean a brush immediately after use. Use plenty of the thinner prescribed for the paint in which the brush was used. See that the thinner is worked well into the brush. Squeeze out as much thinner as possible and repeat the process two or three times. Finally, rinse in clean thinner. Brushes used in emulsion paints should be thoroughly washed in soap and water. Do not throw away solvents and thinners used for cleaning brushes. Store them in *closed* containers for future brush cleaning. After setting in storage, particles of paint will settle to the bottom and the clear portion of the liquid can be poured off and will be clean enough for future brush cleaning.

After cleaning, brushes in daily use should be hung in a slotted, galvanized tank designed so that each slot holds each brush in the proper solution a height of about an inch off the bottom of the tank to prevent distortion of the bristles. The interior of the tank should be compartmented so as to contain a half and half solution of varnish and paint thinner for varnish brushes, a lacquer-alcohol solution for lacquer and shellac brushes, and linseed oil for paint brushes. See Figure 10. Keep the lid closed to prevent evaporation of solvents and reduce fire hazard. Never leave

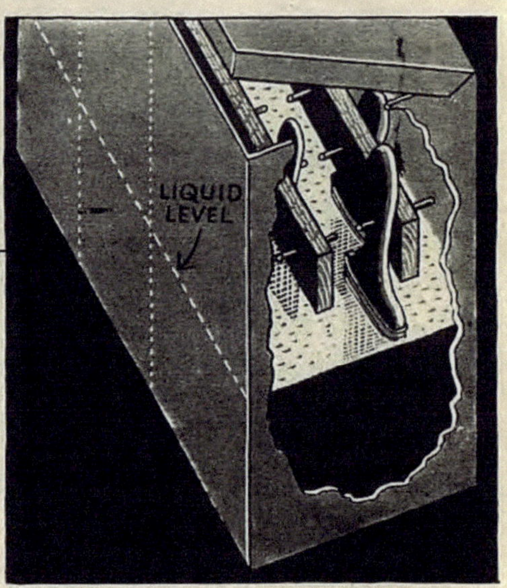

Fig. 10. Multiple brush keeper.

brushes standing on their bristles or in water.

If the regular galvanized tank is not available, a makeshift arrangement can be used as in Figure 11. Drill a small hole through the brush handle such that a stiff wire laid across an empty paint container will support the brush at the proper height off the bottom.

A brush that has become hard with paint can be cleaned. Soak it in linseed oil for about 24 hours, then in paint thinner for the same period. A putty knife and wire brush will prove useful in scraping off loosened paint. Repeated soaking in paint thinner with scraping and cloth wiping should result in a workable brush. A brush that has been reclaimed cannot be expected to give the performance of a well cared for brush. However, it can be used for applying wood preserva-

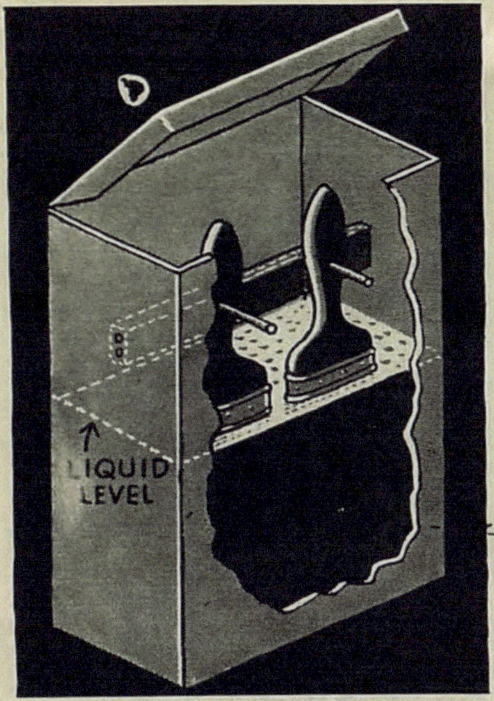

Fig. 11. Small brush keeper.

tive, painting masonry, applying paint removers, and other rough jobs which would ruin a good brush.

Before starting to use a brush for painting, drain the oil from it. Wipe the bristles clean and wash the brush in the paint thinner recommended for the paint to be used. You will find the bristles pliable and the brush easy to work with.

CAUTION: After using a brush, never leave it in an open can of paint or exposed to the air. Good brushes are expensive and hard to get. Clean them immediately after use and store them properly.

1–7–6, STRIPING

A craftsman-like job of striping with a brush is easy to accomplish by either of two methods. Both involve the use of masking tape which can be attached to almost any surface and which can be easily removed without damage to the surface.

The first method applies to surfaces which are finished and the striping is the only painting to be done. Mark the position and width of the stripe and apply masking tape to both sides of the stripe. The striping color is then brushed on. When the paint has set, remove the masking tape.

The second method applies to surfaces which are to receive a complete paint job. Mark the position of the stripe and apply the striping color overlapping the boundary marks slightly. After the striping color has dried, apply masking tape to the exact area of the stripe. Apply the finish coat to the entire area. When dry, remove the masking tape to expose the stripe.

When removing masking tape from a surface, pull it off somewhat diagonally and back upon itself. Do not pull the tape directly away from the surface at right angles.

1–7–7, SKIN IRRITATIONS DUE TO PAINTING

Paints, solvents and thinners are toxic to the skin. The degree of toxicity depends upon the person and the material being used. Persons not accustomed to continual contact with paint materials should wear clothing which protects their arms and should rub a skin lubricant or cold cream into the hands and face before painting or paint mixing.

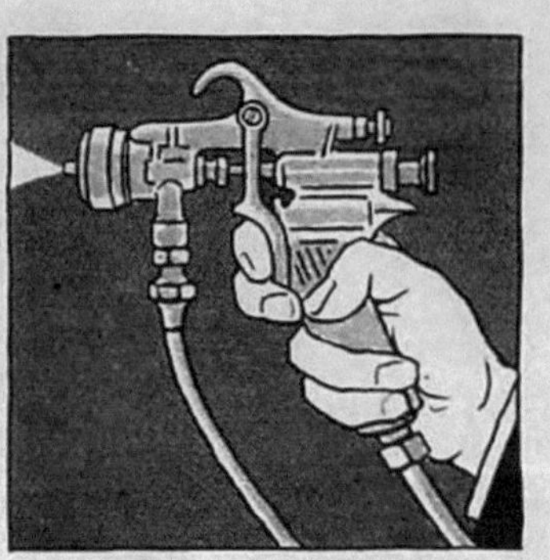

SECTION 1-8

SPRAY PAINTING

Spray painting is widely used throughout the service. This method of application can save many man-hours of labor as compared to brush painting. It is particularly well adapted to painting overheads and areas where a large number of interferences are encountered, such as aboard ship. The principal problem in spray painting is masking to prevent overspray on adjacent areas of contrasting color. Exterior spray painting on windy days will result in larger paint consumption per sq. ft. of surface. The following paragraphs describe paint spray equipment and its use.

1-8-1, PAINT SPRAY EQUIPMENT

A spray painting outfit consists of an Air Compressor to supply air power to the Spray Gun; an Air Transformer which filters out dust and dirt, condenses moisture and oil, and regulates the flow of air; Hoses to conduct both air and paint to the Spray Gun; a Paint Container, either Cup or Tank, to contain the material being applied and meter it out to the spray gun; and finally the Spray Gun itself.

(A) TYPES OF SPRAY EQUIPMENT

Spray equipment is commonly classified according to the type of container used, i.e., it is usually called a Tank Type or a Cup Type. This classification is an indication of the size job the equipment is capable of doing but it gives no information about the Spray Gun

Fig. 12. Separate container type.

Fig. 13. Attached container type.

which is, after all, the heart of the spraying operation, and, therefore, much more important than the container. See Figures 12 and 13.

Three factors determine the type of spray gun.

(a) Method of air supply pressure regulation, i.e., bleeder and non-bleeder.

(b) Method of atomization, i.e., internal mix or external mix.

(c) Method of fluid feed, i.e., gravity suction or pressure.

(B) BLEEDER AND NON-BLEEDER GUNS

Bleeder guns (Figure 14) allow air to leak from some part of the gun when the Trigger is released. They are used with air compressing outfits having no pressure controlling device to eliminate excessive pressure build-up in the hose. Bleeder guns can also be used with compressor outfits having controlling devices. Non-bleeder guns (Figure 15) must always be used with air compressing outfits which have pressure controlling devices, since releasing the Trigger shuts off the flow of air and liquid completely. Air pressure is usually automatically regulated either by the compressor or by a pressure regulator on the tank.

(C) INTERNAL AND EXTERNAL MIX GUNS

External mix guns (Figure 16) mix the air and material outside the air cap.

Internal mix guns (Figure 17) mix the air and material inside the air cap.

(D) PRESSURE SUCTION AND GRAVITY FEED

A Pressure Fed Gun (Figure 18) uses air pressure on the fluid to force it from the container to the gun. Either cup or tank containers are used. Pressure feed gives better results than suction or gravity feed since better control of air to fluid ratio can be obtained through separate controls for each. Either external mix or internal mix guns may be used with pressure feed and they may be of the bleeder or non-bleeder type depending upon whether pressure control is available.

Suction and Gravity Fed Guns are essentially the same, the difference being whether the Container is above or below the gun. Suction and Gravity Feed Containers (Figures 19 and 20) are usually attached to the gun and, due to weight, are necessarily small, usually one qt. or less.

(E) STANDARD PORTABLE PAINT AND VARNISH SPRAY OUTFIT, TYPE A

This spray outfit is available through Catalog of Navy Material, Class 40, and is intended

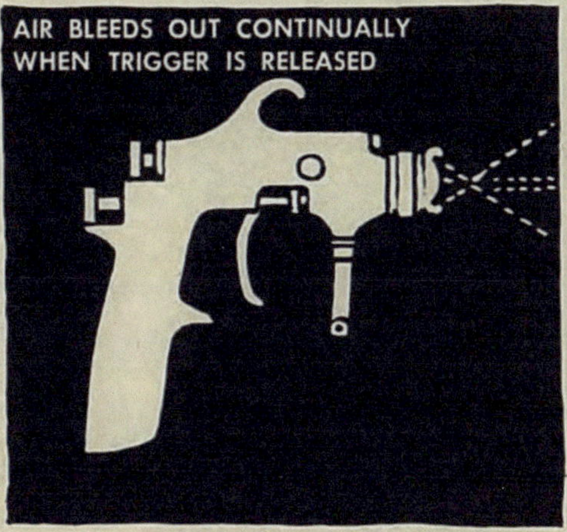

Fig. 14. Bleeder type.

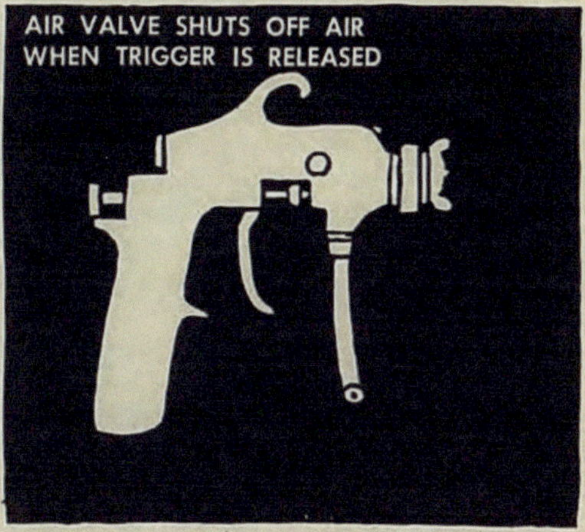

Fig. 15. Non-bleeder type.

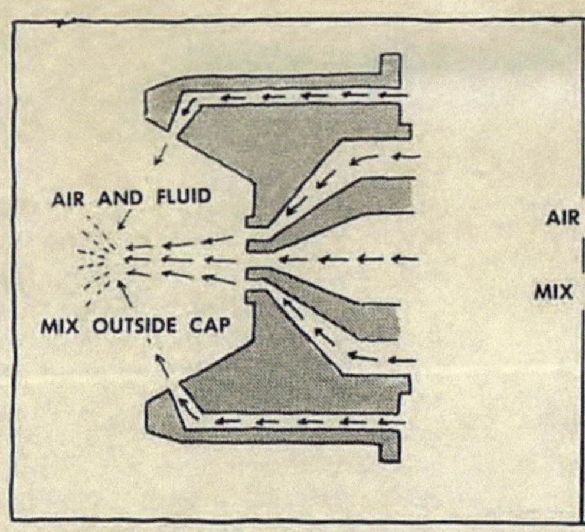

Fig. 16. External mix air cap.

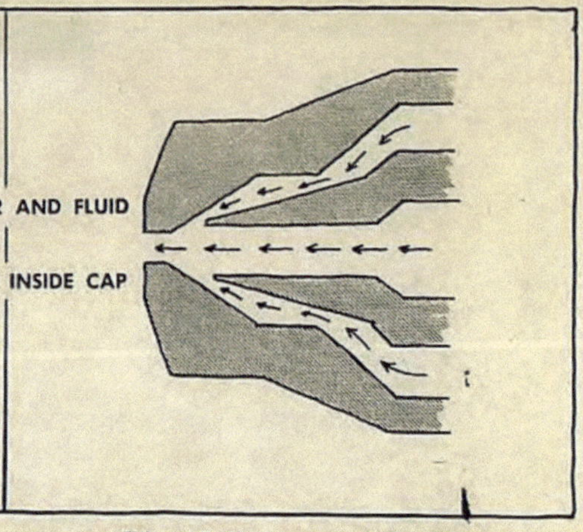

Fig. 17. Internal mix air cap.

for general use where a source of compressed air is available. The outfit includes the following components:

1 Spray Gun complete with Nozzle adjustable for Round or Fan Shaped Spray, as desired, and provided with a removable 1 qt. Pressure Type Cup.

1 five gal. Pressure Tank, for removable paint containers. Tank provided with: Pressure Regulators, controlling both atomizing air pressure for gun and rate of flow of paint from tank to gun; Air Pressure Gauges, indicating atomizing pressure and tank pressure; Safety and Relief Valves; Hose Connections; 1 Removable Cover, equipped with Air Driven Agitator, to prevent settling of paint solids.

1 set of Air and Material (Paint) Hose (two 25-ft. lengths of each).

This outfit is rugged enough for use with a maximum air pressure of 100 lbs. per sq. in. and the Safety Valve is constructed to lift at that pressure. The Regulators can be quickly and easily set at any desired pressure from 5 to 90 lbs. The Removable Paint Container holds 4 gals. of paint and facilitates cleaning after use.

(F) LIGHTWEIGHT, ELECTRIC MOTOR-DRIVEN PORTABLE, TYPE B

This spray outfit is available through Catalog of Navy Material, Class 40, and is intended for miscellaneous work of a limited nature where compressed air is not readily available. The outfit includes the following components:

1 Spray Gun, complete with interchangeable Round and Fan Shaped Spray Nozzles, and Removable 1 qt. Pressure Type Cup provided with a Safety Valve.

1 Air Compressor Assembly, consisting of: An Air Cooled, Motor-Driven Compressor; Electric Motor, either AC or DC (specify which), belt or direct-connected to compressor, with Extension Cord and Plug; an Air Tank, provided with Moisture Separator, Drain Cock, and Safety Valve. The Compressor, Motor, and Air Tank are mounted on a common base, arranged for

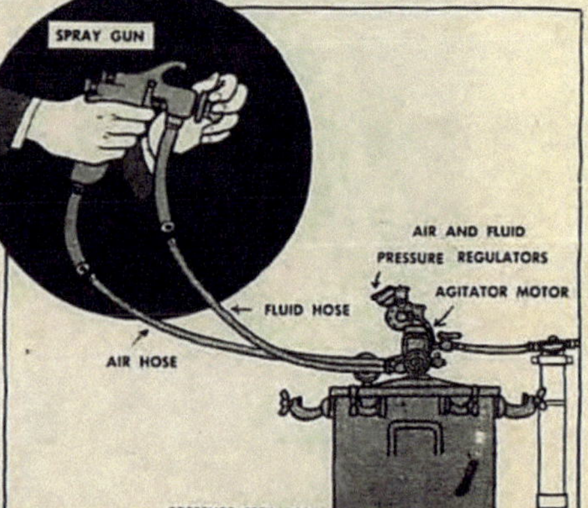

Fig. 18. Pressure feed gun.

ready transportation by one person.

1 Air Hose (15-ft. length).

The Air Compressor is of the single stage type capable of delivering sufficient air for continuous spraying. The Air Tank dampens pulsation in air pressure. This outfit is designed for a maximum air pressure of 100 lbs. per sq. in.

(G) THE PAINT SPRAY GUN

The principal parts of the Spray Gun used with both types of the aforementioned standard stock equipment are shown in Figure 21. Both the Air Valve and Fluid Valve are opened and closed by the pull and release of the Trigger. The fluid Needle Adjustment is a valve which allows more or less material through the Nozzle by controlling the movement of the Fluid Needle. The Spreader Adjustment Valve controls the air to the Spreader Horn Holes. This valve is equipped with a dial for facilitating settings. The Locking Bolt locks the removable Spray Head and Gun Body together. See Figures 22 and 23. Various spray patterns are produced by changing the dial settings. See Figure 24.

The removable Spray Head Assembly consists of the Air Cap, Fluid Tip, Fluid Needle and Spray Head Barrel. It can be quickly removed from the Spray Gun Body (Figure 25). The principal advantages of removable spray heads are:

(a) Quick change from one material or color to another. One Spray Gun Body with several Heads will answer the purpose of what otherwise may require several guns.

(b) Ease of cleaning. The removable Spray Head is the only part of the gun which comes in contact with spraying material.

(c) In case of damage to front of gun, new Gun Body is not required.

(d) An extra Spray Head can be substituted for one being repaired or cleaned.

The Air Cap is that part at the front of the gun which directs the air into the material stream, atomizing the material and forming it into a suitable spray pattern. All Air Caps are of the Internal Mix or External Mix Types and may differ for suction or gravity fed guns. See Figures 26 and 27. The Fluid Needle is actuated by the Trigger and meters the fluid through the Fluid Tip into the air stream. The amount of movement of the Fluid Needle is controlled by the Fluid

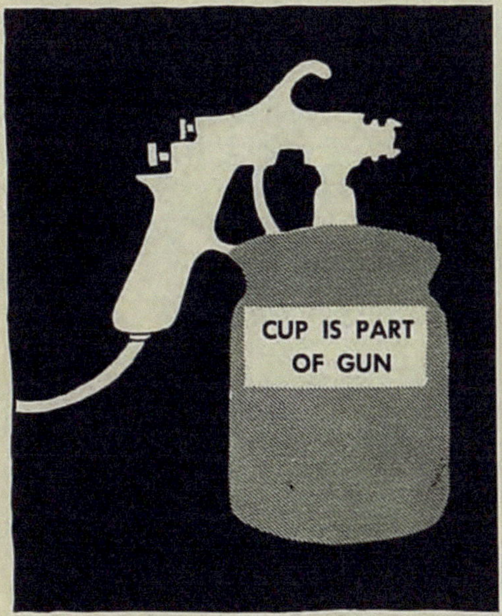

Fig. 19. Suction feed type.

Fig. 20. Gravity feed type.

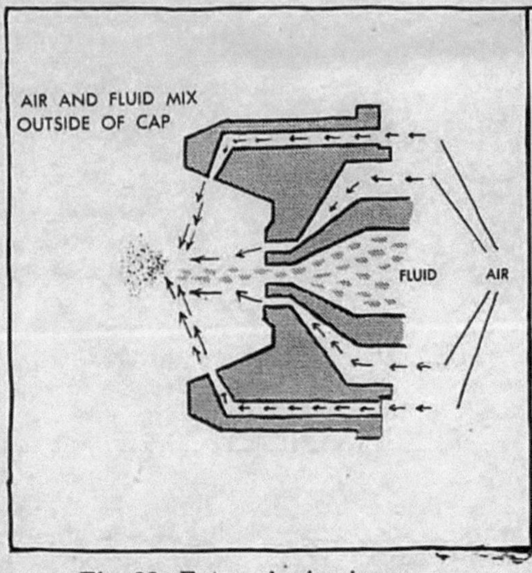

Fig. 22. External mix air cap.

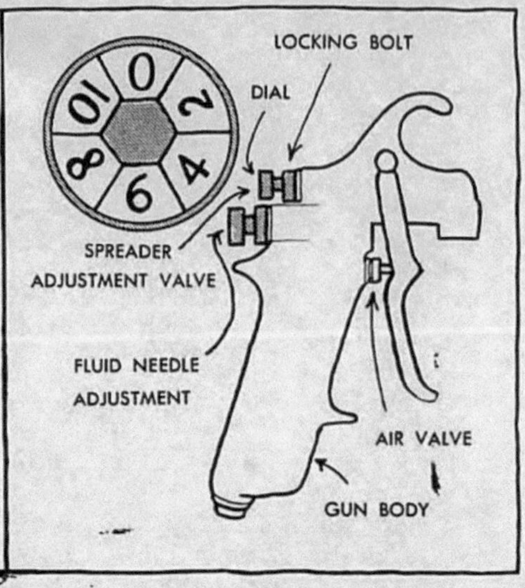

Fig. 23. Spreader dial.

Needle Adjustment. When the Trigger is released the Fluid Needle cuts off the fluid flow. The Fluid Tip fits inside the Air Cap. It meters and directs the material into the air stream. It provides a self-aligning slot for the Air Cap and equalizes the air leaving the center orifice of the cap. The term "Nozzle" refers to the opening in the Fluid Tip. Nozzle sizes are shown in Figure 28. The Nozzle size letter is stamped on the collar of the Fluid Needle and on the outer edge of the Fluid Tip. Sizes E–FF–Fx–F are generally used. Size E is furnished with standard Navy equipment.

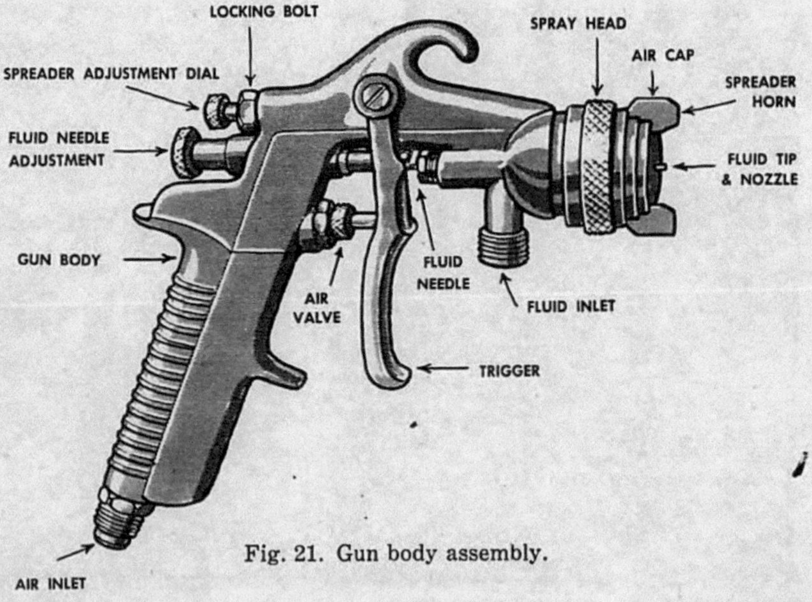

Fig. 21. Gun body assembly.

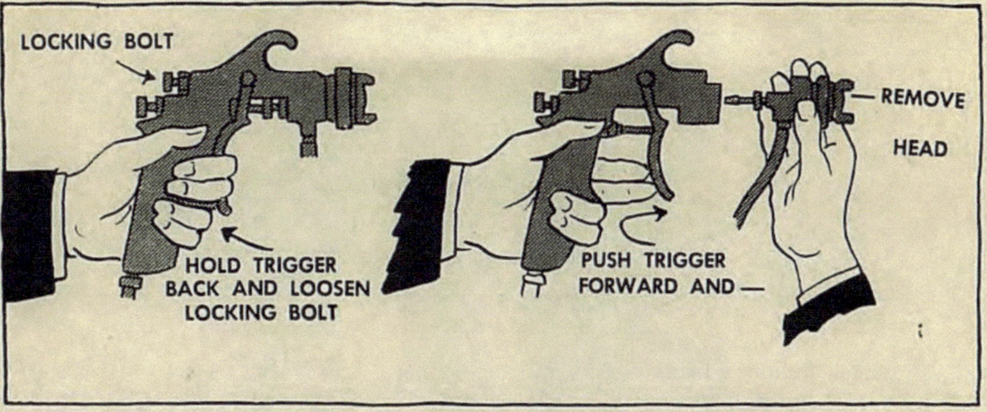

Fig. 25. Method of removing spray head from gun body.

1–8–2, OPERATION OF SPRAY EQUIPMENT

(A) MASKING

Before starting to spray, all surfaces which are not to be painted shall be masked. Large surfaces may be covered with drop cloths or a combination of newspapers and masking tape. Lapping masking tape over onto bulkhead when masking objects such as switch boxes will necessitate touch up by brush. Confine masking to the area to be masked and touching up will be eliminated.

Liquid masking is effective and easy to use on large glass windows. In some cases grease can be utilized effectively as a masking material.

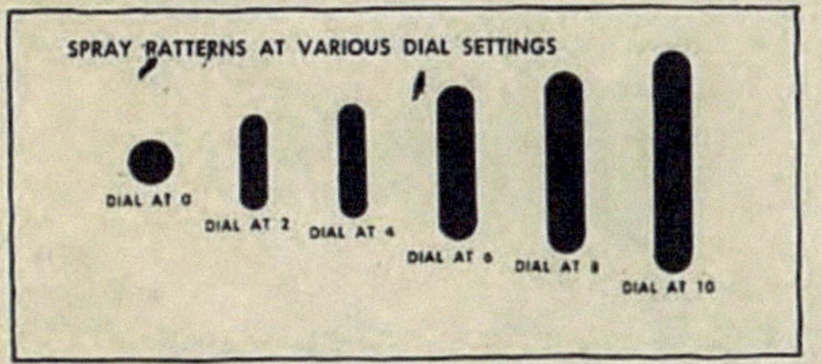

Fig. 24. Patterns at various dial settings.

(B) SPRAY PAINTING TECHNIQUE

The handling of a spray gun is best learned by practice. However, a few pointers on the subject are pertinent. At the beginning, make sure the spray gun is thoroughly clean. Strain paint through screen wire or cheese cloth as it is poured into the spray container. Air pressure to the gun should be

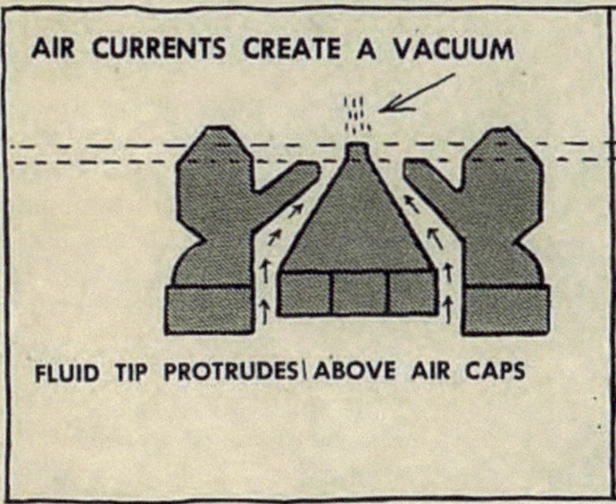

Fig. 26. Suction or gravity feed cap.

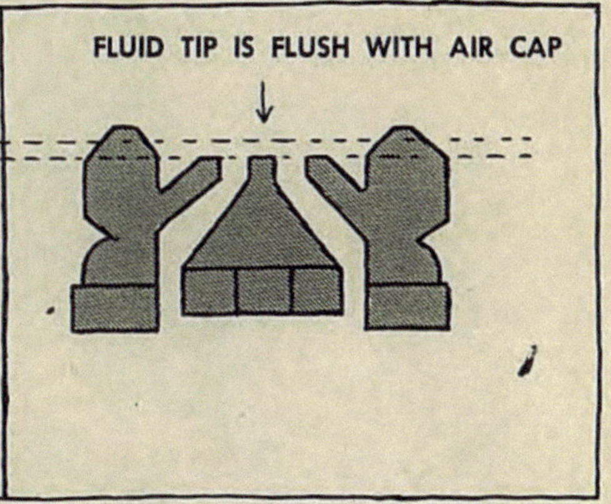

Fig. 27. Pressure feed cap.

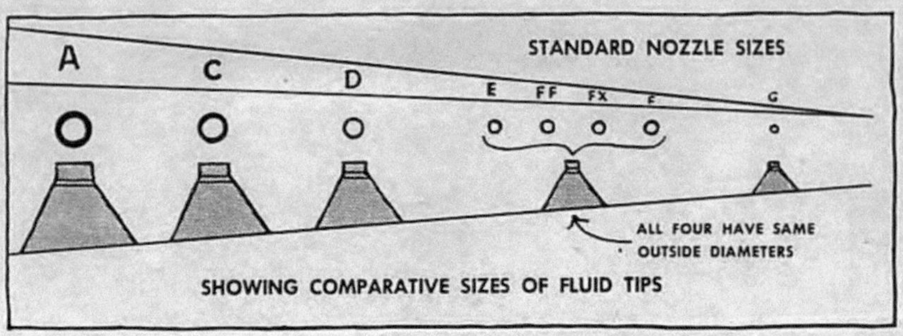

Fig. 28. Fluid tips come in different sizes.

adjusted to approximately 45 or 50 lbs. for lacquer and thin liquids and 60 to 70 lbs. for paint and enamel. Initially, adjust fluid pressure to 10 or 15 lbs. for lacquer and thin liquids and 15 to 20 lbs. for paint and enamel.

Having made the initial adjustments to air and liquid pressures, final adjustments are made by observation of spray patterns. Normal spray patterns will appear as illustrated in Figures 29 and 30.

Imperfect spray patterns are due to one of two basic troubles, i.e., clogging of passages or improper balancing of air and fluid pressures.

Imperfect patterns due to clogged passages will take the following forms:

1. Heavy Top Pattern (Figure 31). Due to:
   - (a) Horn Holes partially plugged.
   - (b) Obstruction on top of Fluid Tip.
   - (c) Dirt on Air Cap Seat or Fluid Tip Seat.
2. Heavy Bottom Pattern (Figure 32). Due to:
   - (a) Horn Holes partially plugged.
   - (b) Obstruction on bottom side of Fluid Tip.
   - (c) Dirt on Air Cap Seat or Fluid Tip Seat.
3. Heavy Rightside Pattern (Figure 33). Due to:
   - (a) Right side of Horn Holes partially plugged.
   - (b) Dirt on right side of Fluid Tip.
4. Heavy Left Side Pattern (Figure 34). Due to:
   - (a) Left side Horn Holes partially plugged.
   - (b) Dirt on left side of Fluid Tip.

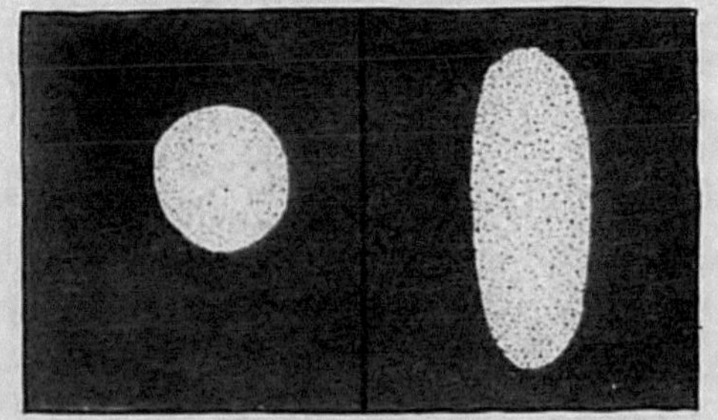

Figs. 29, 30. Normal spray patterns.

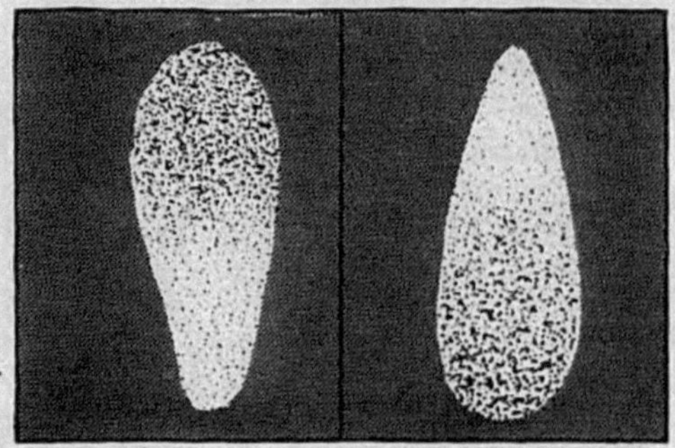

Figs. 31, 32. Imperfect spray patterns.

It should be remembered that the above troubles could be caused by burrs inside orifices. For this reason, care should be exercised in cleaning so as not to damage or alter the shape of orifices. Always use a piece of wood such as a toothpick for this job. Never use a wire or other hard instrument.

Imperfect spray patterns due to improper balance of air and fluid pressures will take one of the following forms:

1. Heavy Centered Pattern (Figure 35). Due to:
   - (a) Too low a setting of Spreader Adjustment Valve.

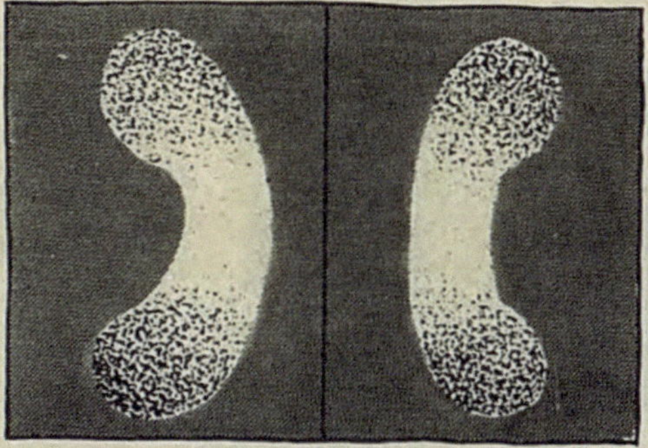

Figs. 33, 34. Imperfect spray patterns.

(b) Too high fluid pressure.
(c) Viscosity of material too great.

It will be noted that adjusting (a) or (b) will accomplish essentially the same thing, i.e., *increase* the pressure differential between air and fluid pressures. Lowering fluid pressure will decrease the rate of flow of the fluid thereby decreasing the rate of application. Increasing air pressure to the Horn Holes will increase the atomization rate. The choice of which adjustment to make will depend upon the rate of application desired. It is important that the manufacturer's recommendations be followed both as to type of thinner and its proportion in the spraying material when making adjustment(c). To change a recommended viscosity merely to make it more sprayable is to invite trouble.

2. Split Spray Pattern (Figure 36). Due to:
   (a) Too high setting of the Spreader Adjustment Valve.
   (b) Too low fluid pressure.
   (c) Material too thin.

Here again adjusting (a) or (b) will accomplish the same purpose, i.e., *decrease* the air and fluid pressure differential. The choice of which adjustment to make again depends upon the application rate desired. Lowering air pressure to Horn Holes by turning the Spreader Adjustment Valve to a lower setting will decrease the atomization rate while increasing fluid pressure will increase the rate of application. Material should never be thinned beyond the manufacturer's recommended proportioning. However, if condition (c) exists, add enough paint as received to the thinned material to restore proper viscosity.

Another common defect in spraying is "Mist" or "Fog," (Figure 37). Due to:

1. Over atomization, caused by—
   (a) Air Pressure to Gun too high.
   (b) Fluid Pressure too low.

Again, the air to fluid balance is off. Reducing the differential as for the split spray pattern above will eliminate the mist unless it is caused by:

2. Improper use of the gun.
   (a) Incorrect stroking.
   (b) Gun held too far from surface.

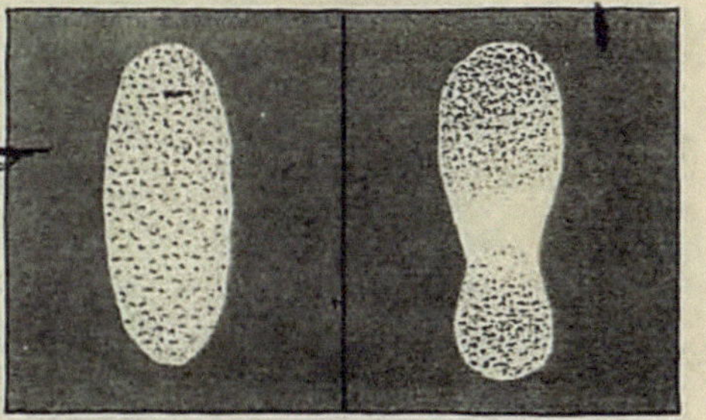

Figs. 35, 36. Imperfect spray patterns.

Correcting these faults lies with the operator of the gun. Carefully study Figure 38 which illustrates the correct method of stroking the gun and the correct distance between

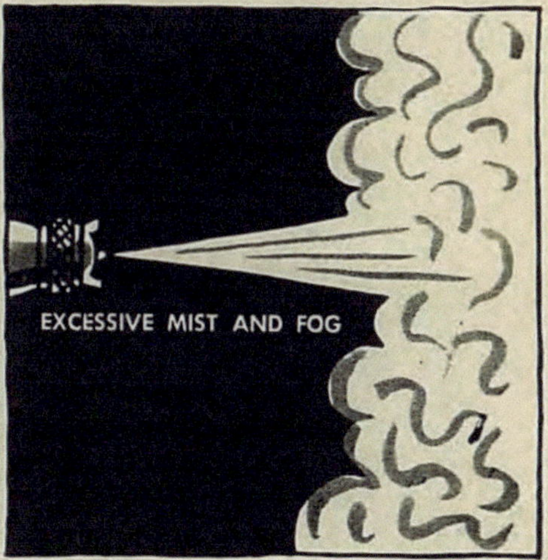

Fig. 37. Improper mist or fog.

Fig. 38. Proper method of spraying.

gun and surface. Improper stroking is shown in Figure 39.

(C) PAINT APPLICATION

A good finished paint film depends as much on proper handling of the gun as it does on correct adjustments of the equipment. Hold the spray gun from 6 to 10 in. away from the surface being painted. Begin the stroke before the Trigger is pulled and release the trigger before the end of the stroke. This prevents "piling up" paint at the beginning and end of each stroke. Always keep the gun at right angles to the surface—never swing the stroke in an arc which would put too much paint in the center of the stroke and too little on the ends. The objective is to obtain adequate coverage with an even application and without runs and sags.

Before applying the overall coat, spray small protruding parts such as hatch casings, door casings, rivet heads and small built-in parts. This will insure adequate coverage and will eliminate bridging and touching up later.

When spraying corners, first spray up to within 1 or 2 in. of the corner; then, turn the gun on its side and hold gun at an angle to corner and, starting at the top, spray downward along the edge so that both sides of the corners are sprayed at once. See Figure 40.

(D) SPRAYED FILM DEFECTS

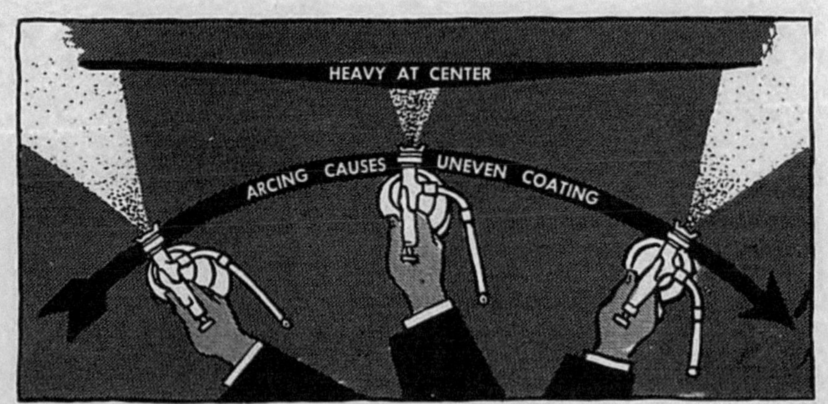

Fig. 39. Improper method of spraying.

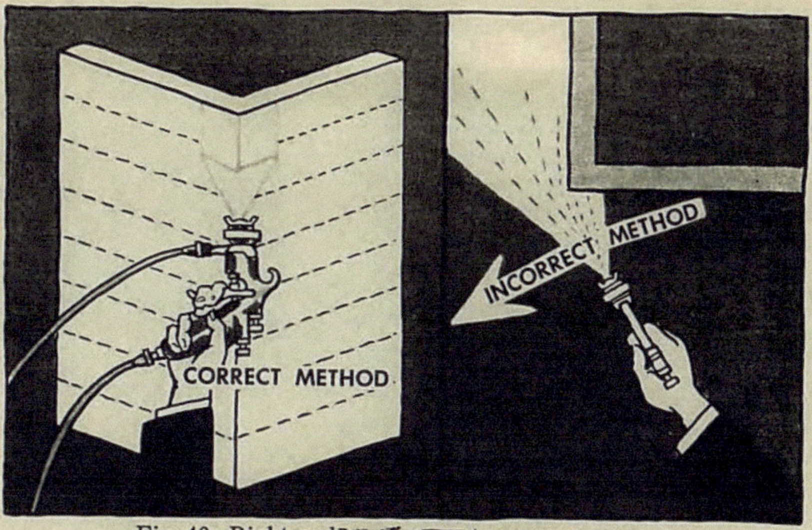

Fig. 40. Right and wrong way to spray corners.

The most common defects in spray paint coatings are: "Orange Peel," Runs and Sags, Pinholes, Blushing, and Peeling. Orange Peel is a general term used to describe a painted surface which has dried with a pebbled texture resembling an orange peel. This may be caused by the use of improper thinners, insufficient atomization, holding the gun too far away from (or too close to) the surface, improper mixture of materials, or low humidity. See Figure 41.

Runs usually result from using material which is too thin. Sags result from too much material. They can also be caused by allowing too big a lap in spraying strokes and by poor adjustment of the Spray Gun or Pressure Tank. Dirty or partially clogged passages for air or fluid will also cause uneven distribution.

Pinholes may be caused by the presence of water or excessive thinner in the paint; or by too heavy application of quick-drying paint. In either case, small bubbles form which break in drying, leaving small holes.

Blushing resembles a powdering of the paint. The cellulose material in the paint separates from its solvent and returns to its original powder form. Water is usually the cause of this—either moisture on the sprayed surface or excessive moisture in the air. When blushing occurs the entire defective area of the coating must be removed because the moisture is trapped within the material and will remain there unless the coating is removed.

Peeling is almost invariably due to careless workmanship in not cleaning the surface properly. Before any spraying is attempted the surface must be absolutely clean.

1–8–3, CARE OF SPRAY EQUIPMENT AFTER USE

Spray equipment should be cleaned immediately after use. First, back off the Fluid Needle Adjusting Screw and release the fluid pressure from the Pressure Tank by means of the Release Valve. Hold a cloth over the Air Cap and pull the Trigger. This forces the paint back into the Tank. Remove the

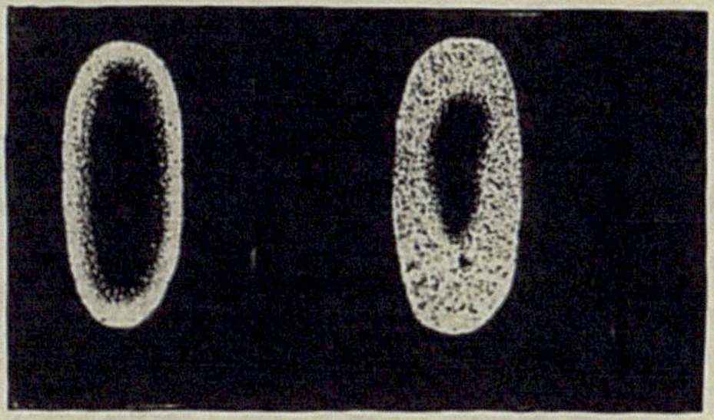

Fig. 41. Good and poor spray patterns.

Fluid Hose from the Gun and run the thinner prescribed for the paint through it. If Air Holes are clogged, use a toothpick to clean them out. Put all clean parts back in place, and the Gun is ready for use again.

To clean an Attached-Container Type Gun, first remove the Container. Place the Pick-Up Tube over a waste bucket. Hold a cloth over the Air Cap and pull the Trigger clearing the gun of paint. Empty the Container and spray in the usual way. This process cleans out all passageways. Clean the Air Cap by soaking in a solvent, then replace. Never immerse the entire Spray Gun in thinner or solvent because all lubricating oil and grease will be removed and the Packing will subsequently dry out hard. See Figure 42.

Fig. 42. Steps in cleaning a container type gun.

Spray guns require an occasional lubrication. This is a small job but one that is well worth while. The Fluid Needle Packing should be removed and softened with oil. The Fluid Needle should be coated with grease or petroleum. All moving and rubbing parts should get a few drops of light oil. Most Spray Guns are provided with Oil Holes for these parts.

For cleaning or repair, or when changing the color of paint used, the Spray Head may have to be changed. Remove the Gun from the Liquid and Air Hose Lines. Hold the Gun in the left hand, pull the Trigger all the way back and loosen the Locking Bolt with a wrench. Push the Trigger forward as far as possible. Then pull the Spray Head forward. To replace the Spray Head, push the Trigger forward and insert the Spray Head. Then hold the Trigger back and tighten the Locking Bolt. See Figure 43.

1–8–4, SAFETY PRECAUTIONS FOR SPRAYING

Fire and Explosion precautions: All paints, formulated solvents and thinners other than water are volatile and flammable and can create an explosive atmosphere in the work area when sprayed. The possibility of an explosion occurring varies in accordance with the flash point of the material and the temperature of the compartment in which the spraying is being done. The hazards for out-

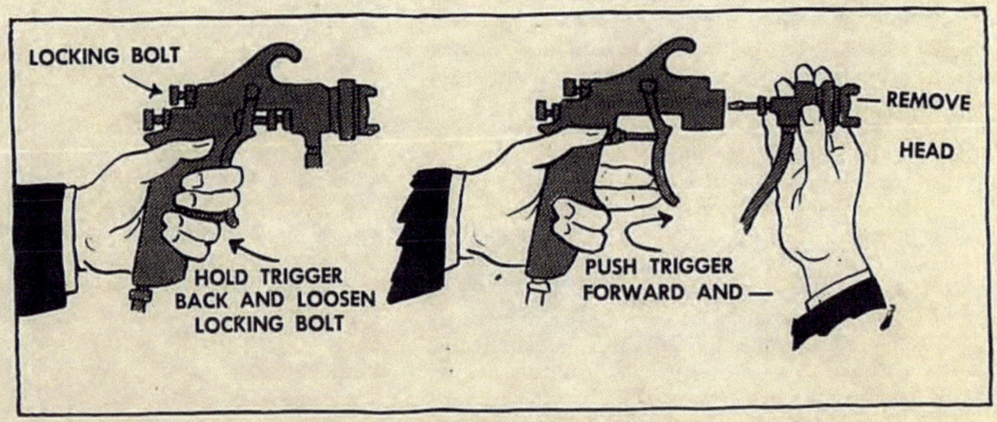

Fig. 43. Removing the spray head.

side painting are less severe and are essentially limited to the fire hazard in the immediate vicinity of the spraying.

In any case where interior spray painting is being done with low flash point materials (flash point below 100° F.), ventilation shall be provided to keep vapor concentrations below the explosive range in all parts of the compartment. This will require roughly 1500 cu. ft. of fresh air per minute for each spray operator. Blower exhausts and intakes must be sufficiently separated to guard against recirculating fumes. The vapors of all paint solvents and thinners, other than water, are heavier than air and tend to accumulate in the lowest portion of a compartment. Ventilation must reach all parts of the tanks or compartments to avoid such accumulations. Blowers should be located outside the area being sprayed. If they are located within the hazardous area, they must be powered by air or explosion-proof motors in safe working condition. Ventilation must continue after spraying is completed until all explosive fumes are purged from the compartment.

No smoking, welding, burning, or other flame or spark-producing operations such as chipping, grinding, etc., shall be allowed within a compartment while spraying is going on or for at least one hour after spraying has ceased.

Health precautions: Spray painting breaks up the paint into a fine spray in which fumes, pigment and vehicle are released in the air. Breathing these fumes and particles or otherwise absorbing them into the body can cause injury. Always wear a respirator when spraying or when in the vicinity of spray work. The most common types are:

(a) The filter respirator, equipped with filter pads, can be used in spraying, grinding, or dust blowing work when dust and fumes are not too severe. See Figure 44.

Fig. 44. A filter respirator.

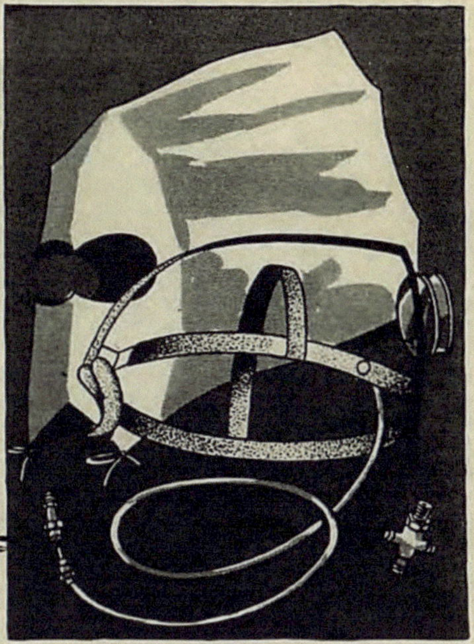

Fig. 45. A head respirator.

(b) The cartridge respirator, designed for more severe conditions than the filter respirator, is equipped with a large purifying cartridge made of chemically treated charcoal, in addition to a filtering pad.

(c) The dust respirator contains a replaceable cartridge and is easy to wear due to its light weight. This respirator may be used for surface preparation operations but it is relatively ineffective for paint spraying and shall not be used for this purpose.

(d) The air supply respirator provides complete protection when working in holds and tanks. This type is supplied with fresh air by a compressed air line, purified by a charcoal cartridge, and then fed to the breathing compartment of the respirator.

(e) The hood respirator consists of a flame-proof cloth hood, a fiber headgear, a metal eye-piece, an air filter, and an air hose. The neck cloth at the bottom of the hood ties snugly around the neck. The opening in front of the hood is the only outlet for a constant flow of air entering through a hose at back of the hood. Foul air cannot enter because pressure inside the hood is slightly greater than pressure outside. See Figure 45.

If air for supplied-air masks or hoods is taken from yard air lines, its purity must be safeguarded by a carbon monoxide alarm on each compressor and no anti-freeze shall be used in the lines.

To reduce the danger of absorbing chemicals through the skin when spray painting, keep well covered. Apply a light coating of petroleum jelly on exposed areas of skin such as the hands. This provides protection and will aid in cleaning up afterwards. Wear goggles when spray painting. Getting paint in the eyes can cause a great deal of suffering and may permanently impair eyesight.

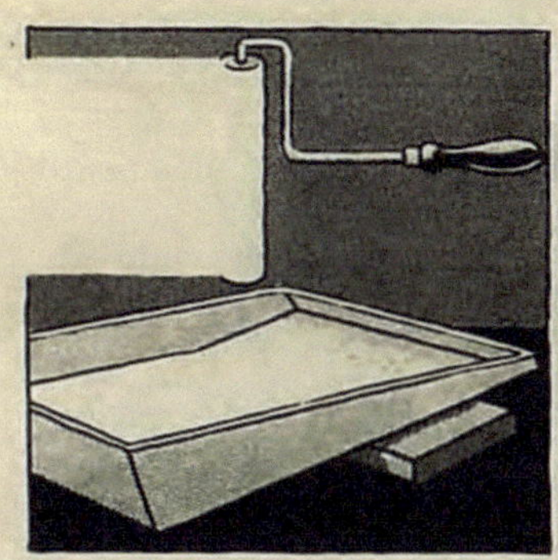

SECTION 1–9

ROLLER COATING

1–9–1, TYPES OF ROLLER COATERS

Roller coating is another method of applying paint which has proved practical. There are three types of hand roller coaters: dip, can, and pressure. All three types consist essentially of a cylinder covered with lamb's wool, flat knit wool or cotton, or short pile carpet material. The cylinder rolls about an axle through its center which is attached at both ends to a two-pronged handle of the necessary length.

(A) DIP TYPE ROLLER COATERS

The dip type requires a flat pan or container into which the roller can be dipped to receive a charge of paint which is simply rolled onto the surface to be painted. This process is repeated somewhat like recharging a brush in brush painting.

(B) CAN TYPE ROLLER COATERS

The can type is essentially the same as the dip type except that the cylinder is a perforated can. Paint is poured into the can and is sucked out through the perforations by the sponge action of the lamb's wool, as it is rolled over a surface.

(C) PRESSURE TYPE ROLLER COATERS

The pressure type is essentially a can type with provisions made for forcing paint from a separate container through a hose into the can. The pressure type containers hold from 2 to 5 gals. and are provided with a hand pump by which air pressure is maintained above the liquid. This type is well suited to large surfaces, since it eliminates the frequent dipping and refilling required by the dip and can type.

1–9–2, USE AND CARE OF ROLLER COATERS

Roller coaters eliminate much of the work involved in rigging staging and scaffolding. Inaccessible areas can be satisfactorily painted by attaching long handles to the roller coaters. This method of paint application has been found very effective in painting the hull of a vessel since all work can be accomplished rapidly from a raft without the use of staging. They are particularly well suited to interior painting at shore stations.

The use of roller coaters is limited, however, to plain surfaces. They cannot be used on irregular surfaces such as piping, beams,

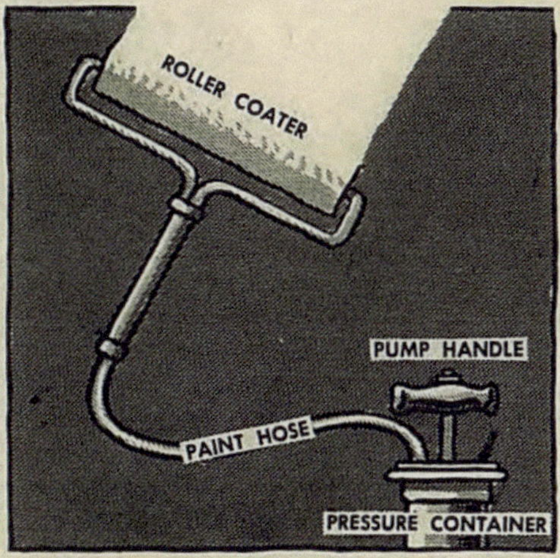

Fig. 46. Pressure type roller coater.

rails, shingles, drop or lap siding and the like.

Roller coaters should be cleaned immediately after use by washing thoroughly in the thinner recommended for the paint which was used. After cleaning with thinner, the cylinder cover should be washed thoroughly with soap and water, rinsed and dried on the roller to prevent shrinkage. Combing lamb's wool while damp will prevent matting. Figure 46 shows pressure type roller coater with connecting pump and tank.

SECTION 1–10

PAINT FAILURES

All paints can fail, sometimes badly, due chiefly to four reasons: improper preparation of the surface prior to painting; improper application of the material—which includes selection of the wrong type of coating for a given surface; use of poor quality paint—not a problem if standard Coast Guard or Navy paints are used; and neglecting to determine if a surface is suitable for the application of paint—for example the moisture content of wood may be too high. The following is a list of the more common paint film defects and failures with notes on how to avoid or remedy the condition.

1–10–1, CHALKING

Chalking is caused by the action of sunlight on a paint film. A layer of oil is present on the surface of a new paint film which is destroyed by the ultraviolet rays in sunlight. This leaves loosely bound particles of paint on the surface which wash off during rain. See Figure 47. Slight chalking is desirable because it makes the surface self cleaning. Heavy chalking is highly undesirable as it washes away rapidly leaving the surface unprotected. Heavy chalking usually occurs on wood surfaces which were not properly primed and did not receive an adequate number of finish coats. The porous wood sucks the oil out of the paint, thus promoting early, heavy chalking. This condition can be avoided by properly applying a primer-sealer followed by a sufficient number of finish coats (usually two). To remedy the condition, remove the heavy chalk by scrubbing or wire-brushing, then repaint.

Chalking which occurs on metal surfaces is generally due to excessive pigment to vehicle ratio and is found in some fire retardant paints. Newer formulations achieve the same degree of fire retarding without the undesirable chalking characteristic. Scrub the surface and recoat.

1–10–2, WASHING

Paint failure of the type known as washing occurs when the outer layer of a paint film washes off. This is evidenced by accumulations of dry pigment at the foot of vertical surfaces such as walls or columns. This washing away of the paint is caused by the formation of water soluble inorganic and organic compounds. The formation of these compounds is due to improper drying of the paint during cold, damp weather or to the presence of sulphur dioxide in the air which reacts with paint. Prolonged moderate temperatures and light are necessary for complete drying of the paint film. Painting should be done during that time of the year when maximum sunlight and fairly constant warm temperatures prevail.

1–10–3, ALLIGATORING

Alligatoring and checking of a paint film exists when the outer layer of paint cracks and presents a pattern similar to alligator leather. This condition occurs when relatively hard finishing coats are applied over relatively soft priming or underlying coats. Undercoats which are too rich in oil, or which have been allowed insufficient drying time, cause this softness. Expansion and contraction of the painted surface where paint coats have unequal flexibility due to soft undercoats and hard outer coats causes alligatoring and checking. To avoid this failure, allow undercoats to dry sufficiently hard before applying the next coat. Undercoat materials should dry harder than top coat materials. Do not add too much oil to priming materials. To remedy this condition, scrape and wire-brush to remove loose scale, sand to smoothness and repaint.

1–10–4, FLAKING, CRACKING AND SCALING

Flaking, cracking and scaling is a condition caused by either painting over wood which is saturated with water (over 25 per cent moisture) or by painting over pitch pockets and knots in pine wood. In the former case, adhesion is poor and shrinkage of the wood upon drying out causes the paint film to crack and flake off. Wait until the weather is dry—at least a week after a rainy period—before attempting to paint exterior wood. In the latter case, resin in the pitch bleeds into the paint making the film too hard and brittle. When this occurs, the paint film lacks the flexibility necessary to withstand expansion and contraction of the wood due to temperature changes. To prevent this condition, seal pitch pockets with Ready-Mixed Aluminum Paint and prime bare wood surfaces with Exterior Wood White Primer.

1–10–5, BLISTERING

Blistering is a term applied to a paint failure where large blisters appear in the paint film—usually the top coats lift leaving the prime coat intact. This condition is the result of moisture being absorbed by the wood. Moisture may come from steamy kitchens or bathrooms, leaks in roofing or plumbing, or leaks around window and door casings. Exposure to heat from the sun causes this moisture to migrate to the surface of the wood and collect under the paint film to form blisters. This condition can be alleviated by either reducing humidity inside the building to safe limits or preventing the moisture from reaching the siding. Humidity inside the building can best be controlled by proper ventilation, repairing leaks in walls, roofs and plumbing, and by operating humidifiers at lower settings. Moisture can be prevented from reaching siding by installing vapor barriers between plaster and insulation in new construction or by the application of moisture-resistant paint to interior sides of exterior walls and ceilings of existing buildings. After the cause has been rectified, scrape and sandpaper blistered areas, then repaint.

1–10–6, GAS DISCOLORATION

Gas discoloration sometimes occurs in white or light-tinted paint films in areas where in-

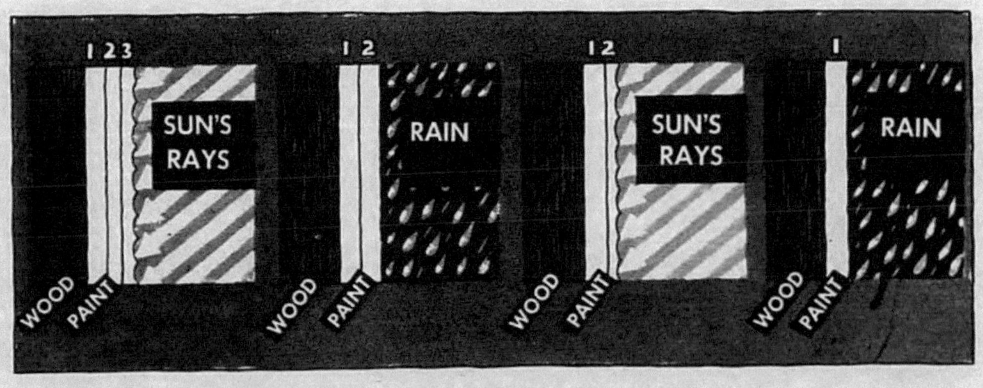

Fig. 47. Progress of chalking with outside paint.

dustrial fumes or sewer gases are prevalent. Hydrogen sulfide reacts with lead in the paint to form gray or black lead sulfide which resembles graphite. This condition is usually found on newly applied soft paint films (3 to 6 months old) which are permeable to the hydrogen sulfide gas. Affected paint films may lighten up during periods of gas-free air. If it is necessary to remove the deposits of lead sulfide immediately, sponge with hydrogen peroxide or a weak solution of acetic acid or muriatic acid, avoiding contact of acid solutions with the skin. Flush the area with fresh water to remove chemicals. Where gas discoloration persists, wash the surface and repaint using a fumeproof paint.

1–10–7, MILDEW

Mildew affects paint films oftenest in southern localities where humidity and warm temperatures are suitable for its growth. Mildew grows to produce a black (aspergillus niger) or green (penicillum glaucum) discoloration. Mildew spores in the air stick to tacky paint films and grow in colonies. Mildew does not grow any better on painted surfaces than on many other surfaces. Use a powerful magnifying glass to determine if mildew is actually present. The soiled appearance may be dirt rather than mildew. Dirt washes off much easier than mildew. Badly mildewed surfaces must be thoroughly scrubbed with soap and water before repainting. Rinse with fresh water and allow a day or two for drying. Add a prepared mildewcide to new paint and mix thoroughly. These compounds may be purchased at paint stores and very small amounts of the order of ½ oz. to 1 gal. of paint are needed. The use of mercury compounds is prohibited due to their toxicity.

1–10–8, SUCTION SPOTTING

Suction spotting is a curious light spotting which occurs usually within 3 to 6 months of application, on light gray and other light tints. This condition is caused by porous areas in the surface of wood which absorb oil from paint, leaving insufficient binder in the paint film to resist weathering. This condition is generally observed in 2 coat systems on wood. Properly applied primer-sealer with a sufficient number of finish coats (usually 2) will prevent such spotting.

1–10–9, DIRT COLLECTION

Dirt collection of a painted surface indicates that either there is excessive dirt in the atmosphere or that the paint film is not chalking properly, or both. In industrial areas where dirt content of air is high, delay new painting until time when minimum winds prevail. Addition of oil to paint cuts down the ratio of pigment to vehicle and aggravates the condition. To alleviate the condition, scrub with soap and water. Rinse with fresh water. Any new painting should be done with a chalking paint. Thin with paint thinner if necessary but do not add oil which inhibits chalking.

SECTION 1–11

SUPERVISION OF

PAINTING

1–11–1, FREQUENCY OF PAINTING

It is of paramount importance to bear in mind that the cost of painting is relatively high. For example, during the calendar year 1951, the value of paint materials used in the Coast Guard exceeded $1,000,000. The cost of application is approximately five times the cost of paint materials. When the labor is taken into account, the 1951 painting costs to the Coast Guard are in excess of $6,000,000.

(A) INTERIOR SURFACES

Interior surfaces, other than those exposed to sweating, are generally painted more frequently than necessary, the painting being more often done for appearance than for protection. It must be borne in mind that too frequent repainting results in paint film failure due to films which are too thick or to films which are incompatible.

Since interior paint films are generally renewed due to yellowing or staining rather than mechanical film failure, the interior paints which have been selected for Coast Guard use are designed to have good scrubability to permit repeated washing. Scrubbing the surface with Paint Cleaner and water will usually result in a reasonably clean, fresh appearance.

Don't waste paint materials on interior surfaces where the paint film is intact and can be cleaned. Eventually, of course, dirt and grease stains will work their way into paint films too deep for removal. When the surface becomes unsightly by reason of these stains, it must be repainted. As a check against too frequent repainting, adequate painting records should be kept and the date of painting stencilled on the surface as described in Article 1–11–7. Interior painting should normally not be done oftener than every three years.

(B) EXTERIOR SURFACES

In exterior painting the most economical practice is to establish a painting schedule. The frequency of painting should be such as to repaint prior to failure of the paint film. Exterior wood surfaces ashore normally will not need painting oftener than every three years. Exterior steel surfaces ashore and those on vessels above the main deck should not normally be painted oftener than every two years. Exteriors of hulls except boottopping should not be painted more frequently than every year. With these generalizations in mind and taking into account previous practice at individual units, each unit should establish and adhere to a painting schedule. Touch-up painting between the scheduled periods will be all that is necessary to protect the surface and impart a satisfactory appearance.

1–11–2, WEATHER CONDITIONS FOR PAINTING

Painting shall only be done under such weather conditions as will insure a dry surface and no precipitation.

In order to insure a dry surface the temperature of the surface must be above the temperature of the dew point. If the temperature of the surface to be painted is at or below the dew point temperature, condensation will occur on the surface. Visual examination of the surface for condensation cannot be relied upon since condensation, although it exists, cannot be detected until it is in an advanced stage. The dew point, of course, depends upon the relative humidity and the air temperature. Below is a table showing the temperature of the dew point corresponding to any given wet and dry-bulb temperature.

After selecting the dew point from the table compare it with the temperature of the surface obtained by taping a thermometer to the surface (in a location where it will not be damaged) for a period of five minutes. *If the temperature is at or below the dew point, do not paint.* It is best to paint only when the temperature of the surface is several degrees above the dew point.

In any case, do not paint when either the temperature of the surface or the air temperature falls to 50° F. or below. Paints dry by the combination of three totally different reactions, i.e., evaporation, oxidation, and polymerization. These reactions must proceed at a given rate with respect to one another in order to obtain a dry film with optimum characteristics. Paints are so formulated that at approximately 70° F. these three drying actions are in balance with one another. The greater the temperature difference between 70° F. and the temperature at which the paint is applied, the poorer performance the paint will give. While no definite demarcation exists between a temperature at which painting can be satisfactorily applied and a temperature at which it cannot be applied, 50° F. is generally accepted by painting authorities as the lower limit at which paint should be applied. Vinyl paints are an exception and since they dry by solvent evaporation only, they may be successfully applied by spray in temperatures as low as 5° F.

During winter, spring and fall the days are short. This delays the start of the paint-

TEMPERATURE OF THE DEW-POINT

Temperature Difference

(Dry Bulb—Wet Bulb)

Dry Bulb Temp.	1	2	3	4	5	6	7	8	10	12	14
90	89	87	86	85	83	82	81	79	76	73	70
85	84	82	81	80	78	77	75	74	71	68	64
80	79	77	76	74	73	72	70	68	65	62	58
75	74	72	71	69	68	66	64	63	59	55	51
70	69	67	65	64	62	61	59	57	53	49	44
65	63	62	60	59	57	55	53	51	47	42	37
60	58	57	55	53	51	49	47	45	40	35	
55	53	51	50	49	47	44	42	40	33		
50	48	46	44	42	40	37	34	32			
45	43	41	38	36	34	31					

To use the table enter the row corresponding to the Dry-bulb thermometer temperature and locate the Dew Point temperature under the column heading which is equal to the difference between Dry and Wet-Bulb temperatures.

ing day until dew and frost have disappeared and the temperature of the surface is above the dew point. The painting day is further shortened by having to stop the work at least two hours before the temperature of the surface is expected to drop below the dew point to allow the paint time to set before being exposed to moisture.

Regardless of the season or temperature, painting should not be done when fog, mist, drizzle, or rain is expected or occurring.

1-11-3, ORDERING PAINTS

Most vessels and shore establishments maintain paint lockers and keep an allowance of paint on hand. These allowances are necessarily low due to the fact that paint deteriorates in the container. Therefore, only enough paint for routine maintenance should normally be kept in the paint locker. Prior to the painting season, larger stocks should be obtained.

When ordering paint, first check Chapter 4 of this Manual to determine what paint is authorized for the intended purpose. State correct name, stock number and amount of the material on requisition and submit to your supply support activity. For large jobs such as the exterior of buildings or ships, requisitions should be made well in advance of the contemplated accomplishment date. At least 30 days should be allowed to insure receipt of the material in time for the job. It should be pointed out that lack of foresight in procurement of paint materials, resulting in so-called "emergency" open market procurement, increases material costs by as much as 300 per cent and usually results in the application of inferior materials. The best guide for determining the amount of paint to order is previous experience. If paint records are properly maintained in accordance with Article 1-11-7, information will be readily available to permit accurate estimation of paint quantities needed for a particular job. Where information is not available, paint quantity needed must be estimated.

Estimating the proper amount of paint for a particular job takes a great deal of experience. The coverage of a paint for a particular job depends upon (a) the type of paint—some paints are absorbed more easily than others and some paints are more viscous than others leaving a thicker surfacing film; (b) the material to which it is applied—there is a wide variation in the degree of absorption by different materials; and (c) the degree of surface irregularity.

In order to take the above three factors into consideration, it is necessary to first estimate the coverage of a flat surface in terms of sq. ft. per gal. This data appears in an accompanying table.

Having obtained the coverage on a flat surface from the table, multiply the flat surface coverage by an arbitrary factor to account for surface irregularity. The resulting coverage figure should be a reasonably accurate estimate of the amount of projected surface area that a gallon of paint will cover. Sample irregularity factors are given herewith.

Surface		*Irregularity Factor*
Vessel,	Overheads	.50
	Bulkhead	.70
Building,	Ceilings	.90
	Walls	.80

1-11-4, ORGANIZING FOR THE JOB

Organization is absolutely necessary. It helps save time and material. It has a good effect on the men because they know what is expected of them. Whatever system is used, it should be carefully planned. No system will work equally well throughout the service. The best system for a particular unit can only be determined by experience. However, the following points will apply universally:

1. Insist on a clean, well conducted paint mixing room and paint locker.
2. Develop a full knowledge of various surfaces and how to prepare them for painting.
3. Gain an understanding of weather conditions and how to avoid dangers of painting when conditions are unfavorable.
4. Learn to gauge the number of man-hours a job will take.
5. Let the men know that you have a personal interest in the job by keeping a close check on them and their work. Insofar as

practicable inspect each phase of the work as it is completed, i.e., surface preparation, priming and finish coat.

6. Instill in the men a sense of personal responsibility in doing the job well.

7. Know the capabilities of your men. Put the right man in the right place.

8. Promote a spirit of carefulness. Impress upon them the importance of keeping their equipment clean and stress the dangers of smoking on the job or storing oily rags or cloths where spontaneous combustion can occur.

9. See that the best painters do not have to waste time fetching paint, preparing materials, or moving equipment. Let someone else do these tasks.

10. Arrange for paint locker personnel to have an early breakfast and to "turn to" ½ hour before the painting crew in order to have all paints mixed, brushes prepared, etc. They should also remain on the job after the paint crew has finished its work in order to clean brushes and leave the paint locker properly secured. Compensatory time off may be allowed paint locker personnel during the day when the workload in the paint locker is light and on rainy days.

11. Arrange for petty officers and leading seamen to also "turn to" prior to scheduled working hours when staging or scaffolding is to be rigged. Paints, paint brushes, spray guns, mechanical and hand cleaning tools, staging and scaffolding should be ready and

COVERAGE IN SQ. FT. PER GALLON ON A FLAT SURFACE

Surface and Material	*1st or Primer Coat*	*2nd Coat*	*3rd Coat*
Siding and Trim			
Exterior Oil Paint	350	400	450
Porch Floors and Steps			
Deck Paint	300	400	450
Asbestos Wall Shingles			
Exterior Paint	150	300	
Shingle Siding			
Exterior Oil Paint	250	300	
Shingle Roofs			
Roof Paint	100	200	
Brick Exterior			
Exterior Oil Paint	150	300	
Cold-Water Cement Powder Paint	75	100	
Cement and Cinder Block			
Cold-Water Cement Powder Paint	75	100	
Exterior Oil Paint	150	180	
Interior Doors and Windows			
Interior Gloss Enamel	350	400	450
Walls, Smooth Finish, Plaster			
Interior Gloss or Semi-Gloss Oil Paint	350	400	450
Interior Synthetic Rubber Emulsion Paint	450	500	
Plasterboard (Sheetrock)			
Interior Gloss or Semi-Gloss Oil Paint		400	450
Interior Synthetic Rubber Emulsion Paint		450	500
Metal			
Pretreatment-Wash Primer	300		
Quick Drying Red Lead Primer		300	
Exterior Enamel			350

waiting by scheduled "turn to" time. Painting should not be secured more than 10 minutes prior to the close of scheduled working hours. This will be adequate time for returning materials to the paint mixing room. Generally speaking, too much time is lost in "standing by to stand by."

By applying these principles in conjunction with practicing painting economy, a supervisor can earn his pay many times over.

1–11–5, SCAFFOLDING

In order to paint high and low places which are normally out of reach, various types of ladders, scaffolds, and staging are used. The boatswain's chair is frequently used and is essentially a one-man stage. The swinging stage is used when two or more men are working. It is suspended by block and tackle and can be raised or lowered as required. Where staging or boatswain's chairs cannot be suspended from above, ladders and scaffolding are employed. Ladders are used at shore establishments where only one workman is required and where the height is relatively small (20 ft.). Where more than one workman is required, scaffolds are constructed which consist of structural timbers, well shored, with wide horizontal boards secured at intervals from which to work.

Workmen on any type of scaffold or stage should be constantly tended by someone on deck. Each man over the side of a vessel shall wear a lifejacket and shall have a lifeline looped around his chest under his arms and secured on deck. These lines will require tending whenever the stage is raised or lowered; also the men will need paint materials at intervals. Scaffolds should be kept clean of loose materials such as tools, scraps of material, and other stumbling hazards. Portable electric wire and lines which may trip or throw a man are also dangerous. Loose clothing such as long sleeves and trousers which are too long or which have loose cuffs are extremely dangerous to wear on staging. Other hazards are improper rigging, falling objects, taking unnecessary risks, and daydreaming.

When working aloft or over the side, all tools shall be secured by a lanyard to either a substantial part of the ship or to the worker, depending on the size and weight of the tool. There is no excuse for dropping tools on people below or for dropping tools over the side. Portable electric tools used for preparing surfaces to be painted shall be checked for grounds at least once each week and shall be grounded while in use.

Ladders, boatswain's chairs, stages and scaffolding materials shall never be painted. Painting such equipment covers any existing defects such as cracks, breaks, and dry rot. Paint drippings should be cleaned up while still wet and shall not be allowed to accumulate. Ladders, boatswain's chairs and stages shall be cleaned to bare wood and coated with boiled linseed oil for protection. Careful inspection of all such equipment before use is mandatory.

1–11–6, CONTROL OF FILM THICKNESS

The protection afforded by a paint film is, within certain limits, directly proportional to the thickness of the film. It is believed that the dry paint film thickness is more important than the number of coats, since after the paint has dried only the dry film remains to protect the surface. The number of coats is of concern only in minimizing the possibility of holidays. It is difficult to estimate just how thick the paint film on a particular surface should be without being cognizant of prevailing conditions. However, with the view in mind that the most economical thickness which will afford the desired protection is the one to use, it has been determined that paint films on exterior steel such as ships' hulls and superstructure should be at least 8.0 mils thick and for interior steel on ships 5.0 mils. Proper application of the painting system specified in Chapter 2 for these surfaces will yield paint films having these minimum thicknesses.

Film thicknesses are criteria for service life of coatings only when the film thickness is built up from the bare surface as a system. Theoretically, once a surface is properly prepared and adequate film thickness is built up as a system, this expensive operation need not be repeated provided a definite inspection and touch-up system is established and that local wirebrushing, spot priming and cover coating is done as soon as inspection shows the slightest indication of localized failures. While this theory may not be one hundred per cent true, it is a fact that the frequency of surface preparation and complete repainting can in most cases be greatly reduced by application of the above principles.

The importance of applying the entire prescribed coating system at one time rather than applying only part of the system, such as applying only one finish coat where two are prescribed, cannot be too strongly emphasized. Applying less than the minimum film thickness at one time and then applying subsequent coats at a later date is an uneconomical practice. Films thinner than the minimum film thickness are too permeable to moisture for the conditions to which they are exposed. The sub-surface will be attacked and breakdown of the paint film will have commenced, even though it appears to be in good condition, before the application of additional coats of paint increases the film thickness to the prescribed minimum for adequate protection. Frequent repainting indicates previous films were too thin for adequate protection and covering failure spots with another thin film is not the answer to the problem. Frequent repainting is also expensive due to the fact that the surface must be prepared to receive the paint prior to each coat. It is estimated that surface preparation constitutes approximately half the cost of a paint job. Therefore, it is of the utmost importance that the job be done properly in the beginning. Prepare the surface properly; build up the minimum film thickness as a system by applying the prescribed number of coats of the specified material; establish a definite inspection system; wirebrush, spot prime and build up film thickness on localized failure spots as soon as the slightest failure is evidenced.

Generally, the first indication of paint film failure will be found over edges, welds and projections. To insure adequate coverage when applying the paint by brush, first coat all edges and sharp corners as a preliminary to applying the overall coat. Apply the overall coat without waiting for the preliminary coat to dry. Follow the system of "laying on" and "laying off" as described in Article 1–7–4, Brushing Technique. The preliminary coating of edges and corners is not counted as a paint coat, but is considered a part of each regular coat. On spray painting the same preliminary treatment of coating edges and corners should be applied. It will be found that a double pass spray coat is about the equivalent of one brush coat. In some cases coverage can be insured by using alternate color coats of the same material, for example, anticorrosive bottom paints which are available in red and brown, to eliminate holidays.

Film thickness on steel can be checked with film thickness gauges after the film has dried. Film thickness gauges are relatively simple devices which measure the strength of a magnetic field which passes from one pole of a permanent magnet through the paint film, across the painted steel surface and back through the paint film to the other pole of the magnet. The strength of this magnetic field which is proportional to the thickness of the paint film, is measured by a meter calibrated in mils of paint film thickness. While these devices act on a very simple principle, they are delicate precision instruments and must be handled with care. These instruments are not available to the field in general but are available to buoy depots, district offices and inspectors.

1–11–7, PAINT RECORDS

In order to eliminate substandard materials, records must be kept which will identify the particular paints that were applied to a surface where early paint failure was experienced. Therefore, records shall be kept on each paint job other than minor touch-up work. Aboard ship this information shall be entered in the Hull Book. At shore establishments, a Painting Log shall be kept. These records shall have separate pages for each compartment or room and for the exterior of superstructures or buildings. The following information is to be recorded whenever complete repainting is done.

1. Date of application.
2. Area painted.
3. Name of finish material used, number of coats. Specification number, name of

manufacturer, date of manufacture and manufacturer's lot number.

4. Name of Primer used and number of coats, specification number, name of manufacturer, date of manufacture, manufacturer's lot number.

5. Number of gals. of each type paint used.

6. Temperature ° F.

In addition to recording the above information, the date of application shall be stencilled on the painted area in an inconspicuous location. This practice will aid in planning future workloads and future material needs by giving visual evidence of how a paint film is holding up. It should also prevent the waste of materials through unnecessary and frequent repainting, since it is a constant reminder of the time interval since the last painting. The stencil will also give a cross-check on the Hull Book or Painting Log.

In any case where paint failure is excessive or where paint supplied is considered to be below par in any respect, a complete report should be forwarded to Commandant (ETD).

1-11-8, CONTRACT PAINTING

In any case where contract painting is done, specifications shall be written to guard against inferior workmanship, materials and other detrimental factors which would shorten the service life of the paint film.

Painting specifications shall be based upon instructions promulgated in the Coast Guard Paint and Color Manual and shall contain all pertinent instructions applicable to the job including weather conditions suitable for painting, surface preparation, the painting systems to be used, paint materials to be used, methods of application, quality of the finished job and the general condition of the premises upon completion.

Paint material may be furnished either by the Government or the contractor. The advantages of having the contractor furnish the paint are principally those of convenience, i.e., it is not necessary to estimate, order or schedule delivery of the paint materials. On the other hand, the furnishing of paint materials by contractors has serious disadvantages. Competitive bidding creates a situation wherein a successful contractor must be able to "cut corners" to enable him to be the lowest bidder and still make a satisfactory profit. It is possible to "cut corners" in paint materials by one or two methods, i.e., using inferior paint products, or excessive thinning of good paint products. Whenever a contractor is required to furnish his own materials, the profit incentive encourages practices which are disadvantageous to the Government.

By contrast, Government furnished paints are manufactured to pass rigid specifications insuring high quality products. In addition, Government furnished paints are normally half the cost of paints of equal quality purchased on the open market. Since the cost of paint materials is only a fraction of the total cost of painting (approximately 1/6 of the total cost) the use of other than high quality paint materials is not economical. High quality can be assured only when Government furnished materials are used. Paint material should, therefore, be furnished by the Government when practicable. Paint material and paint systems are specified in Chapter 2 and cataloged in Chapter 4. Always order paints sufficiently ahead of time to insure receipt prior to the contemplated or contracted commencement date. At least 30 days are required in most instances to process requisitions and make delivery. For estimating paint quantities, consult Article 1-11-3, Ordering Paints.

The contractor may be allowed to furnish paint material when it is impractical for the Government to do so. Contractors shall be required to furnish high quality, first-class products. Since there is no method of correlating commercial paints with Federal specifications, other than through extensive testing of each product, the quality of commercial paints is almost solely dependent upon the integrity of the contractor and the paint supplier. This leaves the Government in a weak position to obtain quality products. Contractor furnished paint materials must be delivered to the job in original, unopened packages bearing the name of the manufacturer and brand.

The following is a sample of a satisfactory specification for the painting of the interior of a plastered building having wood interior trim. The plaster is assumed to be new plaster which has aged the required time as specified in Article 2–2–3 (C).

1. DESCRIPTION: Painting shall include the proper preparation of the surface to be painted, the application, protection and drying of the paint coatings, the protection of all portions of the structure against disfigurement by spatter, splashes and smirches of paint, and the supplying of all tools, tackle, scaffolding and labor necessary for the entire work.

2. WORKMANSHIP AND GENERAL REQUIREMENTS: All painting shall be done in a professional manner such that the resulting finished paint film will be free of sags, runs, pinholes and other blemishes and will give the ultimate service life expected.

All surfaces to be painted shall be smooth, clean and dry. Small cracks and holes in plaster shall be repaired with spackling putty. All nails in wood trim shall be countersunk and all nail holes filled with putty and smoothed off after the first or priming coat has dried. All knots and pitch pockets in Pine Wood shall be scraped to remove excess resin and then sealed with Ready-Mixed Aluminum Paint prior to the application of the prime or first coat.

Inside painting may be done at any time providing the temperature inside the building is kept constantly above 50° F. during and after application until the last coat is thoroughly dry and hard. When outside temperatures are below 50° F., building interiors shall be heated sufficiently prior to paint application to raise the temperature of surfaces to be painted above 50° F. Heat shall not be turned off until paint films are dry. Paint shall be thoroughly mixed before application by hand stirring or by mechanical mixers followed by "boxing" back and forth between containers. Mixing must continue until all of the pigment is in suspension. The paint shall be frequently stirred during application to keep the pigment in suspension.

All paint shall be evenly spread and brushed out to produce a uniform coating. No coat shall be applied until the previous coat is thoroughly dry. The paint shall be worked into all joints, corners and open spaces. Finished surfaces shall be uniform in gloss, finish and color.

3. PAINTING SYSTEMS:

For Plastered Surfaces—

(a) Prepare the surface.

(b) Apply 2 coats of Interior Synthetic Rubber Emulsion Paint.

In sanitary spaces use the following system:

(a) Prepare the surface.

(b) Apply 1 coat Interior Wall Primer-Sealer.

(c) Apply 1 coat Interior White Enamel Undercoat.

(d) Apply 1 coat Interior Gloss Enamel.

For Wood Surfaces—

Interior wood trim including wainscoting, doors, door trim, windows, window trim, baseboards, molding, stair risers and banisters, cabinets and shelving shall be painted according to the following system:

(a) Prepare the surface.

(b) Apply 1 coat Interior White Enamel Undercoat thinned with 1 pt. of Paint Thinner to 1 gal. of the undercoat.

(c) Apply 1 coat of Interior White Enamel Undercoat as received.

(d) Apply 1 coat of Interior Gloss Enamel.

All paint necessary for the completion of this job will be furnished by the Government and any paint in excess of that required shall be left on the premises after completion.

It should be noted that the above painting specifications are not universally satisfactory. Each contract-painting job will have its peculiarities which must be provided for. In every instance, the objective is to obtain a first class paint job by making the contractor cognizant of what is required of him.

CHAPTER 2, COATING SYSTEMS

TABLE OF CONTENTS

SECTION 2–1, INTRODUCTION

Paint films generally consist of a layer of two or more different types of paint such as an anticorrosive paint used with an exterior enamel. For any given surface there is a particular combination of types of paints and film thickness of each type which will produce optimum results. Some paints are even incompatible with others. It is, therefore, necessary to treat painting from a coating system approach; i.e. one in which the materials, the number of coats of each material and the sequence in which the coats are applied are considered as a unit—the coating system.

While a coating system specifies the materials to be used for particular surfaces, it does not include the color of the material to be used nor does it include complete application instructions or ordering information. Where a question of color is involved, refer to Chapter 3, Color Practice. For ordering information and detailed application instructions, refer to Chapter 4, Materials.

The coating systems contained herein are written for coatings to be applied to properly prepared bare surfaces. The preparation for painting of the various types of surfaces normally encountered in the Coast Guard is discussed in Section 1–5, Surface Preparation.

Where painting is to be done to previously painted surfaces which are in good condition throughout, clean the surface thoroughly by wiping oil and grease spots with Paint Thinner followed by washing the entire surface with Paint Cleaner. Surface irregularities are removed by sanding.

If a previously painted surface shows signs of paint failure, the deteriorated area should be "touched up" by removing the defective paint film down to the bare surface. The edges of adjoining areas of paint film should be sanded to fair the paint film into bare surface. Any surface irregularities will be apparent in the finished paint film. Recoat the bare areas with a complete paint system slightly overlapping onto the old paint. Sandpaper the overlap lightly before each subsequent coat to reduce surface irregularities. After completion of "touch-up" painting a uniform appearance may be obtained by following with 1 coat of the finish paint over the entire surface. (On exterior wood surfaces 2 coats may be necessary.) When touching up or repainting over previously painted surfaces, be sure that touch-up material is compatible with the old paint. If doubt exists, test by applying a small amount of the touch-up material over the old paint in an inconspicuous location.

Various modifications of the above practice may be necessary depending on the type and condition of the surface. In all cases the objective is to cover the entire surface with a dry paint film which is the equivalent of the recommended coating system.

SECTION 2-2

COATING SYSTEMS FOR SHORE ESTABLISHMENTS

2-2-1, WOOD SURFACES

(A) EXTERIOR WOOD, GENERAL

New exterior wood in general such as the exterior of wood buildings, trim, shutters, doors, balustrades, porch ceilings and posts, stair risers, guard rails, hand rails, fences, gates, lampposts, towers, water tanks, and similar surfaces shall receive the following paint system:

(a) Prepare the surface. Seal knots and pitch pockets as necessary with Ready-Mixed Aluminum Paint.

(b) Apply 1 coat Exterior Wood White Primer.

(c) Apply 2 coats Exterior Oil Paint, General Purpose. (In areas where discoloration occurs due to gases in atmosphere, such as industrial fumes and volcanic fumes, use Exterior Oil Paint, Special Fumeproof.)

NOTE: When recoating, apply 2 coats of Exterior Oil Paint.

(B) EXTERIOR FLOORS, DECKS, STEPS, BLEACHERS

(a) Prepare the surface. Seal knots and pitch pockets as necessary with Ready-Mixed Aluminum Paint.

(b) Apply 1 coat of Deck and Floor Paint thinned by the addition of 1 quart of thinning mixture to each gal. of paint. Thinning mixture is composed of ⅔ Boiled Linseed Oil and ⅓ Paint Thinner.

(c) Apply 2 coats of Deck and Floor Paint as received.

(C) EXTERIOR WOOD SIGNS

Exterior wood signs of all types including aerial identification platforms and daymarks shall receive the following paint system:

(a) Prepare the surface. Seal knots and pitch pockets.

(b) Apply 1 coat Exterior Gloss Enamel thinned by the addition of 1 pt. of Boiled Linseed Oil per gal. of paint.

(c) Apply 2 coats of Exterior Gloss Enamel.

(d) Apply markings with 2 coats Exterior Gloss Enamel.

(D) WOOD DOCKS, WHARVES, PIERS, RAMPS

Wood pilings, stringers and all structural members of wood docks, wharves and piers shall be pressure treated with creosote prior to installation in accordance with the provisions of the current edition of Federal Specification TT–W–571. Wood deck planking on docks, wharves and piers shall be treated with Tonalith (Wolman-Salt) in accordance with the current edition of Federal Specifications TT–W–571 and TT–W–573.

In cases where deck planking on docks has been previously painted, painting shall be continued, using the following system:

(a) Prepare the surface.

(b) Apply 1 coat of Deck and Floor Paint thinned by the addition of 1 pt. of Boiled Linseed Oil to each gal. of paint.

(c) Apply 1 coat of Deck and Floor Paint as received.

(E) CREOSOTED WOOD

Creosoted wood which is required to be painted, such as poles used for daymarks, shall be given the following paint system after aging for 6 months:

(a) Clean the surface. Remove dirt, oil.

(b) Apply 2 coats Ready-Mixed Aluminum Paint.

(c) Apply 2 coats Exterior Gloss Enamel.

(F) PREVIOUSLY STAINED WOOD SHINGLES AND SIDING

Wood shingle siding and wood shingle roofs which have been previously stained and maintained by staining may continue to be stained. Whenever stained wood siding and shingles are to be painted the surface shall first be given 2 coats of Ready-Mixed Aluminum Paint to seal in the stain. Finish material shall be 2 coats of Exterior Oil Paint for siding and 2 coats of Tile Red Roof Paint for roofs.

(G) INTERIOR WOOD WALLS AND WOOD WAINSCOTING

(a) Prepare the surface.

(b) Apply 2 coats Interior Synthetic Rubber Emulsion Paint.

(H) INTERIOR WOOD TRIM

Interior wood trim including doors, door trim, windows, window trim, baseboards, molding,, stair risers and balustrades, cabinets and built-in furniture which is to be painted shall be painted according to the following system:

(a) Prepare the surface.

(b) Seal knots and pitch pockets with 1 coat of Ready-Mixed Aluminum Paint.

(c) Apply 1 coat Interior White Enamel Undercoat as received.

(d) Apply 1 coat Interior Gloss Enamel.

(I) INTERIOR WOOD FLOORS

Finished Floors—

New interior wood floors and stair treads that are to receive a natural finish shall be machine sanded to a smooth even surface and swept clean of dust. Open grain wood such as oak, chestnut, and ash require a filler coat. Omit the filler for close grain wood such as pine, fir, maple and birch. Use transparent Paste Wood Filler thinned to brushing consistency with Paint Thinner. Apply a thick coat and rub it into the wood with a stiff bristle brush. After about 15 minutes, the excess filler should be removed, wiping across the grain with burlap or excelsior. Finish by rubbing with a clean cloth. If filler dries hard before wiping can be accomplished, dampen wiping cloth with Paint Thinner. Care must be exercised not to remove the filler from the pores of the wood. Allow 24 hours drying time before proceeding.

After filler coat has dried, clean all dust from the floor, preferably with a vacuum cleaner. Apply 1 thin coat of Floor Sealer, allow it to dry, and buff it in. This will seal the pores of the wood, help to keep out dirt, and resist stains.

After floors have been filled and sealed, Liquid Floor Wax should be applied. A number of thin coats will be more satisfactory than 1 thick coat, because a heavy coat makes the floor greasy. Each layer of wax should be allowed to dry thoroughly and should be polished before adding the next layer. Finishing a floor in this manner will give an attractive, satin-like sheen to the wood and a finish that will not be greasy, will not mar, scratch, or flake off—one that can be touched up at worn spots or in heavy traffic lanes without completely refinishing the entire floor.

Gymnasium Floors—

New Gymnasium floors after being sanded and otherwise prepared for finishing should be given 2 coats of penetrating floor sealer of the lacquer type, followed by painting of floor markings and then followed by 2 coats of lacquer type sealer containing resins to reduce slipperiness. Since these materials are not available from Navy or Coast Guard stocks, open market purchase is necessary. There are several well-known proprietary brands on the market for this purpose. This treatment produces a surface of good appearance which withstands hard usage, reduces slipperiness and readily permits the renewal of the finish whenever necessary. For application of the material, the directions of the manufacturer should be followed.

Painted Floors—

Interior wood floors and stair treads which are to be painted shall be given the following paint system:

(a) Prepare the surface.

(b) Apply 1 coat of Deck and Floor Paint

thinned by the addition of 1 qt. of thinning mixture to each gal. of paint. Thinning mixture is composed of $\frac{2}{3}$ Boiled Linseed Oil and $\frac{1}{3}$ Paint Thinner.

(c) Apply 1 coat of Deck and Floor Paint as received.

(J) FURNITURE FINISHING

Wood finishing may be described as the application of transparent or semi-transparent coatings to wood with the object of accentuating the grain of the wood to bring out the natural beauty, or changing the color to make a less expensive wood appear to be a costlier type.

Mahogany, walnut, maple, birch, and oak are usually preferred in natural finishes while softwoods such as fir, pine and cypress may be either finished natural or stained to resemble more expensive woods.

Furniture finishes may be either varnish or lacquer finishes. Either system provides a satisfactory surface. Nitrocellulose lacquer is recommended for wood finishing in the Coast Guard. It dries dust free in a very short time (10 minutes); therefore, rubbing or sanding between coats is unnecessary. Scaling in lacquer finishes is almost unknown. As each succeeding coat is applied it dissolves into the preceding coat and forms one continuous film from filler to finish coat. Lacquer has almost completely replaced varnish in the furniture industry.

The methods and materials of wood finishing are as numerous and varied as the finishes desired. Due to limited space, only one standard method is discussed. For those interested in exploring the subject further, Practical Finishing Methods, Book No. 4543, obtainable from Delta Manufacturing Division, Milwaukee, Wisconsin, price 50 cents, is recommended.

Surface Preparation—

New wood will require very little surface preparation other than sanding. Use No. 3/0 or 4/0 sandpaper on a block of wood and sand with the grain. Remove all pencil marks, finger prints, etc. Sand until the surface is clean, smooth, and free of scratches. In the case of previously painted or varnished surfaces, the old finish must be completely removed in order to obtain a good new job. Follow directions as outlined in Article 1–5–4 (D), Use of Paint Remover. After the old finish is removed, treat the surface as outlined for new wood.

Nail or screw holes should be filled with Plastic Wood colored to match the desired finish since this material cannot be stained after it has hardened. When applying Plastic Wood, do not strike it off flush with the surface because it shrinks slightly upon drying. Leave a slight hump and sand it down smooth after it has hardened. Remove all dust from the surface and pores of the wood immediately before applying the stain coat by wiping with a "tack rag" (tack rags are resin impregnated cloths made to pick up lint and dust and may be purchased on the open market). Dampen the floor to reduce dust.

Stain Coat—

Stain application by the brushing method is common procedure and gives best results. A flat type, fine bristle brush is used. First apply stain to a sample of the wood to be stained. This is important because a stain will give different results on different types of wood. For example, a walnut stain will produce a rich brown color on gum, while the same stain on white pine, basswood or poplar would appear nearly black. Non-grain raising stains shall be used whenever staining is to be done. Stains must dry thoroughly before proceeding. The application of a second coat may be necessary to obtain the desired color. Non-grain raising stains dry to permit filling 1 hour after application.

Filler Coat—

Paste Wood Filler is used to fill the open pores of wood surfaces to prevent undue absorption of lacquer by the wood. Filler is applied following complete drying of stain coat. For natural finish, omit the stain coat and apply Wood Filler directly to the wood.

Paste Wood Filler is thinned with Paint Thinner and applied by brushing across the grain. The filler is allowed to "flatten out" which is characterized by a surface drying appearance, after which it is wiped into the pores with a pad made of burlap. A circular motion is used to insure filling the pores.

The excess is cleaned from the surface with a clean rag wiping across the grain. If filler sets up too hard for easy wiping, moisten the wiping rag with Paint Thinner. If the wiping is accomplished too soon, the filler will be pulled from the pores and the object will not be accomplished. After wiping clean, the filler is allowed to dry for 24 hours, after which the first coat of lacquer is applied.

Lacquer Coats—

Two coats of clear gloss lacquer shall be applied as received in the can. Follow with 1 coat of clear flat lacquer. As noted before, no sanding is necessary between coats. The flat last coat will greatly reduce the rubbing required to produce the desired luster. If flat lacquer is not available, clear lacquer may be used for this third coat. Allow 2 hours drying time between coats and at least 24 hours on last coat before rubbing. Lacquer should be sprayed if possible. If spray equipment is not available, a lacquer formulated especially for brushing must be used. Since the solvents used in lacquers are of a high solvent power, there is a tendency, when applying second and third coats, to lift previously applied coats. To overcome this do not brush out the lacquer as is done with varnish. Use as few strokes as possible to apply a thin even film and then allow the flowing and leveling properties of the lacquer to give you a mirror-like finish. Do not attempt to apply lacquer on a humid day since it may blush, causing the transparent film to become cloudy or opaque.

Rubbing—

The rubbing medium recommended is pumice stone and water. A felt pad about ½ in. thick, 2 in. wide and 4 in. long is used. This felt is mounted on a wood block with the ends turned and securely tacked in place. The rubbing pad should be thoroughly wet when in use.

Rubbing is executed with the grain. Straight and even strokes are necessary. Long full strokes with the same exerted pressure throughout the stroke are desired. Care must be exercised on corners and edges that the rubbing does not cut through the coating. After rubbing to desired sheen, wash all pumice stone off the surface with clear water and dry with a clean cloth. When dry, the surface should be polished with furniture polish or wax.

To summarize the above, the finishing system is as follows:

(a) Prepare the surface.

(b) Apply non-grain raising stain (omit stain if natural finish is desired).

(c) Apply 1 coat Paste Wood Filler (omit wood filler on close grain woods).

(d) Apply 2 coats clear gloss lacquer.

(e) Apply 1 coat clear flat lacquer.

(f) Rub and polish.

It will be noted that some of the specified materials for furniture finishing are not available through Catalog of Navy Material. As previously stated, the open market purchase of paint materials in general is prohibited. However, the open market purchase of the specified furniture finishing materials is permissible. When procuring lacquers it is recommended that an acid, alcohol, and alkali resistant type be used.

(K) REPAIRING FURNITURE FINISHES

Repairing scratches and other damage to furniture finishes is not a difficult job. It will be noted that the quantities of materials needed are small. For this reason, these materials are not carried in the Catalog of Navy Material but may be purchased at most hardware and paint stores.

There are two methods of patching damaged furniture finishes. "Burning-in" is the name given to the application of melted stick shellac to scratches and other indentations by means of a heated knife. "French polishing" is the application of a surface coating of refined and dewaxed shellac (French Polish) by rubbing with a cloth pad. These methods may be used independently or in conjunction depending upon the extent of the damage.

Burning-In—

Stick shellac is available in about seventy-five different colors, making it easy to match almost any finish. Use a translucent stick of the right color for shallow scratches where the stain color is intact. Use an opaque stick of the right color for deep scratches which go down to bare wood.

To avoid the danger of burning the surrounding surface, the scratch should be masked, as shown in Figure 1, with masking tape or cellulose tape.

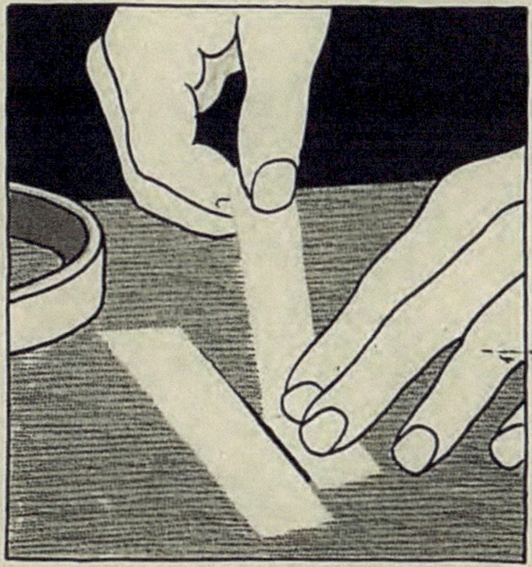

Fig. 1. Use of masking type.

The burning-in knife should be of flexible steel, ground square at the end. A kitchen paring knife makes an excellent tool when ground to shape. The knife is held in the flame of an alcohol lamp, the end of the blade being about an inch beyond the flame. The stick shellac is held in contact with the end of the knife, as shown in Figure 2, the shellac melting and running onto the knife. As soon as a sufficient quantity is melted to fill the scratch, the knife is removed from the flame and quickly drawn along the scratch to deposit the shellac as shown in Figure 3.

After the masking tape is removed, the ridge of shellac is scraped down with a razor blade, as shown in Figure 4. When the patch is as level as the scraper can make it, take a small piece (about 1/8 sheet) of 6/0 finishing paper and back it with a small felt pad, or fold it in thirds and back it with the fingers. Dip the paper in alcohol and then in rubbing oil or water. Rub the patch quickly and firmly, as in Figure 5, using 5 or 6 strokes. Beware of rubbing through the finish

Fig. 2. Use of stick shellac.

around the patch. Dull the patch by rubbing lightly with 3/0 steel wool and the job is done.

Despite care in selecting the stick shellac color, you may find the patch a bit off color. In this case, mix a little alcohol soluble stain and apply it to the patch with a small water-color brush. Selection of the stick shellac color should lean toward a slightly lighter

Fig. 3. Application of shellac to scratch.

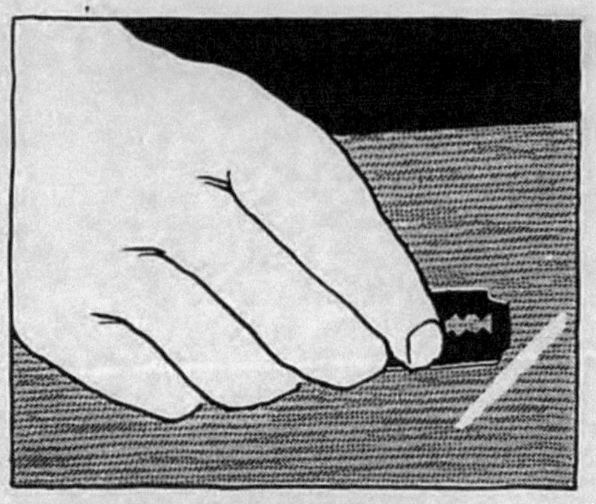

Fig. 4. Patch is scraped level.

color rather than a darker one. The light shellac can always be stained dark, but dark shellac cannot be stained light.

If the finish around the patch is made thin by rubbing, or if the finish is worn in spots, it can be brought back by spraying a coat of clear flat lacquer on the spot as shown in Figure 6, feathering the edges. If the scratch repaired with stick shellac is to be sprayed, water and not oil should be used for rubbing. After spraying spots the entire surface must be rubbed with pumice and rubbing oil.

Fig. 5. Patch is rubbed smooth.

French Polishing—

French polishing is used to work out very shallow scratches, imprints and other defects which do not extend through the finish. The method may also be used for finishing a stick shellac patch. French polish may be a French varnish, ether varnish, or even dilute shellac and is used in combination with lemon oil.

The application pad is made by folding a double thickness of cheesecloth around a small ball of cotton, as shown in Figures 7 and 8. The pad is touched to the wide mouth of the polish bottle (about 2 or 3 tips of the bottle), and then to the small mouth of the oil bottle (1 tip). The lemon oil is necessary as a lubricant to keep the pad from

Fig. 6. Clear lacquer is sprayed.

sticking. Distribute the mixture of oil and polish by patting in the palm of the hand, as shown in Figure 9.

The first rubbing is done with a circular motion, making rings about 3 in. in diameter, as shown in Figure 10. Keep the pad in mo-

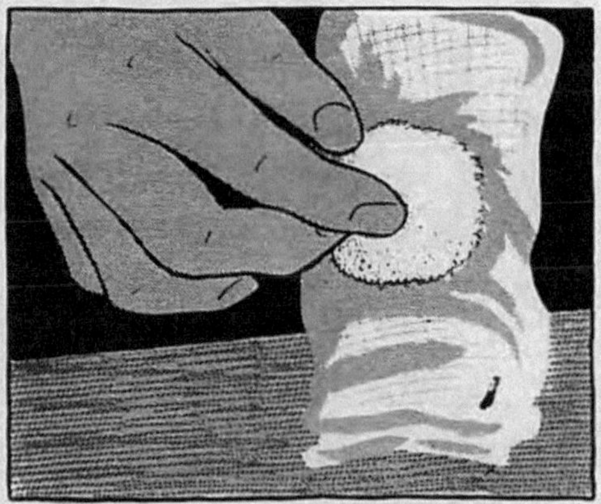

Fig. 7. Cheesecloth is wrapped around cotton.

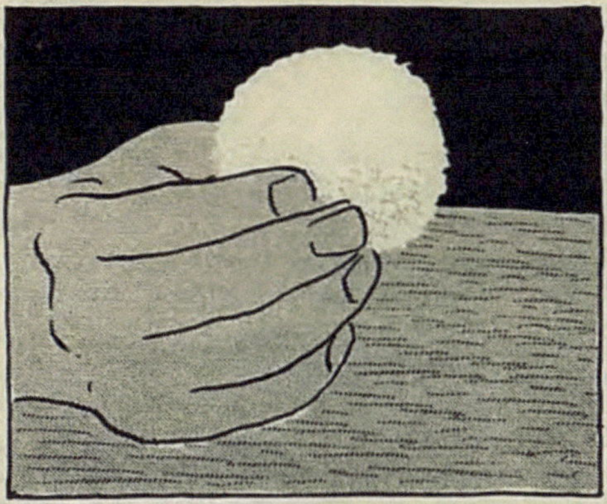

Fig. 8. Finished pad for French polishing.

tion—stopping for even an instant will cause the pad to stick to the surface. Go over the area once or twice with small circles and then change to longer ovals or figure-8 strokes. Finally, rub straight with the grain, beginning the stroke before touching the surface and lifting the pad before ending the stroke, as shown in Figure 11.

After polishing, the area will show a very high gloss which will probably be at variance with the satin polish of the work. To dull the patch, sprinkle with 3-F pumice immediately after rubbing and then scrub the dry pumice 3 or 4 times with a stiff bristle brush, as shown in Figure 12.

Fig. 9. Oil and polish are blended.

(L) WOOD WINDOWS AND DOORS

Sliding surfaces of wood double-hung windows shall not be painted but shall be given 2 coats of Boiled Linseed Oil after fitting. Painting these surfaces will cause sticking and binding of the sash. Unsealed surfaces will swell in damp weather and cause the sash to bind.

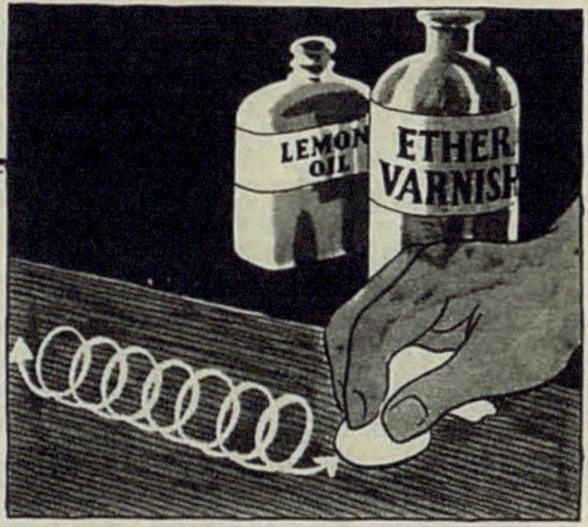

Fig. 10. Circular motion in rubbing.

All edges of wood doors shall be given the same treatment prescribed for the remainder of the door. Particular attention should be given to top and bottom edges where end grain of the wood is exposed, to avoid swelling and subsequent binding of the door in damp weather.

2–2–2, METAL SURFACES

The following paint systems are designed to be applied to properly prepared bare metal surfaces. Methods of surface preparation for the various metals are discussed in the following articles:

1–5–1, Preparation of Metal Surfaces (Except Aluminum and Galvanized Steel).

1–5–2, Preparation of Aluminum Surfaces.

1–5–3, Preparation of Galvanized Steel Surfaces.

Fig. 11. Final rubbing method.

(A) EXTERIOR UNGALVANIZED STEEL

Ungalvanized steel, except those surfaces to receive special treatment as set forth in subsequent paragraphs, shall be painted as follows:

(a) Prepare surface.

(b) Apply 1 liberal coat of Pretreatment-Wash Primer.

(c) Apply 2 coats of Quick Drying Red Lead Primer.

(d) Apply 2 coats of Exterior Gloss Enamel.

(B) EXTERIOR GALVANIZED STEEL AND CORROSION-RESISTING METALS OTHER THAN ALUMINUM

Galvanized steel, corrosion-resisting steel, nickel-copper alloy, brass and other corrosion-resisting metals, except aluminum or

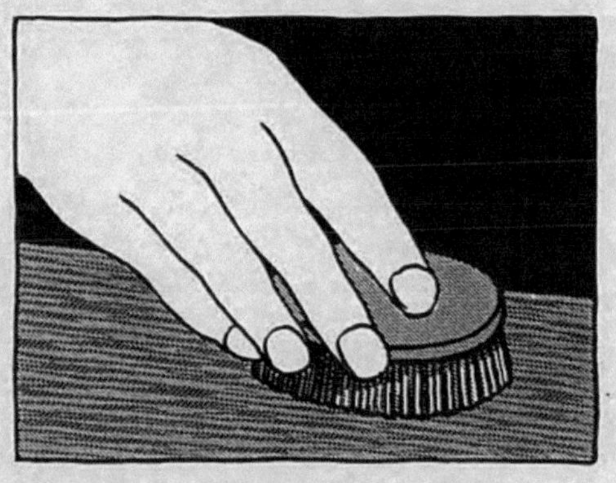

Fig. 12. Surface is brushed with pumice.

those surfaces to receive special treatment as set forth in subsequent paragraphs, shall be painted as follows:

(a) Prepare surface.

(b) Apply 1 liberal coat of Pretreatment-Wash Primer.

(c) Apply 1 coat of Quick Drying Red Lead Primer.

(d) Apply 2 coats of Exterior Gloss Enamel.

NOTE: When a self-washing "house paint" is desired, omit step (d) and substitute therefor 2 coats of Exterior Oil Paint.

(C) EXTERIOR ALUMINUM

Exterior Aluminum surfaces which are to be painted shall be treated as follows:

(a) Prepare surface.

(b) Apply 1 liberal coat of Pretreatment-Wash Primer.

(c) Apply 1 coat of Yellow Zinc Chromate, High Alkyd Type.

(d) Apply 2 coats of Exterior Gloss Enamel.

Aluminum shall be protected from contact with dissimilar metal by use of one of the following Navy standard stock items as appropriate:

Packing, Joint Sealing (Coated Cloth) Oil and Water Stop Items 3437 to 3445, Class 33.

Gasketing, Wool Felt, Adhesive Back, Sheet Items 1008 to 1015, Class 33.

Calking Compound, Oil Type, For Metal Seams and Airports Items 1971 to 1979, Class 52.

Watertight joints between aluminum parts shall be made tight by metal calking rather than the use of gaskets, packing or calking compounds. Where this is not possible the gasket, packing or calking compound material given above may be used.

Asbestos paper or similar absorbent material shall not be used in contact with aluminum.

Threaded parts of aluminum shall be coated, before assembly, with an antiseize mixture of 50 per cent zinc dust and 50 per cent petrolatum by volume.

(D) INTERIOR UNGALVANIZED STEEL

Interior ungalvanized steel except those surfaces to receive special treatment as set forth in subsequent paragraphs shall be painted as follows:

(a) Prepare surface.

(b) Apply 2 liberal coats of Pretreatment-Wash Primer.

(c) Apply 1 coat of Interior Synthetic Rubber Emulsion Paint. (In some cases 2 coats will be necessary to hide the primer.)

NOTE: In sanitary spaces omit step (c) above and substitute therefor 1 coat of Interior White Enamel Undercoat followed by 1 coat of Interior Gloss Enamel.

(E) INTERIOR GALVANIZED STEEL AND CORROSION-RESISTING METALS

Interior galvanized steel, corrosion-resisting steel, nickel-copper alloy, brass, aluminum and other corrosion-resisting metals, except those surfaces not to be painted or to receive special treatment as set forth in subsequent paragraphs, shall be painted as follows:

(a) Prepare surface.

(b) Apply 1 liberal coat of Pretreatment-Wash Primer.

(c) Apply 1 coat of Interior Synthetic Rubber Emulsion Paint. (In some cases 2 coats will be necessary to hide the primer.)

NOTE: In sanitary spaces omit step (c) above and substitute therefor 1 coat of Interior White Enamel Undercoat followed by 1 coat of Interior Gloss Enamel.

(F) GUTTERS, DOWNSPOUTS AND FLASHING

Gutters, downspouts, flashing, roof valleys, roof ridges and similar galvanized or copper or bronze sheet metal surfaces built into or forming part of roofs or walls, if painted, shall be painted as follows:

(a) Prepare surface.

(b) Apply 1 liberal coat of Pretreatment-Wash Primer.

(c) Apply 1 coat of Quick Drying Red Lead Primer.

(d) Apply 2 coats of the finish material used on adjoining surfaces.

Where aluminum is used the following system shall be applied:

(a) Prepare surface.

(b) Apply 1 liberal coat of Pretreatment-Wash Primer.

(c) Apply 1 coat of Yellow Zinc Chromate, High Alkyd Type.

(d) Apply 2 coats of the finish material used on adjoining surfaces.

(G) FURNITURE AND EQUIPMENT

Metal (except aluminum) furniture, steel doors, switch boxes, controllers, switchboards, connection boxes, galley equipment, and miscellaneous equipment in general shall be painted in accordance with the following system:

(a) Prepare surface.

(b) Apply 1 liberal coat of Pretreatment-Wash Primer.

(c) Apply 1 coat Quick Drying Red Lead Primer.

(d) Apply 2 coats Equipment Enamel.

Aluminum furniture and interior aluminum equipment shall not be painted but shall be kept bright.

(H) MACHINERY

Interior Machinery—

Uninsulated, ungalvanized metal surfaces, the operating temperatures of which are 300° F. or less, shall be painted according to the following system:

(a) Prepare the surface.

(b) Apply 1 liberal coat of Pretreatment-Wash Primer.

(c) Apply 1 coat of Quick Drying Red Lead Primer.

(d) Apply 2 coats of Equipment Enamel.

Uninsulated galvanized steel, corrosion-resisting steel, nickel-copper alloy or other corrosion-resisting surfaces, the operating temperatures of which are 300° F. or less, shall be painted according to the following system:

(a) Prepare surface.

(b) Apply 1 liberal coat of Pretreatment-Wash Primer.

(c) Apply 2 coats of Equipment Enamel.

Uninsulated metallic surfaces, the operating temperature of which exceeds 300° F., shall be thoroughly cleaned and coated with 2 coats of Heat-Resisting Paint.

Fabric covered insulation or rough surfaces finished off with Smoothing Cement shall be given 2 coats of Equipment Enamel.

Aluminum surfaces are not to be painted but shall be kept bright.

Exterior Machinery—

The painting system for exterior machinery depends upon the metal used and shall be same as that specified for general exterior use of the metal.

(I) METAL TRAFFIC AREAS

Metal ladders, catwalks, decks, etc. shall be given the coating system prescribed for the particular metal and exposure conditions under consideration, described in Articles 2–2–2 (A) through 2–2–2 (E), except that the finish coat shall be 2 coats of Deck and Floor Paint in lieu of the regularly specified finish coat. Non-skid properties may be added to the Deck and Floor Paint by sweeping washed beach sand on the surface while it is tacky. Sweep the sand well into the paint, allow to dry and sweep off loose sand particles. A final coat of Deck and Floor Paint, thinned by the addition of 1 pt. of paint thinner per gal. of paint should then be applied over the sand.

(J) WATERFRONT AND UNDERGROUND METAL SURFACES

Underground metal piping, interior of floating dry docks, bulkheads, piers, etc. shall be coated with bituminous coatings. These coatings consist of various systems composed of the following materials:

- Coal Tar Primer
- Coal Tar Enamel (applied hot)
- Coal Tar Base Coating (applied cold)
- Bituminous Aluminum Paint
- Petrolatum
- Asbestos Felt
- Reinforced Protective Paper
- Kraft Paper
- Whitewash
- Coupling Compounds

For detail information regarding the use of these materials consult the current edition of Bureau of Yards and Docks Specification No. 34Y.

In the case of piers, piling, bulkheads and other partially submerged structures the economic feasibility of utilizing cathodic protection to protect the underwater surfaces should be investigated. Bureau of Yards and Docks contracts should be employed for cathodic protection surveys and installations.

(K) HOT SURFACES

Hot surfaces (operating temperature above 300° F.) such as boiler fronts, heating furnaces, smoke pipes, hot piping, hot water tanks and hot water heaters shall be given the following paint system:

(a) Prepare surface.

(b) Apply 2 coats of Heat-Resisting Paint.

Where the surface is insulated, Heat-Resisting Paint shall not be used unless the outer surface of the insulation is above 300° F.

(L) WATER TANKS

Feed water tanks and potable water tanks shall be painted with 2 coats of Zinc Dust Paint, over bare metal. Pretreatment-Wash Primer, shall not be used under the Zinc Dust Paint.

To apply the Zinc Dust Paint remove all former paint coatings by sandblasting if possible; power wirebrushing may be used where sandblasting is impractical. Following the cleaning and prior to painting, the Engineer Officer shall inspect the tank to insure that all dirt, sand and rags have been removed from the tank bottom, longitudinal frames and stringers, sounding tubes, vents, filling connections and suctions. Apply first coat of Zinc Dust Paint. Following the first coat of Zinc Dust Paint, unheated air shall be circulated through the tank for 12 hours. The second coat shall then be applied and unheated air circulated through the tank for 12 hours. Tanks shall then be flushed twice with fresh water before placing in service.

CAUTION: Only enough Zinc Dust Paint shall be mixed for daily use. Mixed Zinc Dust Paint shall not be stored. After the material is mixed a gas pressure may build up within closed containers.

(M) INTERIOR OF STEEL TANKS

The interior of steel tanks, such as fuel oil tanks, lube oil tanks and lube oil reservoirs shall be left uncoated except for the oil film. When lube oil tanks or lube oil reservoirs are to be left empty for 30 days or more, the surface shall be coated with Thin-Film Rust Preventive, Grade II. When returning tanks to service, the preservative shall be removed in accordance with the instructions set forth in BuShips Bulletin of Information No. 39, dated 1 July 1939.

(N) INTERIOR PIPING

Insulated metallic piping shall be given 1 liberal coat of Pretreatment-Wash Primer and 1 coat of Quick Drying Red Lead Primer on the exterior surface of the pipe. The fabric-covered insulation shall be given 2 coats of the same material used on adjoining surfaces.

Uninsulated pipe shall be given the same paint system as that prescribed for similar surfaces.

Where the normal operating temperature of the piping exceeds 300° F., 2 coats of Heat-Resisting Paint shall be used in lieu of the conventional system.

2–2–3, MASONRY SURFACES

The material used on masonry surfaces depends primarily on whether the masonry is in a damp or dry location. In the case of concrete floors, abrasion resistance is important. Methods for cleaning, repairing, and pretreatment of masonry surfaces in preparation for painting is discussed in Article 1–5–5, Preparation of Masonry Surfaces.

(A) EXTERIOR WALLS ABOVE GRADE

The following system shall be used whenever the painting of exterior masonry walls above grade is accomplished:

(a) Prepare the surface. Remove defective paint coatings.

(b) Apply 2 coats Cement Powder Paint. (On new masonry, allow this paint 3 months drying time. On old masonry allow this paint 1 month drying time.)

(c) Prepare the surface. Surface must be thoroughly dry prior to the application of oil paint.

(d) Apply 2 coats Masonry Oil Paint.

NOTE: Cement Powder Paint cannot be successfully applied over other paint coatings. Remove old coatings in poor condition. Where old paint coatings are intact, the cement water paint shall be omitted from the above system.

(B) EXTERIOR WALLS BELOW GRADE

New Construction—

The exterior of masonry basement walls below grade which are subject to water penetration shall be waterproofed with a built-up membrane of felt and bituminous compounds constructed in accordance with Bureau of Yards and Docks Specification No. 7 YG.

In dry locations or where excellent soil drainage exists, waterproofing may be accomplished by the application of 2 coats of Cement Powder Paint.

Old Construction—

In most cases, it is impractical to excavate around buried walls for the purpose of waterproofing. Where seepage or leaks occur in basement walls, the causes should be located and rectified before attempting any type of waterproofing. Broken gutters and downspouts and a grade slope toward the building are invitations to leaky basements. Repair all cracks and loose mortar joints in the wall above grade. Treat interior of basement as discussed in paragraph (D) following.

(C) INTERIOR WALLS ABOVE GRADE

New Plaster—

New interior plaster shall be allowed to age at least 2 weeks, or 3 months if possible, and then painted as follows:

(a) Prepare the surface.

(b) Apply 2 coats Interior Synthetic Rubber Emulsion Paint.

Where a glossy paint or a paint with extremely high washability is required, as for sanitary spaces, allow new plaster to age at least 1 month, 6 months if possible, and paint as follows:

(a) Prepare the surface.

(b) Apply 1 coat Interior Wall Primer-Sealer.

(c) Apply 1 coat Interior White Enamel Undercoat.

(d) Apply 1 coat Interior Gloss Enamel.

Interior Masonry Other than Plaster—

Extremely porous masonry, such as cinder blocks, shall be sealed (aging not necessary) with 1 coat of Cement Powder Paint. New masonry and masonry painted with Cement Powder Paint shall be allowed to age at least 2 weeks, or 3 months if possible, before application of the below coating system. Previously painted masonry and wall board may be painted without aging.

(a) Prepare the surface.

(b) Apply 1 coat Interior Synthetic Rubber Emulsion Paint.

(In sanitary spaces, apply 1 coat Interior White Enamel Undercoat followed by 1 coat Interior Gloss Enamel.)

NOTE: In some cases 2 coats of Interior Synthetic Rubber Emulsion Paint may be necessary for adequate coverage.

(D) INTERIOR WALLS BELOW GRADE

Unpainted Basement Walls—

Unpainted basement walls shall be treated as follows:

(a) Prepare the surface.

(b) Apply 1 coat Cement Powder Paint. (Where dampness is present apply a second coat. Rectify possible causes of dampness.)

Previously Painted Basement Walls—

Previously painted basement walls which remain dry shall be treated as follows:

(a) Prepare the surface.

(b) Apply 1 coat Interior Synthetic Rubber Emulsion Paint.

(E) CONCRETE FLOORS ABOVE GRADE

Concrete floors, both exterior and interior, which are not in contact with the ground, shall be treated as follows. New concrete shall be allowed to age for at least 1 month, 1 year if possible, prior to application of the below coating system.

(a) Prepare the surface.

(b) Slick surfaces shall be acid etched by flooding the floor (1 gal. to 100 sq. ft. of floor) with a mixture of 1 gal. of 36 per cent muriatic acid and 3 gals. of water. After 15 minutes, hose off the acid using plenty of water. Allow to dry thoroughly.

(c) Apply 1 coat of Deck and Floor Paint thinned by the addition of 1 qt. of thinning liquid (⅔ Spar Varnish and ⅓ Paint Thinner).

(d) Apply 1 coat of Deck and Floor Paint as received.

(F) CONCRETE FLOORS BELOW GRADE

Concrete floors below grade or in contact with the ground shall receive the following treatment. New concrete floors shall be allowed to age at least 1 month, 2 months if possible, prior to applications of the below rubber-base paint.

(a) Prepare the surface. If a non-rubber base paint was previously used it should be removed with Paint and Varnish Remover and the surface thoroughly scrubbed with soap and water.

(b) Slick surfaces shall be acid etched by flooding the floor (1 gal. to 100 sq. ft. of floor) with a mixture of 1 gal. of 36 per cent muriatic acid and 3 gals. of water. After 15 minutes, hose off the acid using plenty of water. Allow to dry thoroughly.

(c) Apply 1 coat of Basement Floor Paint thinned by the addition of 1 qt. of Synthetic Enamel Thinner per gal. of paint.

(d) Apply 1 coat of Basement Floor Paint as received.

NOTE: Where traffic is heavy an additional coat of the finish paint shall be used.

(G) MASONRY SWIMMING POOLS

New concrete swimming pools shall age for at least 2 months before painting. The pool shall be filled with water during this period in order that the water-soluble salts will be leached out, thus eliminating subsequent blistering of the paint. Any grease or oil spots should be washed with a strong alkaline cleaner and rinsed. New swimming pools with hard slick surfaces should be acid etched by applying a mixture of 1 gal. of 36 per cent muriatic acid added to 3 gals. of water (1 gal. to 100 sq. ft.). After 15 minutes, hose off the acid thoroughly and allow to dry. Apply 3 coats of Paint, Rubber Base (For Swimming Pools), Federal Specification

TT–P–95. The first coat shall be thinned with 1 part Xylene to 4 parts paint (by volume) and thoroughly brushed into the pores. The second and third coats shall be applied as received and applied so as to present a surface free of pin-holes and other imperfections.

While the paint dries tack free within 1 hour, 48 hours drying time is recommended between coats. If the swimming pool is in the interior of a building, provide ventilation while applying paint, and during drying time. If it is necessary to spray the final coat to insure a continuous film, it should be thinned to spraying consistency with Synthetic Enamel Thinner. The paint shall be allowed to dry for at least 7 days before filling the swimming pool with water.

While this paint is intended primarily for painting concrete swimming pools, it may also be used for interior and exterior masonry surfaces which are exposed to water, water vapor, or other severe conditions of moisture. Such structures may include shower rooms, water plants, reservoirs, filter basins, laundries, etc. It is not intended for use over other types of paints.

Xylene and muriatic acid are procurable through Navy supply channels. It will be necessary to procure Rubber-Base Paint (For Swimming Pools), Fed. Spec. TT–P–95 on open market purchase.

2–2–4, COATING SYSTEM FOR DRY WALL CONSTRUCTION

All types of dry wall construction including sheetrock, gypsum board, fiberboard, prestwood, acoustical tile, etc., shall be given the following paint system:

(a) Clean the surface.

(b) Apply 1 coat of Interior Synthetic Rubber Emulsion Paint to seams preliminary to the overall coat.

(c) Apply 1 overall coat of Interior Synthetic Rubber Emulsion Paint.

NOTE: On acoustical tile, apply the paint in a thin coat and brush it out to avoid bridging and clogging of the pores. A small amount of thinning with water is permissible if clogging of pores is experienced. Frequent repainting of acoustical tile shall be avoided.

In sanitary spaces where a glossy paint is required with high scrubbability, the following system shall be used:

(a) Clean the surface.

(b) Apply 1 coat of Interior Wall Primer-Sealer.

(c) Apply 1 coat of Interior White Enamel Undercoat.

(d) Apply 1 coat Interior Gloss Enamel.

2–2–5, COATING SYSTEMS FOR ROOFS

(A) WOOD ROOFS

Wood roofs shall be painted according to the following system:

(a) Clean the surface.

(b) Apply 1 coat Exterior White Wood Primer.

(c) Apply 2 coats Tile Red Roof Paint.

When wood roofs have previously been stained the following system shall be used:

(a) Clean the surface.

(b) Apply 2 coats Ready-Mixed Aluminum Paint.

(c) Apply 2 coats Tile Red Roof Paint.

(B) METAL ROOFS

Metal roofs shall be painted according to the following system:

(a) Prepare the surface.

(b) Apply 1 coat Pretreatment - Wash Primer.

(c) Apply 1 coat Quick Drying Red Lead Primer.

(d) Apply 2 coats Tile Red Roof Paint.

NOTE: Where roofs are used for drinking water, catchments, step (c) shall be omitted and 2 coats applied in step (b).

(C) COMPOSITION SHINGLE ROOFS

Composition shingle roofs on new construction should match the color of Tile Red Roof Paint as nearly as possible. Composition shingle roofs shall not be painted.

(D) SLATE ROOFS

Slate roofs shall not be painted.

(E) BUILT-UP ROOFS

Built-up roofs shall be made of layers of roofing felt and bituminous materials constructed in accordance with Bureau of Yards and Docks Specification No. 7YG.

2-2-6, INTERIOR OF WOOD WATER TANKS

The interior of wood water tanks shall not be painted. Discoloration of water in new Redwood tanks can be avoided by adding Sal Soda to the first filling. Use 2 lbs. of Sal Soda for each 1000 gals. of water. Dissolve the Sal Soda in a bucket of hot water and add the mixture to the tank as it is being filled to insure uniform dispersion of the mixture. Keep the Sal Soda in the tank for two days. Drain and rinse the tank prior to use.

2-2-7, WINDOW AND DOOR SCREENS

Metal screen wire cloth which is new or unpainted shall be protected with 1 coat of Spar Varnish applied to both sides of the screen. Metal screen wire cloth shall be refinished annually with 1 coat of Spar Varnish applied to both sides of the screen. Frequent finishing should be avoided because of the possibility of clogging the mesh.

Surface preparation shall consist of cleaning the mesh with a cloth moistened with paint thinner to remove dust, dirt, oil, and grease. In cases where heavy soiling is experienced, brushing with a stiff bristle brush or broom may be advisable, followed by the paint thinner wash.

Varnish can be easily applied to screen wire cloth, without clogging the mesh, by using a clean chalkboard eraser or similar device as an applicator. If clogging difficulties persist the varnish may be thinned as necessary with Paint Thinner.

Window screen frames and screen door frames shall be painted with the same materials and in the same color as the adjoining window or door trim.

2-2-8, INTERIOR OF REFRIGERATION BOXES

Interior surfaces of refrigeration boxes shall be painted according to the following system:

(a) Clean the surface (surface must be dry and warm).

(b) 1. Metal. Apply 1 liberal coat of Pretreatment-Wash Primer, follow with 1 coat of Quick Drying Red Lead Primer.

2. Wood. Apply 1 coat of Interior White Enamel Undercoat.

(c) Apply 2 coats Exterior Gloss Enamel.

2-2-9, ROAD AND RUNWAY MARKINGS

Road and runway markings may be applied by hand and/or by striping machine. Apply 2 coats of Traffic Paint to dry, clean pavement. Paint Thinner may be used to remove grease and oil from the surface.

2-2-10, BRIGHTWORK

Brightwork shall be polished and coated with lacquer. Polish the brightwork to a shiny surface with metal polish. Wipe off all polish. Remove grease film by wiping with Lacquer Thinner. Apply 1 coat of Brushing Lacquer. This coating should protect interior brightwork for many months and exterior brightwork for at least 1 month. When the surface shows sign of tarnish, remove the old coating by wiping with Lacquer Thinner or Vinyl Paint Thinner, then repeat the above procedure. Do not apply more than 1 thin coat of the lacquer. This will facilitate its removal when necessary. Brightwork that can be removed for spraying should be given 1 coat of Spray-Type Plastic Coating. Do not attempt to spray plastic in humid weather as it will blush.

2-2-11, SURFACES REQUIRING ACID RESISTANCE

Surfaces in darkrooms, battery lockers and laboratories which require acid-resisting protection shall be given the vinyl paint system specified below:

(A) METAL SURFACES

(a) Prepare the surface.

(b) Apply 1 coat of Pretreatment-Wash Primer.

(c) Apply 1 coat of Vinyl Red Lead Primer.

(d) Apply 2 coats of Vinyl-Alkyd Paint.

(B) WOOD SURFACES

(a) Prepare the surface.
(b) Apply 2 coats of Vinyl-Alkyd Paint.
(c) Apply 2 coats Spraying Lacquer, Pigmented.

2–2–12, SHOP HELMETS

(a) Clean surface.
(b) Wipe with Lacquer Thinner to remove grease and oil films.

2–2–13, SURFACES NOT TO BE PAINTED

It is impractical to enumerate all of the surfaces which are not to be painted at shore establishments. However, the following examples are given as a guide.

SURFACES NOT TO BE PAINTED

Applicator Nozzles	Polish
Asbestos-Cement Siding	No Coating
Bells, Bronze	Polish and coat with clear plastic or lacquer
Boatswain's Chairs	2 coats of Boiled Linseed Oil
Bolts, Threaded Parts Exposed to Weather	Chain, Gear and Wire-Rope Lubricating Grease
Brick Walls	No Coating
Brightwork (all types)	Polish and coat with clear plastic or lacquer
Canvas Covers, Removable	No Coating
Cedar Closet Linings	No Coating
Chain, Galvanized	No Coating
Composition Water Ends of Pump	No Coating
Concrete; Walls, Walks, Decks, Roads Previously Unpainted	No Coating
Creosoted Wood Where Appearance is Not Important such as telephone poles, Antenna Masts, Dolphins, etc.	Creosoted only
Fire-Hose	No Coating
Fire-Hose Nozzles	Polish
Floors, Industrial Shop	No Coating
Floors, Linotile or Linoleum	Waxed
Floor Treads, Non-Skid	No Coating
Gaskets, Rubber, for Joints	No Coating
Gears, Exposed	Chain, Gear and Wire-Rope Lubricating Grease
Glands, Packing	No Coating
Gratings, Wood	2 coats of Boiled Linseed Oil
Grease Cups	No Coating
Hose, All Types	No Coating
Insulators	No Coating
Ladders, Portable, Wood	2 coats of Boiled Linseed Oil
Leather Coverings	Oil with Leather Preservative, Neatsfoot
Linoleum	Wax with Liquid Floor Wax
Mahogany and other fine wood paneling	Treat as before
Markers, U. S. Coast and Geodetic Survey Line Markers	Do not paint or disturb. Protect with guard posts or fence if necessary. .
Masonry, Previously unpainted	No Coating
Name Plates	No Coating
Oil Cup	No Coating

Oil Holes	No Coating
Ordnance, Working Parts of	Polished and/or Lubricated
Plaques	No Coating
Plumbing Fixtures, Plated	Polish
Porcelain Enamel Surfaces	No Coating
Spark Proof Floors	No Coating
Stages	2 coats Boiled Linseed Oil
Tile, Asphalt, Rubber, Linoleum, Cork	Waxed
Tile, Ceramic	No Coating
Tree Trunks	No Coating
Valves; stems, glands, threaded and machined parts	No coating on interior valves. Exterior valves may be coated with Chain, Gear and Wire-Rope Lubricating Grease.

SECTION 2–3

COATING SYSTEMS FOR VESSELS

Since the majority of vessels are constructed of steel, coating systems are based primarily on the location of the surface rather than the type of material. The following paint systems are designed to be applied to properly prepared bare surfaces. Methods of surface preparation for the various surfaces encountered in vessel construction are discussed in Section 1–5, Surface Preparation.

2–3–1, PROTECTIVE COATINGS DURING CONSTRUCTION

(A) STEEL VESSELS

Rust and mill scale shall be removed and the steel thoroughly cleaned by acid-pickling, flame-descaling or by an approved blasting process.

Pickling and subsequent cleaning shall be in accordance with the following sequence:

Bath 1, pickling solution: To consist of a 5 per cent (by volume) solution of sulfuric acid maintained at 170°-180° F., with an inhibitor, N. D. Spec. 51–I–2. Acid concentration shall not be permitted to drop below 3.5 per cent (by volume), or iron content to rise above 6 per cent (by weight).

Bath 2, hot water rinse: To consist of fresh circulating water maintained at 170°-180° F. Immersion time shall be about 2 minutes. Combined concentrations of sulfuric acid and iron sulfates in this bath, due to *carry-over* by the pickled steel, shall not exceed 2.0 grams per gal. To prevent rapid contamination of the bath the pickled steel shall be withdrawn slowly from Bath 1, and allowed to drain for one-half minute over pickling tank.

Bath 3, inhibiting solution: To consist of a solution in accordance with the following formula for 100 gals.:

	Pounds
Sodium dichromate, technical grade	6.3
Phosphoric acid, 85 per cent grade	4.9
or	
Phosphoric acid, 75 per cent grade	5.6

Maintain bath at 195° to 205° F. with an assured minimum of 190° F. Two minutes (minimum) immersion is required; 5 to 10 minutes immersion is required for a higher degree of rust inhibition.

During the above processes, steel shall be handled on edge and not laid flat.

Remove grease with nonflammable solvent and wipe surfaces dry. If required by Supervisor, pitted surfaces otherwise acceptable, may be smoothed out with smoothing cement for iron or steel.

Immediately after cleaning the steel, coat all surfaces with 1 coat of Pretreatment-Wash Primer, followed by 1 coat of Quick Drying Red Lead Primer.

On steel vessels of welded construction the cleaning and coating with wash primer may be done either before or after fabricating the ship's structure. When this treatment is given to the metal prior to its being welded into the ship's structure all welds and burned areas occurring in construction shall be cleaned to bare metal and touched up with 2 coats of the wash primer.

Where riveted construction is used, apply the above system to all faying surfaces prior to assembly.

Interiors of tanks, galvanized steel and metals other than steel shall be treated in accordance with the detailed requirements in subsequent paragraphs.

(B) WOODEN VESSELS

Wood vessels under construction shall have all surfaces, except the wearing surfaces of deck planking and surfaces to be oiled or varnished, treated with Copper Naphthenate Wood Preservative. The treatment shall be applied after the structural members have been assembled and prior to planking or decking. Joints, faying surfaces and butt ends shall be thoroughly drenched and particular care shall be given to these surfaces in the bow and stern of the boat. After being planked and calked the hull planking shall be treated, working as much of the preservative into the seams and calking material as possible. Seams in hull planking shall be finished using either Oil Type Calking Compound or an approved commercial seam compound. Calking compound shall be worked well down into the seams and the seams filled nearly flush (slightly concave) with the surface of the planking to allow for swelling after vessel is waterborne.

Deck planking shall be calked with cotton and oakum. The seams shall be filled almost to overflowing with the preservative. Care must be taken to immediately wipe up any preservative which is spilled on the wearing surface of the planking since it will stain. After the preservative has soaked in and allowed to dry, the seams are to be payed with Marine Glue or an approved seam compound.

Countersunk bolt and screw holes in hull and deck planking shall be treated with preservative before plugging. Plugs shall be treated by soaking in the preservative for at least 5 minutes. Butt blocks shall be soaked at least 30 minutes.

2-3-2, BOTTOM PAINTING

(A) STEEL VESSELS IN SALT WATER

Bottom painting (from keel up to bottom edge of boot-topping or bottom edge of color coat if there be no boot-topping) of steel vessels and steel boats regularly kept afloat in salt water shall be accomplished as follows. For painting of sonar domes and transducers see Article 2-3-30, Sonar Domes and Transducers.

(a) Prepare surface.

(b) Apply 1 liberal coat of Pretreatment-Wash Primer.

(c) Apply 1 coat Anticorrosive Shipbottom Primer, Formula 14N.

(d) Apply 1 coat Anticorrosive Shipbottom Primer, Formula 14ND.

(e) Apply 1 coat Anticorrosive Shipbottom Primer, Formula 14N.

(f) Apply 3 coats of Cold Plastic Antifouling Paint, Formula 105.

NOTE: The only difference between formulas 14N and 14ND is one of color.

(B) STEEL VESSELS IN FRESH WATER

The bottoms of steel vessels regularly kept afloat in fresh water (2nd and 9th Districts) shall not be painted below the lower edge of the boot-topping. Sea chests and draft marks shall be painted. However, bottoms of inactive vessels which are to be laid up for more than 6 months at a time in waters contaminated by industrial waste shall be painted. When bottom painting and painting of sea chests is accomplished the following system shall be used.

(a) Prepare surface.

(b) Apply 1 liberal coat of Pretreatment-Wash Primer.

(c) Apply 1 coat Anticorrosive Shipbottom Primer, Formula 14N.

(d) Apply 1 coat Anticorrosive Shipbottom Primer, Formula 14ND.

NOTE: At routine overhaul dockings, unpainted bottoms shall be cleaned by high pressure hoses or scrubbed as necessary to permit inspection of hull plating.

(C) STEEL 40-FT. UTILITY BOATS

(a) Prepare surface.

(b) Apply 1 liberal coat of Pretreatment-Wash Primer (0.5 mils dry film thickness).

(c) Apply 1 coat Vinyl Red Lead Primer (Brown).

(d) Apply 1 coat Vinyl Red Lead Primer (Orange).

(e) Apply 1 coat Vinyl Red Lead Primer (Brown).

(f) Apply 2 coats of Vinyl Antifouling Paint.

NOTE: Combined dry film thickness of (c), (d) and (e) should be at least 4.0 mils. Dry film thickness of (f) should be at least 3.5 mils. Total dry film thickness should be 8.0 mils. On 40-ft. Utility Boats attached to the 2nd and 9th Districts, omit step (f) above.

(D) WOODEN BOATS

The following instructions apply to the bottom painting of wooden boats regularly kept afloat. Wooden boats not regularly kept afloat, such as shipboard boats and boats in launchways at Lifeboat Stations, shall have their bottoms painted with the same paint system as is prescribed for the hull above the boot-top or waterline.

(a) Prepare surface. Treat with Copper Naphthenate Wood Preservative.

(b) Apply 1 coat Vinyl Antifouling Paint thinned by the addition of 1 pt. of Vinyl Paint Thinner per gal. of paint.

(c) Apply 2 coats Vinyl Antifouling Paint as received.

Boats Previously Painted With Conventional Antifouling Paint—

Remove all existing bottom paint to bare wood. Examine underwater body carefully for evidence of marine borer attack. If found, localities or spots of entrance shall be treated by burning with torch or replacing destroyed material as necessary.

(a) Apply 1 coat Vinyl Antifouling Paint thinned by the addition of 1 pt. of Vinyl Paint Thinner per gal. of paint.

(b) Apply 2 coats Vinyl Antifouling Paint as received.

NOTE: On initial application, paint is to be applied by brush. Because of rapid drying and tendency of each coat to soften previous coat, some difficulties may be encountered if workmen attempt to brush paint out as much as conventional paints. Cross work only once.

Boats Previously Coated With Vinyl Antifouling Paint—

Principal growth which will normally be present is grass and algae. These are to be removed by vigorous scrubbing with stiff fiber bristle brush such as a deck scrubber. Remove all oil and grease. Since vinyls are very resistant to alkalies and oils, strong non-abrasive detergents, kerosene or paint thinner may be used. After rinsing and allowing to dry thoroughly, wire brush any areas where paint appears to be peeling, blistering or loosely adhering. Sand all such rough spots smooth. Examine bottom for marine borers and treat as described above.

(a) Touch up bare wood areas with 1 coat of Vinyl Antifouling Paint thinned by the addition of 1 pt. of Vinyl Paint Thinner per gal. of the Vinyl Antifouling Paint.

(b) Apply 2 coats Vinyl Antifouling Paint as received on entire bottom.

NOTE: If spray application is used apply a fourth coat.

Boats With Sheathed Bottoms—

Boats with sheathed bottoms shall have their entire hull, including the area in way of sheathing, treated with Copper Naphthenate Wood Preservative. Copper sheathing shall not be painted.

Boats sheathed with metal other than copper shall have the sheathing painted as described for bottoms of steel vessels.

Marking—

In order to readily identify boats which have vinyl bottom paint, the date of application shall be marked in No. 14 Brilliant Yellow on the starboard quarter not less than 1 foot below normal waterline.

Zincs—

Zincs shall not be used on wooden boats except where steel is used in shafts, rudders or other important exterior appendages and only when the steel is in vicinity of or coupled to bronze propellers or fittings. The use of zincs reduces the effectiveness of antifouling paints and causes accelerated fouling of bronze and copper materials.

Painting Schedule—

All wooden boats regularly kept afloat shall have their underwater bodies repainted with three brush coats or four spray coats of Vinyl Antifouling Paint as follows:

(a) Second and Ninth Coast Guard Districts, once every 2 years.

(b) All other Districts, once each year normally in March or April.

(c) In all Districts other than Second and Ninth, if boat has not been hauled for other reasons and touched up within 6 months after complete bottom painting, the boat shall be hauled for vinyl touch-up prior to being 200 days waterborne.

(d) In all Districts whenever a boat is hauled for any reason, bare and abraded areas shall be touched up with 2 coats of the Vinyl Antifouling Paint.

2–3–3, BOOT-TOP PAINTING

(A) STEEL VESSELS IN SALT WATER

Vessels With White Hulls—

(a) Prepare surface.

(b) Apply 1 liberal coat of Pretreatment-Wash Primer.

(c) Apply 1 coat Anticorrosive Shipbottom Primer, Formula 14N.

(d) Apply 1 coat Anticorrosive Shipbottom Primer, Formula 14ND.

(e) Apply 1 coat Anticorrosive Shipbottom Primer, Formula 14N.

(f) Apply 3 coats Cold Plastic Antifouling Paint, Formula 146.

NOTE: The only difference between formulas 14N and 14ND is one of color.

Vessels With Black Hulls—

(a) Prepare the surface.

(b) Apply 1 liberal coat of Pretreatment-Wash Primer.

(c) Apply 3 coats Quick Drying Red Lead Primer.

(d) Apply 3 coats Red Boot-Topping Paint.

(B) STEEL VESSELS IN FRESH WATER

(a) Prepare surface

(b) Apply 1 liberal coat of Pretreatment-Wash Primer.

(c) Apply 3 coats Quick Drying Red Lead Primer.

(d) Apply 3 coats Exterior Gloss Enamel.

(C) STEEL 40-FT. UTILITY BOATS

(a) Prepare surface.

(b) Apply 1 liberal coat of Pretreatment-Wash Primer.

(c) Apply 3 coats Quick Drying Red Lead Primer.

(d) Apply 3 coats Exterior Gloss Enamel.

(D) WOODEN BOATS

These instructions apply to the painting of boot-topping areas on wooden boats regularly kept afloat. Wooden boats not regularly kept afloat, such as shipboard boats and boats in launchways at Lifeboat Stations, shall have their boot-topping painted with the same system as is prescribed for the hull above the boot-topping.

(a) Prepare the surface. Treat with Copper Naphthenate Wood Preservative.

(b) Apply 1 coat of Exterior Gloss Enamel thinned by the addition of 1 pt. of Boiled Linseed Oil per gal. of paint.

(c) Apply 2 coats Exterior Gloss Enamel as received.

2–3–4, EXTERIOR OF VESSELS ABOVE THE BOOT-TOP—GENERAL

(A) EXTERIOR UNGALVANIZED STEEL

Ungalvanized steel, except those surfaces to receive special treatment as set forth in subsequent paragraphs, shall be painted as follows:

(a) Prepare the surface.

(b) Apply 1 liberal coat of Pretreatment-Wash Primer.

(c) Apply 3 coats of Quick Drying Red Lead Primer.

(d) Apply 2 coats of Exterior Gloss Enamel.

(B) EXTERIOR GALVANIZED STEEL AND CORROSION-RESISTING METALS OTHER THAN ALUMINUM

(a) Prepare the surface.

(b) Apply 1 liberal coat of Pretreatment-Wash Primer.

(c) Apply 2 coats of Quick Drying Red Lead Primer.

(d) Apply 2 coats of Exterior Gloss Enamel.

(C) EXTERIOR ALUMINUM

Exterior aluminum surfaces shall be painted as follows:

(a) Prepare the surface.

(b) Apply 1 liberal coat of Pretreatment-Wash Primer.

(c) Apply 2 coats Yellow Zinc Chromate, High Alkyd Type.

(d) Apply 2 coats Exterior Gloss Enamel.

Aluminum shall be protected from contact with dissimilar metal or wood by use of one of the following Navy standard stock items as appropriate.

Packing, Joint Sealing (Coated Cloth) Oil and Water Stop, Items 3437 to 3445, Class 33.

Gasketing, Wool Felt, Adhesive Back, Sheet, Items 1008 to 1015, Class 33.

Calking Compound, Oil Type, for Metal Seams and Airports, Items 1971 to 1979, Class 52.

Watertight joints between aluminum parts shall be made tight by metal calking rather than the use of gaskets, packing or calking compounds. Where this is not possible the gasket, packing or compound material given above may be used.

Wood in contact with aluminum shall be given 1 coat of Spar Varnish, prior to assembly.

Asbestos paper or similar absorbent material shall not be used in contact with aluminum.

Threaded parts of aluminum shall be coated, before assembly, with an antiseize mixture of 50 per cent Zinc Dust and 50 per cent Petrolatum (by volume).

Label plates, regardless of material, to be installed on aluminum structure or fittings shall be imbedded in the Calking Compound, Oil Type referred to above.

(D) EXTERIOR WOOD

Wood which is to be painted, including hull planking, shall be treated as follows:

(a) Prepare surface.

(b) Apply 2 coats of Copper Naphthenate Wood Preservative. Allow 72 hours drying time, longer if possible, after each coat.

(c) Apply 1 coat of Exterior Gloss Enamel thinned by the addition of 1 pt. of Boiled Linseed Oil to each gal. of paint.

(d) Apply 2 coats of Exterior Gloss Enamel as received.

Wood which is to be varnished such as accommodation ladders shall not be treated with preservative but shall be stained to the shade desired and given 5 coats of Spar Varnish. Undercoats shall be sanded to provide a suitable gripping surface for subsequent coats of varnish.

Wood surfaces where visual appearance is not important and where moist conditions tend to prevail shall be treated with Copper Naphthenate Wood Preservative, but not painted. The wood shall be re-treated periodically to maintain the green color in wood caused by the Preservative.

Exterior wood surfaces exposed to view but not to be painted or varnished such as wood gratings shall be treated with 3 coats of Boiled Linseed Oil.

2-3-5, INTERIOR OF VESSELS—GENERAL

(A) INTERIOR METAL SURFACES NOT SUBJECT TO MOISTURE

Interior metal surfaces (all types of metal) not subject to sweating, steam or moisture shall be painted as follows:

(a) Prepare the surface.

(b) Apply 2 liberal coats of Pretreatment-Wash Primer.

(c) Apply 1 coat of Interior Gloss Enamel.

(B) INTERIOR METAL SURFACES SUBJECT TO MOISTURE

Interior metal surfaces subject to sweating, steam condensation, water leakage or other forms of moisture shall be painted as specified below. This system shall be used for the interior of the skin of the ship, tank boundaries, pantries and galleys:

(a) Prepare the surface.

(b) Apply 1 coat of Pretreatment-Wash Primer.

(c) Apply 1 coat Quick Drying Red Lead Primer.

(d) Apply 2 coats Interior Gloss Enamel.

(C) INTERIOR WOOD

Interior wood surfaces which are to be painted shall be treated as follows:

(a) Prepare the surface.

(b) Apply 2 coats Copper Naphthenate Wood Preservative to all surfaces not coated during construction and subject to moist or humid conditions. Allow 72 hours drying time, longer if possible, after each coat.

(c) Apply 1 coat Interior White Enamel Undercoat.

(d) Apply 1 coat Interior Gloss Enamel.

Wood which is to be varnished shall not be treated with preservative but shall be stained to the shade desired and given 3 coats of Spar Varnish. Undercoats shall be sanded and finish coat polished with pumice and water to give an acceptable finish.

Wood surfaces where visual appearance is not important and where moist conditions tend to prevail shall be thoroughly soaked with wood preservative and left unpainted. This includes such spaces as chain locker, tank compartments, voids, forepeaks, buoyancy spaces, lazarettes, bilges, lockers, etc. Those spaces near the transom and bow and in the area of the turn of the bilge shall be given particular attention. These spaces should be examined frequently and the wood re-treated periodically in order to maintain the green color which indicates that adequate toxic material is present. It will be necessary to examine bilges more frequently than other spaces since the preservative will leach out rapidly due to the washing action of the bilge water.

Wood surfaces exposed to view but not to be either painted or varnished such as wood gratings, shall be treated with 2 coats of Boiled Linseed Oil.

(D) INSULATION

Fibrous glass board, glass cloth and other insulating surfaces shall be given 2 coats of Interior Gloss Enamel.

2–3–6, DECKS AND FLOOR PLATES

Metal decks to be painted shall be treated in accordance with the aforegoing general instructions for the particular location and the particular metal under consideration, except that the finish coat shall be Deck and Floor Paint. Non-skid properties may be added to the deck paint by sweeping washed beach sand on the surface while it is tacky. Sweep the sand well into the paint, allow to dry and sweep off loose sand particles. A final coat of Deck and Floor Paint, thinned by the addition of 1 pt. of Paint Thinner per gal. of paint, shall then be applied over the sand.

Decks to be covered by Mastic, tile or non-slip treads shall be cleaned to bare metal immediately before application of the deck covering.

The top surface of floor plates shall be wire-brushed to a bright surface and coated with a thin film of diesel oil renewed weekly. Edges and bottoms of floor plates shall be cleaned to bare metal, given a liberal coat of Pretreatment-Wash Primer and 2 coats of Quick Drying Red Lead Primer.

Wooden decks shall be calked with oakum and cotton. The seams shall be filled almost to overflowing with Copper Naphthenate Wood Preservative. Care must be taken to immediately wipe up any preservative which is spilled on the wearing surface of the planking since it will stain. After the preservative has soaked in and allowed to dry, the seams shall be payed with Marine Glue or an approved seam compound.

In general, wooden decks are to be left unpainted with the exception of small work craft such as buoy boats, motor cargo boats and barges. Decks which are to be painted shall be treated as follows:

(a) Prepare the surface.

(b) Apply 1 coat of Deck and Floor Paint thinned by the addition of 1 qt. of thinning mixture to each gal. of paint. Thinning mixture is composed of ⅔ Boiled Linseed Oil and ⅓ Paint Thinner.

(c) Apply 2 coats of Deck and Floor Paint as received.

2–3–7, WATER TANKS

Feed water tanks and potable water tanks shall be painted with 2 coats of Zinc Dust Paint over bare metal. Pretreatment-Wash Primer shall not be used under the Zinc Dust Paint.

To apply the Zinc Dust Paint remove all former paint coatings by sandblasting if possible; power wirebrushing may be used where sandblasting is impractical. Following the cleaning and prior to painting, the Engineering Officer shall inspect the tank to insure that all dirt, sand and rags have been removed from the tank bottom, longitudinal frames and stringers, sounding tubes, vents, filling connections and suctions. Apply first coat of Zinc Dust Paint. Following the first coat of Zinc Dust Paint, unheated air shall be circulated through the tank for 12 hours. The second coat shall then be applied and unheated air circulated through the tank for 12 hours. Tanks shall then be flushed twice with fresh water before being placed in service.

CAUTION: Only enough Zinc Dust Paint shall be mixed for daily use. Mixed Zinc Dust Paint shall not be stored. After the material is mixed a gas pressure may build up within closed containers.

2-3-8, FUEL OIL TANKS

Fuel oil tanks whether fitted for ballasting or not shall be left uncoated except for the oil film.

2-3-9, LUBE OIL TANKS AND RESERVOIRS

Lube oil tanks and lube oil reservoirs in machinery shall be left uncoated when in use. When lube oil tanks or lube oil reservoirs are out of service for 30 days or more the surface shall be coated with Thin-Film Rust Preventive, Grade II. When returning tanks or reservoirs to service the preservative shall be removed in accordance with BuShips Bulletin of Information No. 39, dated 1 July 1950.

2-3-10, VOIDS AND COFFERDAMS

Clean to bare metal and treat with a liberal coat of Pretreatment-Wash Primer. Prime with 2 coats Quick Drying Red Lead Primer. Finish with 2 coats of Interior Gloss Enamel, White.

Voids which are normally dry but occasionally flooded shall be given 2 coats of Bituminous Emulsion Coating over bare clean steel.

2-3-11, CHAIN LOCKER

Clean to bare metal and apply 2 coats of Bituminous Emulsion Coating.

2-3-12, BILGES

Clean to bare metal. Apply a liberal coat of Pretreatment-Wash Primer. Apply 3 coats of Quick Drying Red Lead Primer.

2-3-13, INACCESSIBLE SPACES

Inaccessible steel surfaces such as bilge wells, peak spaces, machinery and crane foundations, areas under built-in furniture and bulkheads behind switchboards shall be cleaned to bare metal and coated for greater permanence as follows:

(a) Apply 1 liberal coat of Pretreatment-Wash Primer.

(b) Apply 3 coats of Quick Drying Red Lead Primer.

Inaccessible steel surfaces such as the interiors of rudders, bilge keels, docking keels, and skegs shall be liberally coated with Thin-Film Rust Preventive, Grade II. The rust preventive may be applied by brush, spray or filling and draining, whichever process gives most thorough application in the particular case under consideration.

2-3-14, SPACES SUBJECT TO HEAVY SWEATING

Spaces subject to heavy sweating shall be given an antisweat treatment as follows:

(a) Clean to bare metal.

(b) Apply 1 liberal coat of Pretreatment-Wash Primer.

(c) Apply 2 coats of Quick Drying Red Lead Primer.

(d) Apply 1 liberal spray coat of Anti-sweat-Coating Binder.

(e) Apply 1 coat of Expanded Vermiculite.

(f) Apply 2 coats of Interior Gloss Enamel.

2-3-15, DUKW RUBBING RAILS

Rubbing rails on DUKWs shall be cleaned by burning off or removing the end plates for access. Fit filling, drain and vent holes in each member by welding threaded pipe bosses into the member permitting closure by pipe plugs. These pipe bosses are to be installed on the tops or bottoms of the members and not on the rubbing faces. Fill and drain each member with Thin-Film Rust Preventive, Grade II. Filling and draining shall be done annually to insure adequate protection against corrosion.

2–3–16, PIPING

(A) INTERIOR PIPING

Interior uninsulated piping, the operating temperature of which is 300° F. or less, shall be thoroughly cleaned and coated as follows:

(a) 1 liberal coat of Pretreatment-Wash Primer.

(b) 2 coats Quick Drying Red Lead Primer.

(c) 2 coats Interior Gloss Enamel.

Piping below the level of the lower level floor plates shall be given an additional coat of Quick Drying Red Lead Primer (3 coats total) in lieu of the Interior Gloss Enamel.

Uninsulated metallic piping, the operating temperature of which exceeds 300° F. shall be cleaned thoroughly and given 2 coats of Heat-Resisting Paint.

Insulated metallic piping, the operating temperature of which does not exceed 300° F. shall be given 1 liberal coat of Pretreatment-Wash Primer and 1 coat of Quick Drying Red Lead Primer on the exterior surface of the pipes. The fabric covered insulation shall be given 2 coats of Interior Gloss Enamel.

Insulated metallic piping, the operating temperature of which exceeds 300° F. shall not be painted prior to application of insulation. Exposed flanges and valve bonnets shall be given 2 coats of Heat-Resisting Paint.

(B) EXTERIOR PIPING

Exterior piping shall be painted in accordance with the preceding instructions for painting exterior surfaces.

2–3–17, MACHINERY

(A) INTERIOR MACHINERY

Uninsulated metallic surfaces, other than aluminum surfaces, the operating temperature of which is 300° F. or less, shall be thoroughly cleaned and coated as follows:

(a) 1 liberal coat of Pretreatment-Wash Primer.

(b) 2 coats Quick Drying Red Lead Primer.

(c) 2 coats Equipment Enamel.

Uninsulated aluminum surfaces, the operating temperature of which is 300° F. or less, shall be thoroughly cleaned and coated as follows:

(a) 1 liberal coat of Pretreatment-Wash Primer.

(b) 2 coats Yellow Zinc Chromate, High Alkyd type.

(c) 2 coats Equipment Enamel.

Uninsulated metallic surfaces, the operating temperature of which exceeds 300° F. shall be thoroughly cleaned and coated with 2 coats of Heat-Resisting Paint.

Fabric covered insulated surfaces or rough surfaces finished with Smoothing Cement shall be given 2 coats Equipment Enamel.

(B) EXTERIOR MACHINERY

Exterior Machinery shall be painted in accordance with the preceding instructions for painting exterior surfaces.

2–3–18, FURNITURE AND EQUIPMENT

(A) INTERIOR

Furniture, joiner doors, switch boxes, controllers, switchboards, gauge boards, connection boxes, laundry machinery, galley equipment and miscellaneous equipment of all types shall be painted as follows:

(a) Prepare surface.

(b) Apply 1 liberal coat of Pretreatment-Wash Primer.

(c) Apply 1 coat Quick Drying Red Lead Primer.

(d) Apply 2 coats of Equipment Enamel.

Equipment presently installed having a lacquer or baked enamel finish in a color other than the prescribed color shall not be repainted for color purposes alone. When repainting is accomplished the above system shall be followed.

(B) EXTERIOR

Exterior equipment shall be painted in accordance with the preceding instructions for painting exterior surfaces.

2–3–19, ELECTRIC CABLE

When installing new cables, interior or exterior, the support brackets shall be painted before installing cables. Cables shall be painted after they have been pulled but prior to being secured in position.

(A) INTERIOR CABLES

Armored cables shall be painted as follows:

(a) Apply 1 liberal coat of Pretreatment-Wash Primer.

(b) Apply 2 coats of Yellow Zinc Chromate, High Alkyd type.

(c) Apply 2 coats of Interior Gloss Enamel.

(B) EXTERIOR CABLES

Exterior armored cables shall be painted as follows:

(a) Apply 1 liberal coat of Pretreatment-Wash Primer.

(b) Apply 3 coats of Yellow Zinc Chromate, High Alkyd type.

(c) Apply 2 coats of Exterior Gloss Enamel.

Tape armored and rubber covered cables shall not be painted except with Electrical Insulating Varnish, Grade BA.

2–3–20, RADIO AND RADAR ANTENNAS

Radio direction finder loops, and housings and pedestals for other electronics equipment shall be painted as follows:

(a) Apply 1 liberal coat of Pretreatment-Wash Primer.

(b) Apply 3 coats of Yellow Zinc Chromate, High Alkyd type.

(c) Apply 2 coats of Exterior Gloss Enamel.

Care shall be taken to avoid painting the gasket on the direction finder antenna as well as all electrical contact points, ceramic insulators, rubber shock mounts, and other insulating material on electronic equipment, by properly masking all such parts before painting.

Instructions for painting radar antennas, radomes and similar electronic radiators are found in the Equipment Maintenance Bulletin Series.

Whip and wire antennas shall be left unpainted.

2–3–21, ANCHOR CHAIN

The painting of anchor chain serves no useful purpose in so far as prolonging the life of the chain is concerned. Regardless of the paint system employed it is impossible to keep paint on the points of wear. Thus, those areas which will fail first due to wear and corrosion cannot be protected by painting. Painting of anchor chain serves primarily to improve the appearance of the chain. For this reason the following low-cost paint systems shall be used on vessel anchor chain:

(a) Prepare the surface.

(b) Apply 1 liberal coat of Pretreatment-Wash Primer.

(c) Apply 1 coat of Asphalt Varnish.

Color markings shall be applied over the above coat. On light vessels that portion of the anchor chain which normally is below water when the vessel is on station shall not be painted.

2–3–22, PROPELLERS

Propellers, other than cast iron or cast steel, shall not be painted but shall be cleaned and polished bright.

Cast steel and cast iron propellers shall be thoroughly cleaned, the pits filled with Hydraulic Cement and painted with the same coating system prescribed for steel hulls in either salt or fresh water as the case might be.

2–3–23, PROPELLER SHAFTING

(A) OUTBOARD

Rubber covered shafting of vessels used in *salt water* shall be given 3 coats of Cold Plastic Antifouling Paint. Rubber covered shafting of vessels in fresh water shall not be painted.

Steel shafting shall be painted with the same paint system as is applied to bottom of the hull.

(B) INBOARD

Inboard shafting shall be thoroughly cleaned and painted as follows:

(a) Apply 1 liberal coat of Pretreatment-Wash Primer.

(b) Apply 2 coats of Quick Drying Red Lead Primer.

(c) Apply 2 coats of Equipment Enamel.

2-3-24, RUNNING RIGGING

Running rigging shall be treated with Chain, Gear and Wire-Rope Lubricating Grease, Item No. 533-547, Class 14, Catalog of Navy Material.

2-3-25, STANDING RIGGING

Standing rigging shall be swabbed as necessary with Thin-Film Rust Preventive, Grade I. Rigging insulators shall not be painted.

2-3-26, WOOD MASTS AND SPARS

Wood masts and spars shall be treated as follows:

(a) Large checks shall be soaked with Copper Naphthenate Wood Preservative, calked with oakum (only enough material being used to provide a backer) and sealed with Oil-Type Calking Compound.

(b) Small checks and holes shall be soaked with Copper Naphthenate Wood Preservative and filled with Oil-Type Calking Compound.

(c) Surfaces in way of fittings and partners shall be treated liberally with Copper Naphthenate Wood Preservative.

(d) Mast heel shall be soaked in Copper Naphthenate Wood Preservative for a period of 16 hours.

(e) Allow preservative to dry for 72 hours and apply 1 coat of Exterior Gloss Enamel thinned by the addition of 1 pt. of Boiled Linseed Oil per gal. of paint.

(f) Apply 2 coats of Exterior Gloss Enamel as received.

2-3-27, WOOD LADDERS, GANGPLANKS, STAGING AND BOATSWAIN'S CHAIRS

Wood accommodation ladders shall be given 5 coats of Spar Varnish.

Wood pilot ladders, staging and boatswain's chairs, shall be given 3 coats of Boiled Linseed Oil. At least 24 hours drying time shall be allowed between coats.

Wood gangplanks and brows may be given either 3 coats of Boiled Linseed Oil or 5 coats of Spar Varnish.

None of the above items shall be painted in order that cracks and splits may be more easily detected.

2-3-28, SMOKE PIPES

Outside of inner stack casing, inside of outer stack casing, gratings and flats between the inner and outer stack casings shall be cleaned to bare metal and given 2 coats of Heat-Resisting Paint. The exterior surfaces at the top of the stack within the outer casing shall be cleaned to bare metal and given 3 coats of Bituminous Emulsion Coating.

2-3-29, ZINCS

Zincs shall be imbedded in zinc oxide paste and the edges calked. Care must be taken to insure good metallic contact between the zincs and the hull of the vessel. Outside surfaces of zincs shall be left unpainted.

Zincs shall not be used on wooden vessels except where steel is used in shafts, rudders or other important exterior appendages and only when the steel is in vicinity of, or coupled to, bronze propellers or fittings. The use of zincs reduces the effectiveness of anti-fouling paints and causes accelerated fouling of bronze and copper materials.

2-3-30, SONAR DOMES AND TRANSDUCERS

Article 14.2 Supplement No. 19, Sonar Bulletin, NAVSHIPS 900,025 A contains the basic instructions for the care and painting of sonar domes and transducers. The instructions contained herein supersede Article 14.2 in so far as they relate to painting. The Sonar Officer is responsible for the maintenance of sonar domes and transducers and shall be familiar with both that portion of Article 14.2 which relates to care of the equipment other than painting and with the instructions contained herein relative to the painting of the equipment.

(A) INSPECTION AND FREQUENCY OF PAINTING

All surfaces in contact with sea water should be inspected each time the vessel is drydocked and treated as necessary. At each regularly scheduled availability all surfaces should be cleaned, damaged areas touched up and antifouling paint renewed. The number of antifouling coats to be applied will be dependent on the condition of the surfaces. The desired result is to reproduce the original paint film.

(B) CLEANING AND PREPARING SOUND-TRANSPARENT SURFACES

Instruction manuals should be consulted to determine the location of the various types of surfaces to be encountered in order that particular care may be taken in removing fouling from the critical areas, i.e. rubber and corrosion-resistant steel.

Remove the fouling with wooden scrapers and non-metallic non-abrasive brushes as soon after drydocking as practicable to take advantage of the soft, wet conditions of the fouling. Play a water hose on the wet sea growth to help keep it soft during the process of removal. Once the calcium encrustation arises and hardens it defies removal, which greatly increases the chances of injury to transducer diaphragm and sonar dome window.

Removal of fouling can be greatly facilitated by swabbing the affected parts with a solution of 1 volume of commercial nitric acid to 5 volumes of fresh water. Apply the solution with long-handled swabs. Allow the solution to soak a few minutes. Wipe with a rag or non-abrasive brush. Repeat the application if the surface retains some carbonate film. When satisfactorily cleaned, wash the swabbed areas with hot soapy water (on rubber surfaces use only hot water) and flush with fresh water. If the surface shows any traces of oil or grease residue after the soapy water washing, Paint Thinner may be used to dissolve and remove such persistent traces.

Inspect corrosion-resistant, sound-transparent window for pits and holes. Particular attention should be given to welded seams and spot welds. Solder such holes with a 50 per cent lead and 50 per cent tin solder. Commercial strength phosphoric acid may be used as a flux. Remove all excess solder and wash off all excess flux.

(C) PAINTING OF CORROSION-RESISTANT, SOUND-TRANSPARENT WINDOWS

Exterior Surface—

(a) Apply 1 liberal coat of Pretreatment-Wash Primer.

(b) Apply by spray 1 liberal coat of Zinc Chromate Primer, Spec. MIL–P–6889, thinned by the addition of 2 parts of Synthetic Enamel Thinner to 1 part of the Zinc Chromate. Caution: Yellow Zinc Chromate, High Alkyd Type, is not satisfactory for this purpose.

(c) Apply by spray 8 coats of Antifouling Sonar Dome Paint, thinned by the addition of 3 parts of Synthetic Enamel Thinner to 4 parts of the Antifouling Paint. Allow 1½ hours drying time between coats and 24 hours between last coat and undocking.

Interior Surface—

Follow procedure set forth above for painting the exterior surface of corrosion-resistant, sound-transparent window except apply only 4 coats of the antifouling paint.

(D) PAINTING OF SOUND-TRANSPARENT RUBBER SURFACES OF SONAR DOMES AND THE DIAPHRAGMS OF SOUND TRANSDUCERS

Apply by spray 4 liberal coats of Rubber Antifouling Paint, thinned by the addition of 1 part of Paint Thinner to 5 parts of Antifouling Paint.

(E) PAINTING OF NON-SOUND-TRANSPARENT SURFACES OF SONAR DOMES AND TRANSDUCERS

Non-sound-transparent surfaces shall be painted in the same manner as the bottom of steel hulls. Sound-transparent surfaces should be masked when shipbottom paint is applied.

NOTE: It is recommended that District Engineering Divisions keep 1 or 2 gals. of the below listed materials on hand and issue to ship's Sonar Officer the required quantity just prior to vessel's drydocking. The approximate quantities of each material needed by a single vessel are:

Zinc Chromate Primer, MIL-P-6889, 1 pt.

Antifouling Sonar Dome Paint, M-559, 1-2 qts.

Rubber Antifouling Paint, ½ pt.

2-3-31, BRIGHTWORK

Brightwork shall be polished and coated with lacquer. Polish the brightwork to a shiny surface with metal polish. Wipe off all polish. Remove grease film by wiping with Lacquer Thinner. Apply 1 coat of Brushing Lacquer. This coating should protect interior brightwork for several months and exterior brightwork for several weeks. When the surface shows signs of tarnish, remove the old coating by wiping with Lacquer Thinner or Vinyl Paint Thinner then repeat the above procedure. Do not apply more than 1 thin coat of the Lacquer. This will facilitate its removal when necessary. Where brightwork can be removed for spraying, it should be given 1 coat of Spray-Type Plastic Coating in lieu of the Brushing Lacquer. Do not attempt to spray the plastic in humid weather as it will blush.

2-3-32, SURFACES REQUIRING ACID RESISTANCE

Surfaces in darkroom, battery rooms and laboratories which require acid-resisting protection shall be given the vinyl paint system specified below:

(A) METAL SURFACES

(a) Prepare the surface.

(b) Apply 1 coat of Pretreatment-Wash Primer.

(c) Apply 1 coat of Vinyl Red Lead Primer.

(d) Apply 2 coats of Vinyl-Alkyd Paint.

(B) WOOD SURFACES

(a) Prepare the surface.

(b) Apply 2 coats of Vinyl-Alkyd Paint.

2-3-33, SURFACES NOT TO BE PAINTED

It is impractical to enumerate all the surfaces which are not to be painted. However, the following examples are given as a guide.

SURFACES NOT TO BE PAINTED

Annunciators, Brass	Polish and coat with clear plastic or lacquer
Accommodation Ladders, Wood	5 coats Varnish
Applicators Nozzles	Polish
Boat Hook Staffs	3 coats Boiled Linseed Oil
Boat Booms, Wood	5 coats Varnish
Boatswain's Chairs	2 coats Boiled Linseed Oil
Brightwork	Polish and coat with clear plastic or lacquer
Bells, Bronze	Polish and coat with clear plastic or lacquer
Bell Pulls, Brass	Polish and coat with clear plastic or lacquer
Bolts, Threaded Parts Exposed to Weather	Chain, Gear and Wire-Rope Lubricating Grease
Brows	5 coats Varnish or 3 coats Boiled Linseed Oil
Canvas Covers, Removable	No coating
Composition Water Ends of Pump	No coating
Chain, Galvanized	No coating
Drain Holes	No coating
Deck Treads, Non-Skid	No coating
Decks, Stainless Steel	Bright

2-3-33

Decks, Wood	No coating
Decks, Linotile or Linoleum	Wax with Liquid Floor Wax
Dogs, for Watertight Doors, Working Surfaces	No coating. Grease working parts.
Fire-hose, Nozzles	Polish
Gaskets, Rubber, for Joints	No coating
Gaskets, Rubber for Watertight Doors, Hatches and Airports	Pulverized graphite
Gratings, Wood	3 coats Boiled Linseed Oil
Grease Cups	No coating
Gypsy-heads Whelps	No coating
Guns, Working Parts of	Polished and/or lubricated
Glands, Packing	No coating
Gears	Lubricated
Gangplank	5 coats Varnish or 3 coats Boiled Linseed Oil
Hose, All Types	No coating
Heat Exchange Surfaces	No coating
Insulators	No coating
Joint Faces	Uncoated or coated with suitable joint compounds
Knife-edges of WT Door and Hatches	Sandpaper
Ladders, Accommodation	5 coats Varnish
Ladders, Pilot	3 coats Boiled Linseed Oil
Lubrication Fittings	Uncoated
Leather Coverings	Oil with Leather Preservative, Neatsfoot Oil
Linoleum	Wax with Liquid Floor Wax
Masts, Wood	5 coats Varnish
Name Plates	No coating
Oil Cup	No coating
Oil Holes	No coating
Ordnance, Working Parts of	Polished and/or lubricated
Plaques	No coating
Railings, Wood	5 coats Varnish
Release Mechanisms	Chain, Gear and Wire-Rope Lubricating Grease
Searchlight Shutter	No coating
Stages	2 coats Boiled Linseed Oil
Strong Backs, Wood	5 coats Varnish
Springs	No coating, or slosh with Chain, Gear and Wire-Rope Lubricating Grease
Strainers	No coating
Threaded Parts	Chain, Gear and Wire-Rope Lubricating Grease
Valves; Stems, Glands, Threaded and Machined Parts	No coating on interior valves. Exterior valves may be coated with Chain, Gear and Wire-Rope Lubricating Grease
Zincs	No coating

SECTION 2–4

COATING SYSTEMS

FOR BUOYS

2–4–1, EXTERIOR OF METAL BUOYS

Metal buoys shall not be painted more often than every 2 years. The following paint system shall be used when painting metal buoys except those which have high loss probability.

(A) ABOVE WATERLINE

(a) Prepare the surface. Wet or dry sandblast to bare metal.

(b) Apply 1 liberal coat of Pretreatment-Wash Primer. Dry film thickness 0.5 mils.

(c) Apply sufficient coats of Vinyl Red Lead Primer to give a dry film thickness of 4.0 mils. Use orange colored primer for initial and final prime coats and brown for alternate coats. Total average dry film thickness 4.5 mils.

(d) Apply sufficient coats of Vinyl-Alkyd Paint to give average dry film thickness of 3.0 mils. Total average dry film thickness 7.5 mils.

(e) Identification marking. Apply 2 coats of Exterior Vinyl-Alkyd Paint.

(f) Stencil on a suitable horizontal surface the day, month and year of completion of paint application.

(B) BELOW WATERLINE

Follow the system set forth above in 2–4–1 (A), Above Waterline, except that the finish material shall be sufficient coats of Vinyl Antifouling Paint to give average dry film thickness of 3.5 mils of antifouling paint. Total average dry film thickness for the system 8.0 mils.

Second and Ninth Districts paint buoy as specified in 2–4–1 (A), Above Waterline.

2–4–2, INTERIOR OF METAL BUOYS

The interior of buoys whose access openings are not large enough to permit access by a man, shall not be coated with paint, oil or preservative. Those buoys with access openings large enough to permit access by a man shall be painted in accordance with the below systems:

(A) UNLIGHTED AND ACETYLENE BUOYS

(a) Prepare the surface. Chip, wirebrush or sandblast.

(b) Apply 2 coats of Slow Drying Red Lead Primer to all interior surfaces.

Average dry film thickness to be not less than 3.0 mils.

(B) ELECTRIC BUOYS

Follow the system given above in 2–4–2 (A), Unlighted and Acetylene Buoys, except that in tank pockets sufficient coats of Vinyl-Alkyd Paint shall be applied over the primer to give a dry film thickness of 3.0 mils or a total dry film thickness in tank pockets of 6.0 mils.

2–4–3, METAL BUOYS WITH HIGH LOSS PROBABILITY

In some districts, particularly the 2nd, a high percentage of certain class buoys are lost each year. Whenever the expected loss of a given class of buoy exceeds 50 per cent of the buoys of *that class* placed on station, the following low-cost painting system shall be followed for that class of buoys only.

(a) Prepare the surface. Wirebrush or

sandblast the above-water portion only.

(b) Primers, none.

(c) Apply 1 coat of Exterior Gloss Enamel to above-water portion only.

(d) Lettering or numerals, 2 coats of Exterior Gloss Enamel.

(e) Underwater part of buoy, clean surface, apply no paint or preservative.

(f) No paint, oil, or preservative shall be applied to the interior of the buoys.

2-4-4, WOODEN SPAR BUOYS

Wooden spar buoys shall not be painted more often than once a year. The following system shall be used:

(A) ABOVE THE WATERLINE

(a) Apply 2 liberal coats of Copper Naphthenate Wood Preservative (applicable only where paint has been removed to bare or to new wood). Allow 72 hours drying time.

(b) Apply 1 coat of Exterior Gloss Enamel thinned by the addition of 1 pt. of Paint Thinner or Synthetic Enamel Thinner per gal. of paint.

(c) Apply 2 coats of Exterior Gloss Enamel as received.

(d) Lettering or numerals, use 2 coats of Exterior Gloss Enamel.

(B) BELOW THE WATERLINE

(a) Apply 2 liberal coats of Copper Naphthenate Wood Preservative (applicable only where paint has been removed to bare or to new wood). Allow 72 hours drying time before finishes are added.

(b) Apply 1 coat Vinyl Antifouling Paint thinned by the addition of 1 pt. of Vinyl Paint Thinner per gal. of paint.

(c) Apply 2 coats of Vinyl Antifouling Paint as received.

NOTE: Second and Ninth Districts, paint entire buoy as prescribed for above the waterline.

2-4-5, COLOR RENEWAL ON STATION

It is anticipated that some buoys may, because of bird deposits, require renewal of identification color before the buoy needs repainting for protection. Such color renewal is to be accomplished on station by tenders without effort to remove bird guano. One coat of Exterior Gloss Enamel shall be used for such color renewal.

SECTION 2–5

COATING SYSTEMS FOR VEHICLES

2–5–1, NEW VEHICLES

New vehicles shall be painted by the manufacturer using the manufacturer's enamel or lacquer in the color which most nearly matches the specified color for the vehicle.

2–5–2, REPAINTING OF VEHICLES

When repainting is necessary it may be accomplished either by contract or Coast Guard Personnel. Since automotive painting requires some experience and special equipment for best results, it will generally be more economical to contract for such service. When painting is accomplished by contract, standard commercial materials may be used provided they meet Federal Specifications for the Materials Listed in the paint system below. See Article 1–11–8, Contract Painting. Colors shall match the Federal Color Numbers specified in Chapter 3. Lacquers produce better finishes than enamels but at a disproportionately higher cost; therefore, automobile enamels shall be specified in contracts for painting of vehicles.

When painting is accomplished by service personnel the following system shall be used:

(a) Prepare the surface in a manner similar to other metal surfaces. Particular care shall be taken not to nick or scratch the surface.

(b) Apply 1 liberal coat of Pretreatment-Wash Primer.

(c) Apply 1 coat of Quick Drying Red Lead Primer.

(d) Apply 2 coats Exterior Gloss Enamel.

In general the following procedure is used when painting vehicles. Surface preparation is the most difficult part of the job and will take approximately two days for one man. First, mask off all areas not to be painted. To do a good job of masking, the following sizes of masking tape are necessary: ¼ in., ½ in., ¾ in., 1 in., 1½ in. and 2 in. Large areas are masked with wrapping paper. Newspapers are not good for this purpose because they are easily torn and may have pin holes. Do not allow masking tape to lap over onto the surface to be painted because when it is removed, a strip of new paint may come with it. Any bright parts which are easily removed such as hub caps, hood ornaments and exterior rear-view mirrors should be removed from the vehicle. Be sure to remove the spare tire from the trunk of passenger vehicles before masking so that the spare wheel can be painted. Bright parts which cover areas to be painted such as wraparound bumpers must be removed from the vehicle. All rust must be removed to bare metal. When machine sanding is done, numbers 60 to 80 sandpaper should be used on rust and rough places. For disc sanders use number 24 close regular sandpaper and smooth up with number 50 regular. For removing paint with a disc sander use numbers 24 to 36 open sandpaper to avoid clogging. For hand sanding never use sandpaper rougher than number 100 for dry sanding.

When all rust spots and rough places have been sanded down, wash the vehicle with Synthetic Enamel Thinner. Sand the entire surface with number 220 wet sandpaper followed with number 320 wet sandpaper fairing edges of old paint. Allow the paper to soak in water for several minutes before using. Hand wirebrush door edges, and door frames to remove rust and dirt. Wash thoroughly with Synthetic Enamel Thinner to remove grease from hinges and door locks. Dispose of rags used on these areas because they may contaminate the body of the vehicle. Sand smooth with number 220 wet sandpaper followed with number 320 wet sandpaper.

Major body work should be accomplished before starting surface preparation. Small irregularities such as rust pits and small dents should be primed and then smoothed out with body putty. This material should never be applied more than 1/8 in. thick. If thicker films are needed, apply in thin coats with adequate drying time between coats. When body putty is dry, sand down smooth with number 220 wet sandpaper followed with number 320 wet sandpaper.

Wash the entire vehicle with a cloth dampened in Synthetic Enamel Thinner and allow it to dry. From this point on, the areas to be painted must not be touched with the bare hands. No matter how clean the hands may be, there is always some dirt, grease, oil or perspiration which will cause blistering of the paint, rust and poor adhesion.

Apply 1 spray coat of Pretreatment-Wash Primer to bare metal areas. When dry, apply 1 coat of Quick Drying Red Lead Primer to bare metal areas. When the primer has dried hard, sand lightly with number 320 wet sandpaper to remove burrs. Wipe with a cloth dampened in Synthetic Enamel Thinner and allow to dry.

Apply 2 coats of Exterior Gloss Enamel thinned with 1 pt. of Paint Thinner per gal. of paint. Paint door edges and door frames first. Open the trunk and hood and paint around edges of trunk lid and hood. Spray the top of the vehicle, the hood, front fender, doors, rear fender, trunk, rear fender, doors, front fender and wheels in that order. To remove dust particles, use a tack rag (procurable on the open market) on each panel immediately before spraying.

Allow the first coat to dry overnight before applying the second coat in the same manner as the first coat. Very light sanding with number 320 wet sandpaper may be necessary if dust has settled on the first coat. Use a tack rag on each panel immediately prior to spraying the second coat.

Allow the finish to dry overnight and rub with rubbing compound for enamel (procurable on the open market). Rubbing compound for lacquer is too coarse for enamel and should not be used for rubbing enamel. Wash off the compound with clean cold water and polish the vehicle by buffing. Allow about 30 days for the finish to harden before waxing. Periodic washing and polishing will prolong the useful life of vehicle finishes.

2–5–3, UNDERCOATING

Undercoating, consisting of a suitable underbody and chassis protective coating, shall be applied to all new Coast Guard vehicles including passenger cars, trucks and, where practicable, trailers. (Statutory limitations on purchase price of passenger cars shall be kept in mind with reference to cost of undercoating.)

Specifications for purchase of automobiles, station wagons, light general duty trucks (less than 10,000 lbs., gross vehicle weight) and carryalls do not include underbody coating due to manufacturers' high cost of application and delay in delivery. Therefore, in the case of these vehicles it is necessary to have the undercoating applied after purchasing the vehicle. Specifications for buses, special purpose trucks, and all trucks over 10,000 lbs. G.V.W. will require that the underbody coating be applied by the manufacturer.

Where the underbody coating is furnished in accordance with specifications or as a standard equipment item, it shall be inspected to assure that adequate protection is provided. Vehicles to be operated in beach areas or outside the Continental United States shall have the underside of the body and chassis completely coated. The coating shall be applied to a minimum thickness of 1/16 in. on the underside of the hood, cab, body and chassis and to a minimum thickness of 1/8 in. on the underside of the fenders.

Undercoating compounds will not be purchased by Coast Guard units and will not be stocked by Coast Guard Supply Depots. The material is to be supplied and applied by local garages and paint shops having the special equipment needed for its proper application. The compounds approved for use on Coast Guard vehicles are as follows:

Fendix Hiflash	Davison Chemical Corp.
Witcote EC 5465	Witco Chemical Corp.
Hollingshead Underbody Coating No. 70951A	Hollingshead Corp.
Flintkote Underbody Protective Coating No. 11003 AC	Flintkote Corp.
Utica Car Coat	Equipment Mfg. Co.
Compound Chassis Coating No. E2013	J. W. Mortell Co.
Soundproof 99C Hiflash	Non-Rust Chemical Corp.
Underbody Coating No. 7834	Anderson Prichard Oil Corp.
Nokorode Army Coat 827	Lion Oil Co., Refining Div.

SECTION 2–6

COATING SYSTEMS FOR AIRCRAFT

Coating systems and materials for aircraft are prescribed in Specification MIL–F–7179, Finishes and Coatings; General Specifications For Protection of Aircraft and Aircraft Parts. This specification covers the methods and materials required for surface treatment and the application of finishes and protective coatings to parts and surfaces such as fuselages, wings, cowls, struts, empennage, and floats.

For painting fabric surfaces of aircraft, refer to the current edition of Specification NavAer SR–70, Navy Aeronautical Specification For Application of Protective Coatings to Fabric Surfaces of Aircraft. The above references will provide full instruction.

CHAPTER 3, COLOR PRACTICE

TABLE OF CONTENTS

SECTION 3–1, INTRODUCTION

Chapter 3 contains detail color specifications for the Coast Guard. It will be noted that in the case of shore establishments considerable choice of color is offered. This choice however is limited to the district level. District Commanders should select the color principles desired, where options are offered, and inform the units affected. Headquarters units are authorized to make their own selections of color principles where options are offered. Units shall record selections of options by underlining the option selected and deleting options not to be used in each copy of the Manual furnished the unit. Notation shall be made in the margin as to the date and authority for selection or deletion of an optional color. In making this selection the information contained in Section 3–3, General Color Principles for Shore Establishments, shall be consulted in addition to the sections covering types of shore units.

Two interior color schedules are given for vessels over 65 ft. in length. The extent of use of each of these schemes will be governed by separate directives or amendments to this Manual.

Section 3–16, Safety Color Code, is effective for shore units only. Sections other than the Safety Color Code are effective for both shore establishments and vessels where applicable. (Since the use of color for safety purposes aboard vessels presents a unique problem, the color markings to be used on vessels for safety purposes have been incorporated into the sections dealing with vessel colors.)

Chapter 3 contains color specifications only, Chapter 2 specifies the paint coating system to be applied to particular surfaces, and Chapter 4 gives ordering information and application instruction pertaining to the materials to be used.

Discretion must be exercised in applying these color specifications to surfaces not normally painted. As an example, decorative wood paneling and ceramic tile floors should be left in their natural color, whereas linoleum should be procured in a color most nearly matching the prescribed color.

SECTION 3–2, SCIENTIFIC PRACTICE IN THE USE OF COLOR

3–2–1, USE OF COLOR FOR SHORE ESTABLISHMENTS, VESSELS AND VEHICLES

The principles of color adopted for the Coast Guard have been based on extensive research and case history. Above all, purely emotional opinions in the use of color have been set aside for a more factual approach. This has been made possible by taking full advantage of technical studies in color, brightness and illumination and by referring to medical research on the visual and physiological effects of color.

So that the technique of functional color may be better understood, here are a few points regarding its application. Good vision is essential to the efficient conduct of almost countless tasks. Abuse of the eyes rapidly leads to fatigue and to a reduced capacity to concentrate and work.

For example, glare will tend to fog vision and to prevent clear visibility, particularly of dark objects. Extreme contrasts in light and dark will likewise cause the pupil of the eye to open and close alternately and to make seeing difficult.

Vision is usually at its best where brightness in the field of view is relatively uniform. This applies not only to ceilings, overheads, walls and bulkheads, but to floors, decks and equipment, all of which should not conflict with each other. Where possible, surrounding brightness should equal task brightness; the poorest situation is that in which the environment is exceptionally bright and the task dark. The reason for this is that the eye is quick in adjusting itself to brightness and slow in adjusting itself to darkness. Thus, exposed lighting fixtures, white walls and bulkheads (which usually hold little significance) will impair an ability to work comfortably at desks, machinery or special equipment having a dark finish.

By controlling the colors of walls, bulkheads and work surfaces, the requirements of clear visibility may be met. Coast Guard color standards have been carefully selected to agree with modern technical practice. In general, ceilings and overheads should be white for high light reflection and diffusion and to minimize shadows. Wall and bulkhead reflectances should be between 50 and 60 per cent where fairly difficult seeing tasks are performed. Paler colors may be used in large working areas and in spaces devoted to storage or incidental traffic.

Floors, decks, and working surfaces should reflect above 25 per cent where possible and should not be dark except where low illumination levels are required (as in Timer Rooms at Loran Stations and Combat Information Centers aboard ship).

In addition to a proper control of brightness, color serves the added function of offering psychological relief from extremely high or low temperature conditions. Certain

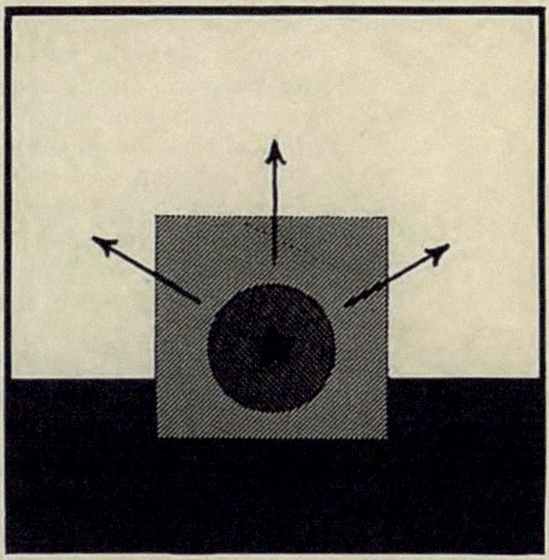

Strong contrasts tire the eye, distract from tasks, lower human efficiency and produce fatigue.

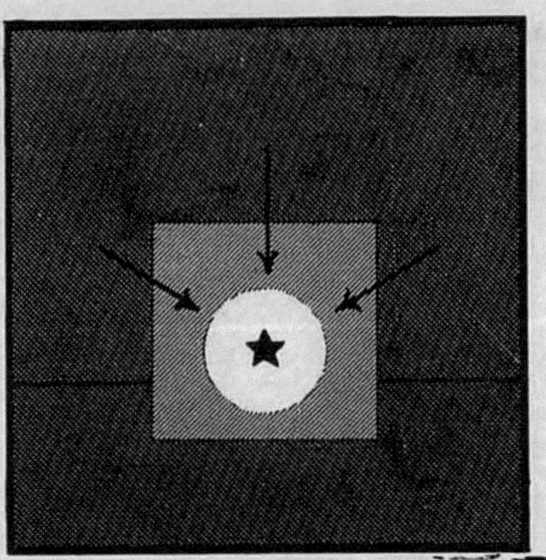

Smooth brightness in equipment and surroundings makes seeing easy and is a fundamental law of good vision.

principles in this Manual recommend the use of cool tones of green and blue for areas in which average temperature may be above normal, particularly for south and west exposures. Similarly, tones of yellow and sand are suggested for facilities having a cooler average temperature, particularly for north and east exposures.

On building exteriors, the use of white or a pale color will, under intense sunlight, reduce interior temperatures as much as ten degrees by reflecting heat rays. This will add to comfort in personnel facilities as well as reduce fire hazards in sheds used for the storing of flammable liquids and paints.

Where safety is important, a special color code has been designed. This code is standard in all Naval shore establishments and is widely employed throughout American industry. Colors of high visibility, such as yellow and orange, are used to mark points of danger; red is reserved for fire protection and damage control; green for the identification of first aid equipment; and blue for electrical controls in general.

The color standards accepted for the Coast Guard are twenty-four in number and are shown in two charts at the end of this Manual. Interior finishes have been checked for ideal reflectance under the conditions in which they will be used. It will be noted that many of these are soft in tone both to avoid needless emotional distraction and to resist soiling and abuse. Machinery and equipment finishes in gray will assure a clean appearance, reflect more light than the conventional dark gray of the past, and will also be easy to maintain and "touch up." In exterior colors, pigments and finishes able to resist prolonged outdoor exposure have been chosen. White, No. 11 Sun Tan, and No. 9 Pearl Gray for walls and siding, No. 26 Tile Red for roofs, and No. 12 Spruce Green for trim, will have excellent permanence under all climates. The safety code colors are as brilliant as modern chemical formulation will permit.

Painting is a costly maintenance expense. Where formerly paint finishes have been used for the preservation of surfaces and materials, they today have added functions that concern human efficiency and well being. Through the use of this Manual, which brings together the most advanced thinking in color science, it is expected that the facilities of the Coast Guard will be improved and that better service will result.

3–2–2, USE OF COLOR IN SEARCH AND RESCUE AND AIDS TO NAVIGATION

While many electronic advances have been made, visual search and visual sighting of objects will continue to play an important part in Search and Rescue and Aids to Navigation.

In order to arrive at a logical use of color, it is well to review some of the fundamental facts associated with the visibility of objects. (Visibility of lights and lighted objects is not a part of this discussion.)

One of the first facts to be considered is the resolving power of the human eye, i.e. what size objects can be seen. In general, the human eye cannot distinguish an object when the object subtends an angle of less than one minute of arc at the eye. Since one minute of arc is the smallest angle that can usually be resolved, it follows that in the actual sighting of objects the probability of detection will be very low until the distance has decreased so that the object subtends from 3 to 5 minutes of arc or more. In a few special situations such as tall poles against the sky and long narrow wires against the sky, the eye resolves angles as small as .12 minutes of arc. However, these special situations should for the most part be excluded from planning concerned with visibility because they are unusual situations. When binoculars are used the net effect is to enlarge the angle subtended by the object so that the eye can resolve objects which might otherwise be invisible.

Perhaps the most important single fact connected with practical visibility problems is the contrast ratio, i.e. the ratio of the brightness of the object to the brightness of the background. The above angular relationships were determined under maximum contrast conditions such as dull black lines on a white background. When the contrast ratio is low, visual detection ranges are relatively short so that a high contrast ratio is desirable for ease of sighting.

Although two objects may each have the same brightness, the color of one object may provide chromatic contrast with its background and thereby be more visible. For reasons discussed below, chromatic contrast is of more importance at short than long ranges.

In all practical visibility problems the scattering effect of the atmosphere, commonly called haze, must be considered. The effect of haze is to lower the contrast ratio and thus reduce the visual range in proportion to the amount of haze present. The effect of haze multiples rapidly with distance, thus becoming a very important factor at long visual ranges. The effect of haze on white and black is to make the whites a shade of light gray and to lighten the blacks to a shade of dark gray. As distance is increased the whites become darker gray and the blacks lighter gray until the object brightness matches that of the background at which time the object is "lost in the haze."

The effect of haze on color is of particular importance to Search and Rescue and Aids to Navigation. At long visual ranges or in heavy haze all colors appear as some shade of gray and are not recognizable as colors. At shorter ranges and even in quite clear weather the effect of haze on color is present and is of considerable importance. Consider the effect on two colors commonly employed in search and rescue, yellow and International Orange. The first effect to be noted as distance from the objects is increased or as the haze between observer and the objects increases is that the yellows take on a greenish cast, thus decreasing chromatic contrast with wooded or grassy backgrounds, while no effect on the International Orange is apparent. As the range is increased, the graying effect of haze is observed. At ranges when objects can still be distinguished clearly, the yellows will be reported as white and the International Orange as black due to the fact that on losing color significance because of haze, the yellow appears as a light gray associated with white objects, while the International Orange appears as a dark gray associated with black objects. See Figure 1.

From the above fundamental facts it is possible to arrive at some general conclusions of a practical nature. Although the visual problem in the use of such aids to navigation as day marks involves the same fundamental facts, the practical application of these facts differs somewhat from the

Fig. 1. Effect of Haze: left square is International Orange. Right square is orange yellow. Film and filter selected to show brightness contrast as seen by human eye at a distance. At distances International Orange appears black and orange yellow appears white.

search and rescue problem. For this reason discussion of the two end applications of the above fundaments will be presented separately.

(A) SEARCH AND RESCUE

It becomes evident that on a large object, such as a ship, color is of relatively low importance since the size of the object in terms of angle subtended at the eye and the contrast ratio of the object with its background will permit a high probability of detection at ranges so great that under most conditions chromatic contrast would be entirely obscured by the graying effect of haze. On a large object, since we cannot change the size, the principal consideration is the contrast ratio, for this can be changed by varying the reflectivity of paint used. The lowest visibility will be obtained with gray at sea, since the intermediate reflectance of gray reduces the average contrast ratio. For high visibility either white or black may be used on large objects. For daylight use white has a slight edge over black when all backgrounds at sea are considered. At night a black ship is more visible than a white one when no moon or searchlight or navigation lights are involved, since it provides a greater contrast with the illumination present in the night sky. In these cases the size of the object is great enough so that no danger of confusion with whitecaps exists.

On small objects the use of color becomes more important, since due to their small size the maximum range at which they can be resolved by the eye is small and at small ranges the effect of haze is lessened. To demonstrate the importance of color contrast in the practical search problems, tests were carried out on visibility of standard dark gray kapok life jackets and the same jacket made of International Orange material. Even though the collar of the jacket was the only visual target, it was shown that under a considerable range of daylight conditions the probability of sighting the orange jacket was of the order of four times as great as the gray jacket.

Although there is general agreement on the desirability of using chromatic contrast to aid in sighting small objects, there has been some difference in opinion as to whether the orange yellow, common to many aeronautical equipment items, or International Orange was the better color.

In the detection of objects in water from aircraft and ships by day and by night, International Orange offers distinct and practical advantages over the orange yellow. Even though ranges are generally small, any haze that is present affects the yellow color before the effect is noticeable on the orange, so that the yellow appears white *before* the orange color has ceased to be useful. While in a flat calm the whitening of the yellow does not affect the sighting adversely, at the first onset of whitecaps the small yellow object becomes lost in the whitecaps. Not only does the International Orange retain its chromatic contrast under more adverse haze condition, but when the color is obscured the object still remains detectable as a *black* object which is not easily confused with either the sea or whitecaps. At night the chances of detecting a small object without artificial illumination are negligible, except when a bright moon on a calm night is considered, under which situation the yellow with its higher reflectance may offer some advantages. However, under searchlight or star shell illumination the yellow object is nearly invisible if any whitecaps are present. During the night phase of the Bermuda Sky Queen operation, great difficulty was encountered in keeping track of a yellow rubber raft drifting on a line from the rescue vessel, since at ranges of the order of 100 yards the raft became indistinguishable from the sea under searchlight illumination.

The advantages of International Orange over orange yellow for use at sea are very real and suggest that yellow or orange yellow is advantageous only under the limited condition of fair weather and flat seas.

Over land, orange yellow has been used because of its higher conspicuity against a dark forest. However, recent studies[1] show that when the whole range of natural background is considered (sky, water, snow, sand, forests, grasslands, agricultural fields, in spring, summer, winter and autumn), the International Orange provides greater average conspicuity than orange yellow even though the yellower and lighter orange yellow remain more conspicuous against dark green backgrounds.

It can be seen from the above fundamentals that increase in the contrast ratio or increase in the chromatic contrast will increase the visual detection range. Thus the utility of the use of daylight fluorescent pigments can be predicted fairly accurately. These pigments sometimes called phosphors have the ability to absorb invisible ultraviolet radiation from the sun or sky and reradiate this energy in a visible wavelength. The effect is twofold in that both the brightness of the object is increased because more light of all wavelengths is reflected and the color is intensified, since the reradiated energy is usually confined to the visual color of the pigment, thus increasing chromatic contrast. In some situations, such as viewing an object against a sky background on a sunny day, the fluorescent pigments lose part of their effect, since the increase in brightness against a bright sky decreases the contrast ratio. However, the increase in chromatic contrast through intensification of the color usually results in a slight gain in visibility in sunny weather against a bright sky. In cloudy weather or against dark backgrounds the gain in visual range is most pronounced, since the increase in brightness increases the contrast ratios. The increased contrast ratio together with increased chromatic contrast give rise to most favorable results against dark backgrounds and in cloudy weather. While presently well adapted for some temporary uses such as signal markers, most of the daylight fluorescing pigments are not very stable and gradually lose this property on long exposure to the weather. Many search and rescue objects are protected from weather when not in use and can employ these pigments effectively. Continuing efforts are being made to produce more permanent daylight fluorescing pigments whose general utility will increase in proportion to the degree of permanence achieved.

(B) AIDS TO NAVIGATION

Whereas in Search and Rescue color is of importance primarily in connection with detection of an object, in Aids to Navigation color is normally used primarily as a means of identification of the object. Color can be used to improve the daylight visual ranges of Aids to Navigation, particularly of shore aids. However, the identification usage together with widely accepted practices must always be carefully considered.

As for Search and Rescue, color, i.e., chromatic contrast, is more important in the case of the small structure and at small visual ranges (under 5 miles). At long visual ranges associated with large structures there appears to be little justification for the use of colors other than black and white, since atmospheric haze will in general obscure the coloration.

For an object of given size the most important factor in the visibility of an aid to navigation is the contrast ratio, i.e., the brightness of the aid as compared with the brightness of the background. Consideration of this factor will generally lead to the use of white or black. While white will give the highest average contrast ratio, further gains in visibility can be obtained by devices which increase the brightness of the white surface. Selection of white paints of highest practical reflectance and proper maintenance of the surface are desirable. Although no designs are now in service in this country, it is possible to design white structures with greater brightness by taking advantage of the fact that the average vertical component of daylight is more intense than the horizontal component.

For example, at noon a sloping white roof has a higher brightness than a vertical billboard. This higher brightness can be used to increase the contrast ratio of the object with its background. Mariners who have made landfalls in sections of the world where white roofs are used to reduce heat absorption are familiar with the fact that the first object sighted through the haze is usually a white sloping roof whose high brightness and resultant high contrast has produced increased visual range. See Figure 2. Various arrangements to more effectively use the vertical components of daylight are possible. One of the simplest is to increase the reflectivity of horizontal surfaces. For example, walkways which frequently surround daymarks whose visual elements are in vertical planes, may have their horizontal surfaces painted white.

Generally black is the most effective color for those aids which are silhouetted against the sky, since it provides the highest possible brightness contrast ratio. White is generally most effective where the aid is not silhouetted but is viewed against a background of land, provided the land, hills and rocks are

Fig. 2. Utilization of more intense vertical component of daylight increases contrast and visibility. Surfaces of both targets are flat white.

dark in color. Against sand and snow a white color is obviously ineffective. It is apparent that seasons can cause marked changes in the visibility of an aid. The white tower which is so easily seen against the summer forest becomes almost invisible with the first snow. Two solutions can be used to overcome the seasonal changes in contrast ratio. One is to use chromatic contrast, i.e., color, and the other is to use bold black and white patterns.

In general the use of chromatic contrast or color will be most effective with relatively small structures which are visible at fairly short ranges (under 5 miles). International Orange (Federal Color 1205), which gives the highest average conspicuity over a wide range of backgrounds, is the logical choice for situations requiring chromatic contrast. It is important to consider that chromatic contrast should be used only where black or white will not meet the problem. While International Orange gives improved visibility over white against snow by reason of the combined effect of color (chromatic contrast) and darkness (brightness contrast ratio), it may actually decrease the summer time visibility to some extent. For example, assume that a small structure is to be viewed against a background of pine trees during summer. The reflectance of a pine tree forest is about 15 per cent, while International Orange paint is also about 15 per cent, so that the brightness contrast ratio is unity, i.e., the painted structure and the background have the same brightness. The visibility of the structure will depend entirely on color contrast which is difficult to evaluate numerically. However, on hazy days the effect of color contrast will be greatly reduced and the structure will become invisible in relatively light haze conditions. A white structure against the pine tree background would have a contrast ratio of 5.3 to 1 and would have about the same visibility as the orange structure on clear days and better visibility on hazy days.

If the structure is small the maximum visual range will be relatively small, so the effect of haze is minimized and the use of International Orange to produce chromatic contrast is justified.

Fig. 3. Resolving Power: both squares have black and white stripes. Stripes in right hand square are 1/3 the width of left hand square. Small pattern is not resolved and target appears light gray.

On large structures the use of bold black and white patterns to overcome seasonal variations of background brightness appears preferable to the use of chromatic contrast. The larger visual range associated with the large structure increases the effect of haze in neutralizing chromatic contrast. Considerable care should be taken to avoid small patterns which cannot be distinguished at the maximum visual range. See Figure 3. While a bold pattern can increase the contrast ratio between parts of the pattern and the background, the net effect of a pattern too small to be resolved at maximum visual range is that the contrast ratio of the structure with its background is reduced. Further consideration of angular relationships and the effect of haze, leads to the conclusion that the minimum dimensions of the pattern (for example the stripe width in a striped pattern) should not be less than 1/4 the maximum dimension of the structure. In other words it appears that normally not more than four stripes or pattern components should be used (two white, two black).

Often, when patterns are employed, colors other than black or white are used. Although in some special cases special color combinations may be warranted, in general the use of color combinations other than black and white reduces contrast ratios and decreases visual range. This is particularly true on large structures. If a pattern is used it is assumed that it is desired that the pattern itself as well as the structure be visible at maximum ranges. Consideration of two possible color schemes versus the contrast ratio of black and white will illustrate this point.

For example, compare the contrast ratio of International Orange and white with black and white. See Figure 4. International Orange reflects 15 per cent of the light incident upon it. White paints reflect about 80 per cent of the incident light. Under any lighting conditions the contrast ratio will be equal to the ratio of the reflectances, in this case a ratio of 1 to 5.3. If black is used in lieu of International Orange, the black paint will have a reflectance of about 4 per cent resulting in a contrast ratio of 1 to 20. Thus the black and white pattern will normally be visible at a greater distance than the same pattern in International Orange and white.

If for example a black and yellow pattern were under consideration, the reflectances are 4 per cent for black and 60 per cent for yellow, giving a contrast ratio of 1 to 15. Thus, the black and yellow pattern would normally be less visible than the black and white pattern by a small amount, but both would be materially better than the Interna-

Fig. 4. Contrast Ratio of Pattern: left pattern black and white. Right pattern International Orange (Fed. Color No. 1205) and white. Film and filter selected to show brightness contrast as seen by human eye at short ranges.

tional Orange and white.

In some cases brown enamel has been used on aids to navigation structures. The brown color used is rather dark with a reflectance of about 10 per cent. The use of a dark color would indicate that such structures are viewed against a lighter background. Brown offers almost no chromatic contrast with natural backgrounds and can be distinguished as a color at only very short ranges under most lighting conditions. When the effect on contrast ratio is considered it is evident that against a light background or as a silhouette the use of brown reduces the contrast ratio over what would be obtained with black. Thus, the sole effect of the use of brown paint would appear to be a slight reduction in the visibility of the structure.

---

[1] "A Study of The Conspicuity of Orange Surface Colors," by W. E. Knowles Middleton, National Research Council of Canada, Illuminating Engineering, Vol. XLVII, No. 2, Feb. 1952.

SECTION 3–3, GENERAL COLOR PRINCIPLES FOR SHORE ESTABLISHMENTS

3–3–1, GENERAL EXTERIOR COLOR PRINCIPLES FOR SHORE ESTABLISHMENTS

Nine color standards have been established for outdoor buildings and structures. These are illustrated on the opposite page and are shown in various air view drawings which precede the discussion of the different types of Coast Guard shore establishments. In certain cases the structures are left unpainted. As an example masonry structures shall be left unpainted where appearance or identification is not important. Also aluminum roofs or entire aluminum structures shall be left unpainted.

WHITE. As in the past, white is standard for most types of painted structures. This would apply to wood dwellings and buildings, stucco and concrete where painted. Outbuildings, galvanized steel sheds (not industrial) and Quonset huts should also be white.

White is the prescribed trim color for red brick buildings, being applicable to window sash, storm doors and windows, gutters, eaves, doors, porch and stair balustrades. It should be used for wood picket fences. Lighthouses should remain as at present and be painted in accordance with the requirements of the Light List.

Flag poles should always be painted white whether of wood or metal.

No. 12 SPRUCE GREEN. This is a Coast Guard color which has been given a wide number of uses for outdoor structures. It is a durable chromium oxide finish that has excellent permanence under all weather conditions.

On white dwellings and buildings No. 12 Spruce Green should be used for exterior trim, window sash, doors, shutters, wood lattice under porches, roof vents, weather vanes, storm doors and windows. It may also be specified for the trim on buildings used for industrial purposes at Bases and Depots. It should be used for metal fences, metal railings, catwalks, ladders, lampposts, metal clothes poles, wind instrument staffs and outdoor waste receptacles. Ornamental iron, however, may be painted black as an alternate.

On tall stationary structures No. 12 Spruce Green should be applied to steel structures supporting water tanks. The tank itself, roof included, should be white. Radio antenna poles and towers, flag towers, range light towers, storm warning towers, lookout towers, should also be No. 12 Spruce Green. (Wood structures on towers should be white.) It is specified for bridges and trestles, large shipways, tall dry dock cranes, hammerhead cranes.

Storage tanks (more than 1,000 gals. capacity) should also be No. 12 Spruce Green. However, small gasoline and diesel oil tanks should be painted No. 14 Brilliant Yellow, with black lettering.

Note: where all structures are near airfields, these must be painted No. 18 International Orange and white as required by Civil Aeronautics Administration. (See CAA bulletin, Obstruction Marking.) Tall structures used as Aids to Navigation shall be painted to conform to the Light List.

No. 11 SUN TAN. This color is acceptable as an alternate for white in tropical regions or where sunlight glare is to be reduced. Trim for No. 11 Sun Tan dwellings and buildings should be white.

No. 19 LIGHT GRAY. This is the color specified for industrial types of buildings and warehouses, particularly those located at Bases and Depots where white may be impractical. It may also be used for painted chimneys (aluminum heat-resistant paint would be an alternate where surfaces of high temperature exist). Where industrial buildings are No. 19 Light Gray, trim should be No. 12 Spruce Green as described above.

No. 19 Light Gray should be used in the

WHITE. For trim on red brick buildings.

NO. 12 SPRUCE GREEN. A trim color for white buildings, poles, metal fences, etc.

NO. 11 SUN TAN. For dwellings and buildings in tropical regions to reduce glare. ~~White~~ should be used for trim.
No. 8 Terra Cotta

NO. 19 LIGHT GRAY. For industrial buildings. NO. 20 MEDIUM GRAY. For equipment such as railway cars.

NO. 18 INTERNATIONAL ORANGE. Used with white for tall structures near air fields.

NO. 13 FIRE RED. NO. 33 SEAL BROWN. For lighthouses as prescribed by the applicable Light List.

form of a deck paint as a standard on all traffic areas such as porch floors, stairs, ramps, platforms and bleachers, both interior and exterior. No. 20 Medium Gray may be used on traffic areas where maintenance is difficult, such as floats, launchings, loading docks and shop floors.

No. 20 MEDIUM GRAY. In addition to being an alternate for No. 19 Light Gray on traffic areas, No. 20 Medium Gray is the prescribed color for outdoor shore equipment such as winches, small movable cranes, truck cranes, derricks, barges, railway cars and locomotives.

No. 26 TILE RED. This color is for painted roofs to blend with the existing red roofs found at most Coast Guard units.

No. 18 INTERNATIONAL ORANGE. This color is designed for high visibility at long ranges. It gives the highest average conspicuity of all colors, against the wide variety of sea and land backgrounds. It should be used with white for tall structures near air fields as required by Civil Aeronautics Administration. (See CAA bulletin, Obstruction Marking.)

No. 13 FIRE RED and No. 33 SEAL BROWN. These two colors are specified for lighthouse structures and should be applied as directed by the Light List.

No. 29 BRIGHT BLUE. The colors of blue and white have long been associated with Coast Guard activities. When this color combination is used to represent the Coast Guard the blue color shall conform to No. 29 Bright Blue.

No. 8 TERRA COTTA

Used as an alternate trim when buildings and dwellings in high glare regions have been painted No. 11 Sun Tan.

ciated that the specification of color for the widespread facilities of the Coast Guard involves many complications. Due to problems of supply and storage it is necessary to standardize on a limited number of colors. These must serve a multitude of uses and be formulated in brightness and hue to agree with the best of modern scientific practice. Different paints of the same color are frequently required for indoor and outdoor use, for metal surfaces, wood, plaster and concrete. Added colors, therefore, increase the stocking problem.

A great deal of research, study and case history has gone into the organization of the principles included herewith. An exhaustive check has been made on almost countless major and minor items found at shore units. It is expected that the Manual will be convenient to use and will lead to improved facilities.

For the sake of simplification, several uniform practices are recommended. For example, *all ceiling and overhead areas should be white.* There are practically no exceptions to this other than in certain control towers and special interiors where operations are unusual.

Again the common practice of having a special trim color for every wall color has been avoided. This would mean double inventories for many standards and add to costs for paint supplies and labor. Interior trim in family quarters may be the same color as adjacent walls or white. Trim in all other shore facilities should be uniformly white or No. 19 Light Gray, except in some industrial applications where dadoes are recommended. Here, however, no special paints are required other than those adopted for other standard purposes. Except for family quarters, uniform white trim is standard for doors, door frames, baseboards and window sash in living and operational facilities at Light Stations, Lifeboat Stations, Radio and Loran Stations. Where deeper trim finish would be more practical, such

White is recommended as a uniform trim color on doors, door frames, baseboards and window sash in average facilities. Deeper colors such as No. 19 Light Gray may be used for trim in industrial facilities, depending on the principle employed.

form of a deck paint as a standard on all traffic areas such as porch floors, stairs, ramps, platforms and bleachers, both interior and exterior. No. 20 Medium Gray may be used on traffic areas where maintenance is difficult, such as floats, launchings, loading docks and shop floors.

No. 20 MEDIUM GRAY. In addition to being an alternate for No. 19 Light Gray on traffic areas, No. 20 Medium Gray is the prescribed color for outdoor shore equipment such as winches, small movable cranes, truck cranes, derricks, barges, railway cars and locomotives.

No. 26 TILE RED. This color is for painted roofs to blend with the existing red roofs found at most Coast Guard units.

No. 18 INTERNATIONAL ORANGE. This color is designed for high visibility at long ranges. It gives the highest average conspicuity of all colors, against the wide variety of sea and land backgrounds. It should be used with white for tall structures near air fields as required by Civil Aeronautics Administration. (See CAA bulletin, Obstruction Marking.)

No. 13 FIRE RED and No. 33 SEAL BROWN. These two colors are specified for lighthouse structures and should be applied as directed by the Light List.

No. 29 BRIGHT BLUE. The colors of blue and white have long been associated with Coast Guard activities. When this color combination is used to represent the Coast Guard the blue color shall conform to No. 29 Bright Blue.

3–3–2, GENERAL INTERIOR COLOR PRINCIPLES FOR SHORE ESTABLISHMENTS

In using this Manual it should be appreciated that the specification of color for the widespread facilities of the Coast Guard involves many complications. Due to problems of supply and storage it is necessary to standardize on a limited number of colors. These must serve a multitude of uses and be formulated in brightness and hue to agree with the best of modern scientific practice. Different paints of the same color are frequently required for indoor and outdoor use, for metal surfaces, wood, plaster and concrete. Added colors, therefore, increase the stocking problem.

A great deal of research, study and case history has gone into the organization of the principles included herewith. An exhaustive check has been made on almost countless major and minor items found at shore units. It is expected that the Manual will be convenient to use and will lead to improved facilities.

For the sake of simplification, several uniform practices are recommended. For example, *all ceiling and overhead areas should be white.* There are practically no exceptions to this other than in certain control towers and special interiors where operations are unusual.

Again, the common practice of having a special trim color for every wall color has been avoided. This would mean double inventories for many standards and add to costs for paint supplies and labor. As illustrated below, *trim in all Coast Guard shore units should be uniformly white or No. 19 Light Gray*, except in some industrial applications where dadoes are recommended. Here, however, no special paints are required other than those adopted for other standard purposes. Uniform white trim is standard for doors, door frames, baseboards and window sash in typical living and operational facilities at Light Stations, Lifeboat Stations, Radio and Loran Stations. Where a deeper trim finish would be more practical, such

White is recommended as a uniform trim color on doors, door frames, baseboards and window sash in average facilities. Deeper colors such as No. 19 Light Gray may be used for trim in industrial facilities, depending on the principle employed.

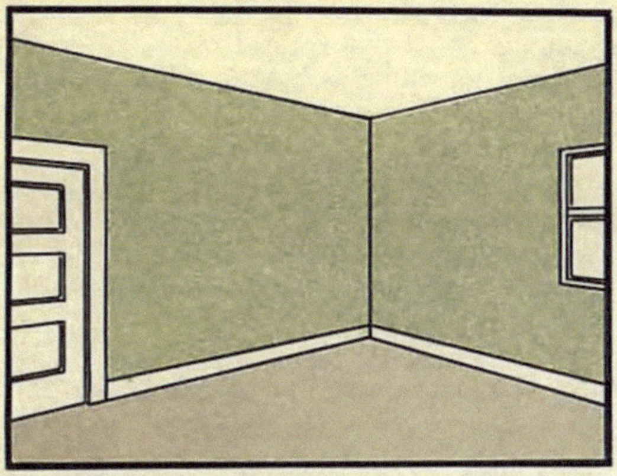

PRINCIPLE A, No. 1 Light Green. A cool color for living spaces, sleeping quarters, offices.

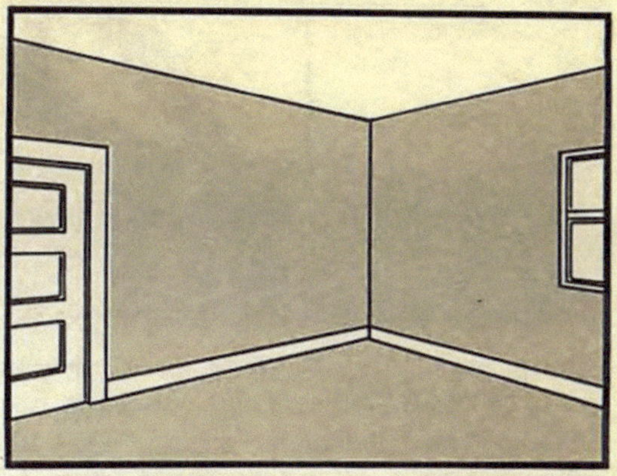

PRINCIPLE B, No. 9 Pearl Gray. For living spaces, offices, and areas where difficult visual tasks are performed.

PRINCIPLE C, No. 25 Beach Sand. A warm color for living spaces, sleeping quarters, mess halls and general offices.

PRINCIPLE D, No. 4 Soft Yellow. A high-reflectance color for passageways, and areas deprived of good light.

PRINCIPLE E, No. 10 Light Blue. A cool color for living spaces, sleeping quarters and mess halls.

PRINCIPLE F, No. 6 Ivory. A general purpose color of high reflection. For passageways or areas needing light.

Ceilings in practically all instances should be white. Wall colors should extend to the bottom of roof beams or trusses. Wood or concrete beams on a flat ceiling may be white. However, steel beams and trusses may be No. 19 Light Gray.

as at Air Stations, Bases, Supply Depots and other large industrial establishments, No. 19 Light Gray may be used. This principle will simplify the task of painting as well as help to tie all facilities together for a neat and orderly appearance. Radiators may be in the wall or trim color.

Wall colors should as a rule extend to picture molding or ceilings. In industrial areas having high bays, wall colors should extend to a line level with the bottom of roof trusses. All areas above, including sash in monitors and skylights, should be white. Ceilings should never be tinted unless specific instructions are given to the contrary.

(A) NON-INDUSTRIAL SCHEMES

Eighteen principles have been devised for interiors from a limited number of color standards. Variety is thus assured with a restricted paint inventory.

PRINCIPLE A, No. 1 Light Green. This color may be widely used for ward rooms, living spaces, sleeping quarters, mess halls, recreation rooms, offices, sick bays. It has a reflectance of 55 per cent and will assure a glare-free environment having a cool and fresh appearance. It is suitable for most climates and particularly well adapted to areas in which personnel are concentrated.

PRINCIPLE B, No. 9 Pearl Gray. This cool, neutral color is appropriate for living spaces, ward rooms, offices. Its best purpose is served in providing a subdued, non-distracting environment for the execution of confining visual and mental tasks. It is not proposed for climates having a low average temperature.

PRINCIPLE C, No. 25 Beach Sand. This color is warm in tone and will help compensate for low average temperatures. It is recommended for living spaces, sleeping quarters, mess halls, rest rooms, general offices, sick bays.

PRINCIPLE D, No. 4 Soft Yellow. This high reflectance color is an ideal standard for passageways and stairways and for other such areas deprived of good natural light. It is suitable for living spaces, sleeping quarters, large general offices. Because of its high brightness, however, it is not recommended for areas in which difficult visual tasks are performed. It is excellent for cold climates.

PRINCIPLE E, No. 10 Light Blue. This is a cool color proposed in limited applications for regions having a generally high temperature. It is suitable for living spaces, sleeping quarters, mess halls. It is not recommended for offices because of the fact that blue light, reflected or otherwise, is not conducive to clear visibility.

PRINCIPLE F, No. 6 Ivory. This is a general purpose color for passageways and stairways, large spaces such as gymnasiums. Its high light reflection will be appropriate to areas having poor natural light. It is not suitable for small offices, nor for living spaces where a more colorful environment would be superior.

PRINCIPLE G, No. 2 Medium Green. This soft color should be given restricted application. It is suitable for living rooms and mess halls in dwellings, for reception rooms and women's rest rooms where a quiet and restful atmosphere would be comfortable, relaxing and appropriate. It should not as a rule be applied to work spaces, except where dim lighting conditions are prescribed as in certain facilities at Radio and Loran stations.

PRINCIPLE H, No. 8 Terra Cotta. This unusual color effect has very limited use. It may be specified for living rooms and mess halls in dwellings where regional temperatures may be on the cool side. It is otherwise suitable for reception rooms and women's

PRINCIPLE G, No. 2 Medium Green. Limited application for living rooms, reception rooms, women's rest rooms.

PRINCIPLE H, No. 8 Terra Cotta. Limited application for living rooms, mess halls, women's rest rooms.

PRINCIPLE I, White. For areas devoted to storage, lockers. For galleys, heads, examining rooms in infirmaries.

PRINCIPLE J, No. 9 Pearl Gray with No. 2 Medium Green end wall. For training rooms, offices, drafting, shops.

PRINCIPLE K, No. 9 Pearl Gray with No. 4 Soft Yellow end wall. For training rooms, offices, having dim light.

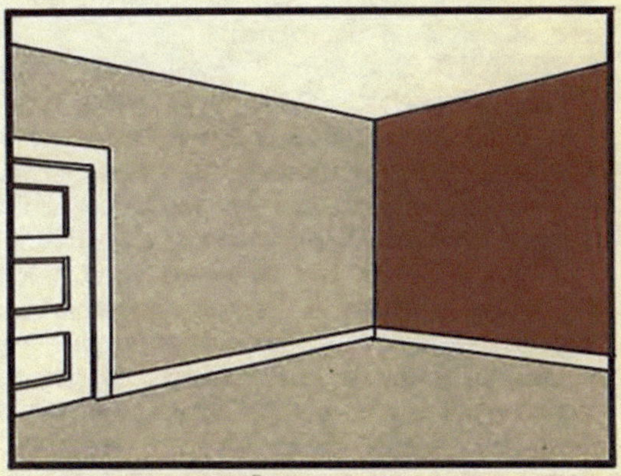

PRINCIPLE L, No. 9 Pearl Gray with No. 8 Terra Cotta end wall. For reception, recreation, women's rest rooms.

rest rooms. A soft color like this requires good natural or artificial light to prevent it from having a dim and shadowy appearance.

PRINCIPLE I, White. Although white is a clean, bright finish, it may produce glare. This principle is recommended for unimportant spaces little occupied, such as storage spaces, closets and locker rooms. To assure cleanliness and high standards of housekeeping it is also prescribed for sanitary spaces such as galleys, pantries, sculleries, laundries and heads. Because of high light reflection it may also be used for passageways and for examining and treatment rooms in infirmaries.

PRINCIPLE J, No. 9 Pearl Gray, with No. 2 Medium Green. In using end wall effects, the No. 9 Pearl Gray should be applied to three walls, with the No. 2 Medium Green restricted to one wall—preferably faced by personnel and not having windows. Thus the principle is suitable for training rooms, reception rooms, general offices, drafting rooms, laboratories, special shops where difficult seeing tasks are performed and where the end wall will serve a functional purpose in removing glare from direct line of sight.

PRINCIPLE K, No. 9 Pearl Gray, with No. 4 Soft Yellow. This principle is applicable to rooms deprived of good natural light and where personnel may be concentrated. For the best effect the No. 4 Soft Yellow end wall should be at the far side of the room away from windows. The scheme is suitable for training rooms, reception rooms, general offices, drafting rooms.

PRINCIPLE L, No. 9 Pearl Gray, with No. 8 Terra Cotta. Three gray walls combined with one end wall in No. 8 Terra Cotta is effective for reception rooms, women's rest rooms, training rooms in relatively cold climates. Its use should always be limited.

PRINCIPLE M, No. 25 Beach Sand, with No. 8 Terra Cotta. This color scheme has very limited use. It is a good combination for recreation rooms in cool climates, having a warm and luminous quality. It is also suitable for women's rest rooms and for reception rooms. It is an alternate for Principle C and may be specified where a deeper trim color is desirable in lieu of white or light gray.

PRINCIPLE N, No. 10 Light Blue, with No. 29 Bright Blue. This is perhaps the most colorful of all Coast Guard principles. Its chief use is for recreation rooms where its cool, clean appearance will be both restful and cheerful—and particularly appropriate where average temperatures are on the warm side. It may also be prescribed for exchanges. Blue is the preferred color of most men and therefore has universal appeal.

(B) INDUSTRIAL SCHEMES

The following principles are for areas having relatively severe usage, for shops and industrial plants where a deep trim color is most practical. The trim color may in all cases be applied to doors, door frames and dadoes. Window sash is best in the trim color if metal, and in the wall color if wood. Radiators should be preferably in the deeper trim color.

In industrial plants having high bays, overhead steel trusses and beams may be No. 19 Light Gray. Otherwise all ceilings and overhead, wood and concrete beams, should be white.

PRINCIPLE O, No. 1 Light Green, with No. 2 Medium Green. This is perhaps the most useful of all principles for working areas. It is recommended for industrial facilities and workshops where personnel is concentrated, and for offices having heavy traffic or occupancy. No. 1 Light Green has an ideal reflectance for relief from glare and ease of seeing, and it is restful and cool in quality. No. 2 Medium Green for dado and trim will withstand abuse and simplify maintenance. Principle O is suitable for almost all climatic conditions.

PRINCIPLE P, No. 4 Soft Yellow, with No. 19 Light Gray. This industrial color scheme is proposed for areas deprived of good natural light or for windowless or below ground working spaces. It is suggested for large workshops, boat houses, hangars, garages. Being warm and luminous in tone, yellow is suitable for cool climates and for large spaces where artificial heating may be none too adequate during winter months.

PRINCIPLE Q, No. 10 Light Blue, with No. 19 Light Gray. This color scheme has been designed specifically for areas and

PRINCIPLE M, No. 25 Beach Sand with No. 8 Terra Cotta dado and trim. Limited application in cool climates.

PRINCIPLE N, No. 10 Light Blue with No. 29 Bright Blue dado and trim. Limited application for special purposes.

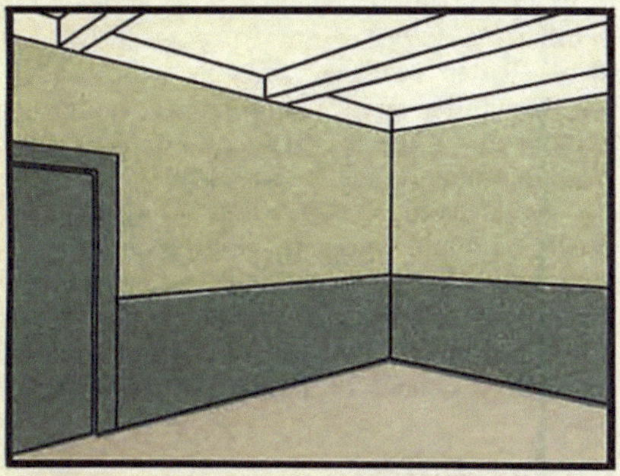

PRINCIPLE O, No. 1 Light Green with No. 2 Medium Green dado. For shops and offices having high occupancy.

PRINCIPLE P, No. 4 Soft Yellow with No. 19 Light Gray dado. For large areas and shops having poor natural light.

PRINCIPLE Q, No. 10 Light Blue with No. 19 Light Gray dado. For working areas exposed to high temperatures.

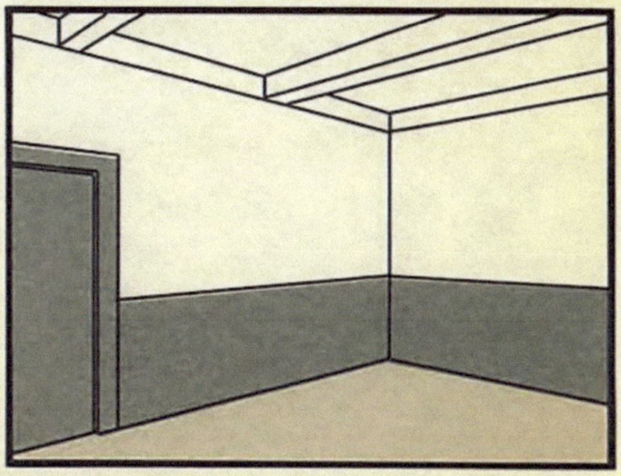

PRINCIPLE R, White with No. 19 Light Gray dado. For storage, boiler rooms, engine rooms, emergency power.

operations involving high temperatures. This would include heat treating departments, spaces devoted to welding and the like. (Boiler and furnace rooms, however, are preferred in white. See Principle R.)

PRINCIPLE R, White with No. 19 Light Gray. This simple combination is recommended for relatively unimportant areas where occupancy is light and where difficult seeing tasks are not performed. Included would be storage and stock rooms. In boiler rooms, engine rooms, furnace rooms, emergency power rooms, the addition of No. 26 Tile Red on floors shall be used.

(C) FLOOR COLORS

Natural wood floors shall be cleaned, sealed and protected with a floor wax rather than painted. Where floors are painted, four standard floor colors are available: No. 19 Light Gray, No. 20 Medium Gray, No. 26 Tile Red, and No. 31 Deck Green. No. 19 Light Gray is the preferred finish for average floors, being practical and easy to maintain. It should be used for sleeping quarters, offices, training areas and laboratories. The deeper No. 20 Medium Gray should be used for industrial areas, workshops, garages, boat houses. No. 26 Tile Red is prescribed for engine rooms, boiler rooms, furnace rooms, generator rooms and the floors of emergency power plants. No. 26 Tile Red or No. 19 Light Gray may be used for passageways, mess halls, recreation areas, heads and locker rooms. No. 31 Deck Green may be used as an alternate for No. 19 Light Gray and applied to the floors of ward rooms, living quarters, sleeping rooms, first aid departments, sick bays.

(D) EQUIPMENT

In non-industrial facilities, fixed equipment against walls, such as cabinets, should be in the wall or trim color. Shelving, bins, racks and movable equipment such as tables, benches, desks should be No. 30 Equipment Gray for the sake of uniformity and adherence to one standard finish.

In industrial areas, fixed equipment against walls may also be in the dado or trim color. As above, No. 30 Equipment Gray should be used as a standard for miscellaneous equipment such as benches, bins, racks, shelving and wire screening. However, where equipment is subject to severe usage, it may be painted No. 20 Medium Gray as an alternate. Heated surfaces such as ovens, heat treating units, metal boiler fronts should be painted with a heat-resistant aluminum paint. Insulated boilers should be white. Coal stokers should be black. Vent and exhaust ducts should be white where they run against ceilings and may be kept relatively clean. In those cases where vent ducts extend down into the interior, that portion of duct below the ceiling should be painted No. 19 Light Gray.

(E) MACHINERY

For the application of color to industrial machinery see Section 3–17, Machinery Colors.

3-3-3, ILLUMINATION

Lighting is essential to vision; proper control of it is necessary to assure comfortable and efficient seeing. Although lighting systems throughout the Coast Guard vary considerably, certain well defined points may be set forth.

As to natural daylight, unless blackout conditions prevail (as during a war) the best window shade materials are a translucent (semi-opaque) shade cloth in pearl gray or ivory. Dark colors such as green and blue are not recommended.

The light colored shade cloth holds two advantages. During daytime it will admit and diffuse light, scattering it usefully throughout the rooms. At night pearl gray or ivory will reflect artificial light in good amount and will help to build up higher illumination as well as minimize unfavorable contrast with walls. Where greater darkness is wanted during the day—such as in sleeping quarters or training rooms where projection equipment is used—fully opaque shades in pearl gray or ivory are available and may be specified.

If the more costly Venetian blinds are employed, these also should be finished pearl gray or ivory; they should not match wall colors such as green in order to avoid distorted light reflections.

Illustrations on these pages show three excellent lighting principles. With incandescent light, semi-indirect fixtures are usually best. Light bulbs will be shielded and the room will have good distribution in all corners.

With fluorescent sources, the semi-indirect type is again preferred for average areas such as living quarters and offices.

In industrial areas, however, which may have a high overhead, direct type RLM units may be installed.

With fluorescent tubes, 3500-degree white is most efficient and has a slightly yellowish quality which is in the high visibility region of the spectrum.

Semi-indirect fluorescent lighting fixtures create a bright overhead. Illumination is both direct and indirect—assuring clear visibility of three-dimensional objects.

Semi-indirect fluorescent units should use 3500-degree white tubes. General light levels from 25 to 35 footcandles are ample. Localized light may be added over difficult tasks.

Where ceilings are high, efficient illumination may require direct light with the type units illustrated. Photographs, courtesy General Electric Company.

SECTION 3–4, COLORS FOR LIGHT STATIONS

The below instructions are also applicable to Light Attendant Stations and Fog Signal Stations

3–4–1, EXTERIOR COLORS FOR LIGHT STATIONS

The exterior painting of light stations, including all structures identified in Light List descriptions, shall follow the color scheme prescribed by the applicable Light List. Where the color scheme given in the Light List for a particular structure is revised to improve operational characteristics or to meet changed conditions, consideration shall be given to Section 3–2–2 on the functional use of color in regard to visibility and to the typical color schemes set forth in this Section. The below color scheme applies in all instances where there is no conflict with the Light List.

(A) WOOD, METAL AND PAINTED MASONRY BUILDINGS

The exteriors of buildings shall be painted white, including siding, eaves, gutters and downspouts (where painted), wood porch enclosures, wood porch railings and wood stair structures and balustrades. Trim shall be No. 12 Spruce Green. This includes window sash, shutters, doors, trellis under porches, roof vents, weather vanes, storm doors and windows, metal porch railings, metal stair structures and metal balustrades. Ornamental iron may be black.

In tropical regions buildings may be painted No. 11 Sun Tan to reduce glare. Where this standard is used, trim as above using white with the exception that metal porch railings, metal stair structures and metal balustrades shall be No. 12 Spruce Green.

(B) RED BRICK BUILDINGS

White shall be used as a trim color on window sash, shutters, doors, eaves, gutters and downspouts (where painted), wood columns, wood balustrades, wood ornamentation, storm doors, storm windows and trellis under porches. Metal porch railings and metal balustrades shall be No. 12 Spruce Green.

(C) UNPAINTED STONE AND MASONRY BUILDINGS

Trim with No. 12 Spruce Green in the same manner as for white buildings. It is best to avoid the painting of masonry where possible.

(D) LIGHT TOWER AND LANTERN HOUSE (wood, steel or masonry)

Unless otherwise specified in Light List, tower and lantern house shall be white. Trim with No. 12 Spruce Green including catwalks, balustrades, exterior lens frames, window frames and sash.

(E) ROOFS AND FOUNDATIONS

Roofs, where painted, shall be No. 26 Tile Red. Concrete foundations, if painted, are to be No. 19 Light Gray.

(F) MISCELLANEOUS STRUCTURES

Miscellaneous structures, towers and details such as lampposts, fence posts and rails, clothes poles, waste cans, etc. shall be painted white if a wooden surface and No. 12 Spruce Green if a metal surface. Flag poles are an exception and shall always be painted white.

(G) TANKS

Water tanks (metal and wood) whether on ground level or atop a structure shall be painted white on both the sides and the roof. Wood structures supporting a water tank shall be white; metal structures shall be No. 12 Spruce Green.

Large storage tanks (over 1,000 gals. capacity) containing flammable liquids are to be painted No. 12 Spruce Green and shall have the name of the contents painted in large No. 14 Brilliant Yellow letters on the

exterior of the tank in a conspicuous location. Smaller containers for flammable liquids or gases shall be painted No. 14 Brilliant Yellow overall with the contents conspicuously indicated in large black letters.

Large flammable liquid storage tanks in tropical and semi-tropical areas may be painted aluminum to reduce heat absorption.

l)

eads

ll be

No. 19 Light Gray. Railings and balustrades attached to buildings shall be painted white if of wood and No. 12 Spruce Green if of metal. Railings and balustrades not attached to a building shall be painted No. 19 Light Gray. No. 20 Medium Gray may be used as an alternate for No. 19 Light Gray where maintenance is unusually difficult.

(I) OUTDOOR MACHINERY AND EQUIPMENT

Winches, cranes and other outdoor machinery and equipment shall be painted No. 20 Medium Gray. See Section 3–13.

3–4–2, INTERIOR COLORS FOR LIGHT STATIONS

★ Note:

(A) CEILINGS

Ceilings are to be white throughout.

(B) TRIM

Trim including baseboards, window frames, window and door sash and doors shall be white or No. 19 Light Gray. Fixed equipment against walls such as cabinets shall be in either the wall or trim color. White is preferred for trim especially in dwellings. No. 19 Light Gray may be used as an alternate where maintenance is difficult as in generator rooms or shops. Where the No. 19 Light Gray is used for trim, window sash shall be in the wall color.

(C) SLEEPING QUARTERS

In bedrooms and sleeping quarters Principle A (No. 1 Light Green) and Principle E (No. 10 Light Blue) are recommended for hot climates or for areas having south or west exposure. Principle C (No. 25 Beach Sand), Principle D (No. 4 Soft Yellow) and Principle F (No. 6 Ivory) are recommended for cold climates, for spaces deprived of good natural light or for north and east exposure.

(D) LIVING ROOMS

Living rooms, wardrooms, day rooms and parlors are allowed a maximum number of color schemes to provide for wide latitude in functional and decorative effect. Principle A (No. 1 Light Green) and Principle E (No. 10 Light Blue) impart a cool and fresh atmosphere. These colors are best in warm or moderate climates or for south or west exposures. Principle C (No. 25 Beach Sand) and Principle D (No. 4 Soft Yellow) are warm in tone and will help compensate for low average temperatures or for north or east exposures. Principle G (No. 2 Medium Green) and Principle H (No. 8 Terra Cotta) are quiet and restful and should be restricted to casual areas in which difficult eye tasks are not performed. Principle F (No. 6 Ivory) is a general purpose color scheme suitable in all environments and where natural or artificial light is weak. Principle J (No. 9 Pearl Gray with No. 2 Medium Green end wall), Principle K (No. 9 Pearl Gray with No. 4 Soft Yellow end wall) and Principle L (No. 9 Pearl Gray with No. 8 Terra Cotta end wall), have a unique decorative effect. In each instance the colored wall should be at the end or far side of the room and never on a wall with windows.

(E) OFFICES, OPERATIONS ROOMS AND RADIO ROOMS

Offices, operations rooms and radio rooms may take a variety of color effects: Principle A (No. 1 Light Green) or Principle B (No. 9 Pearl Gray) for south and west exposures; Principle C (No. 25 Beach Sand) or Principle F (No. 6 Ivory) for north and east exposures.

(F) DINING ROOMS, MESS HALLS, AND RECREATION ROOMS

For dining rooms, mess halls and recreation rooms Principle C (No. 25 Beach Sand) is good as it provides a cheerful and appetizing effect. For a cool effect Principle A (No. 1 Light Green) may be used. Principle G (No. 2 Medium Green) and Principle E (No. 10 Light Blue) may also be applied.

★ NOTE: Shore Units use Interior Synthetic Rubber Emulsion Paint except in sanitary spaces.

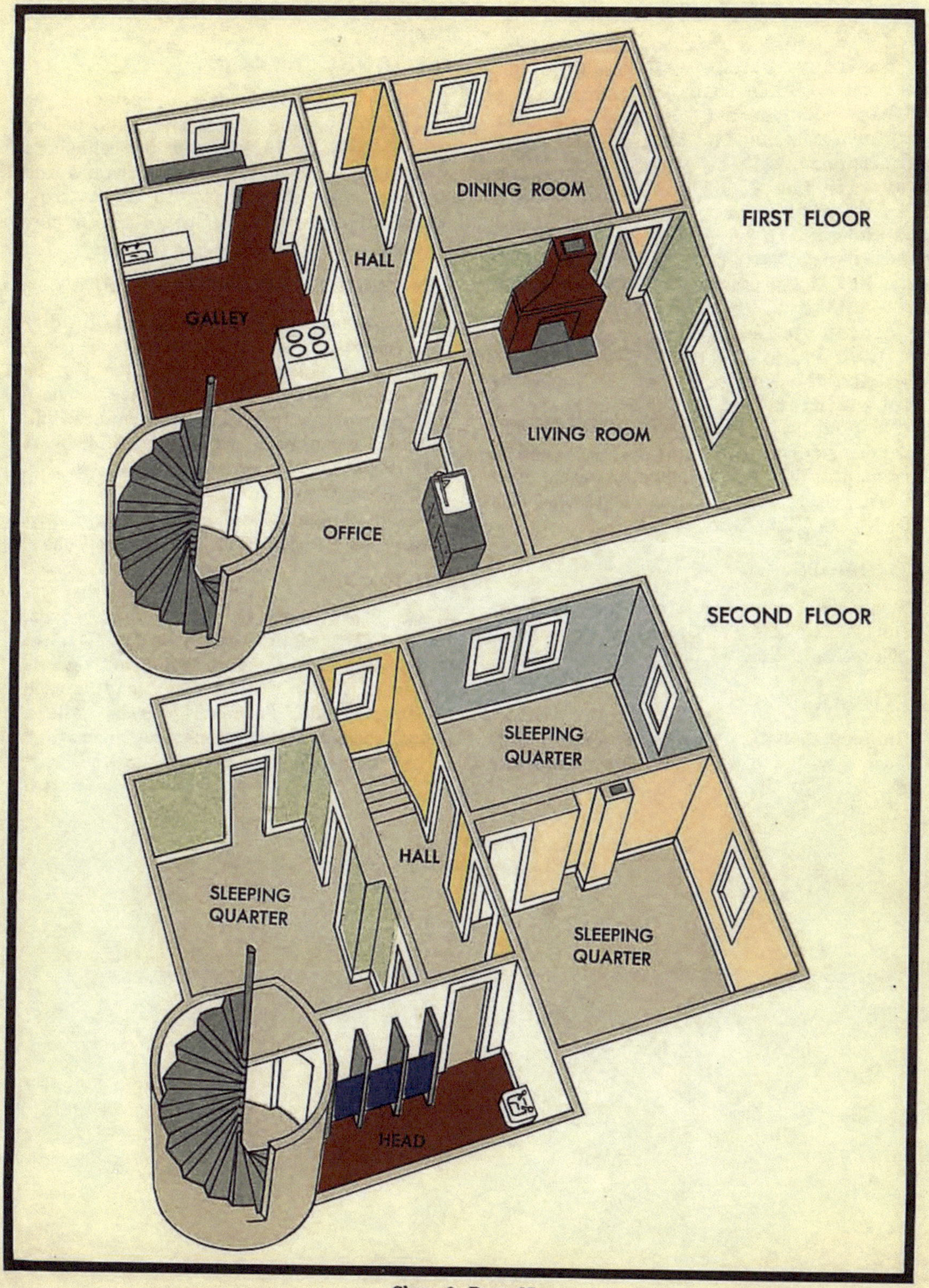
FIRST FLOOR
DINING ROOM
HALL
GALLEY
LIVING ROOM
OFFICE
SECOND FLOOR
SLEEPING QUARTER
HALL
SLEEPING QUARTER
SLEEPING QUARTER
HEAD

(G) SANITARY SPACES

The use of Principle I (white) in galleys, pantries, sculleries, heads, and shower rooms will insure cleanliness and high standards of housekeeping. In white heads, toilet stall partitions shall be No. 19 Light Gray with No. 29 Bright Blue doors. As an alternate to white, heads and showers in barracks may be painted Principle O (No. 1 Light Green with No. 2 Medium Green dado and trim). Galleys, kitchens and sanitary spaces in family quarters may be painted Principle C, F or D, depending upon the color scheme adopted by the district.

...ither ...ple F (No. 6 Ivory) or Principle I (white). Soffits (under side) of stairs and landings shall be painted white. For practical reasons, stair treads, risers, stringers and balustrades may be No. 19 Light Gray.

(J) SHOPS

Small general purpose shops shall be Principle A (No. 1 Light Green) or Principle D (No. 4 Soft Yellow).

(K) GARAGES

Garages shall be Principle P (No. 4 Soft Yellow with No. 19 Light Gray dado and trim). Storage sheds shall be Principle I (white).

(L) TOWER INTERIOR

The light house tower interior, including the lens room, shall be white. The metal ladder shall be No. 19 Light Gray. Miscellaneous metal railings, metal sash, lens frames and supports shall also be No. 19 Light Gray. As an alternate No. 12 Spruce Green may be used.

(M) EQUIPMENT AND MACHINERY

Most equipment and machinery at Light Stations including desks, filing cabinets, benches, bins, racks, shelving, etc. shall be No. 30 Equipment Gray. For equipment where rougher work is involved, maintenance difficult and cleanliness not essential, such as battery racks, tool racks and bins, use No. 20 Medium Gray.

For the finishing and highlighting of machinery see Section 3–17, Machinery Colors.

(N) FLOORS

Wood floors shall be sealed and waxed. Painted floors shall normally be No. 19 Light Gray. Where maintenance is difficult, such as in shops, No. 20 Medium Gray may be used as an alternate. Painted floors of galleys, heads, furnace rooms, generator rooms and locker rooms shall be No. 26 Tile Red.

exterior of the tank in a conspicuous location. Smaller containers for flammable liquids or gases shall be painted No. 14 Brilliant Yellow overall with the contents conspicuously indicated in large black letters.

(H) TRAFFIC AREAS (where painted)

Traffic areas such as porches, stair treads and risers, platforms and catwalks shall be No. 19 Light Gray. Railings and balustrades attached to buildings shall be painted white if of wood and No. 12 Spruce Green if of metal. Railings and balustrades not attached to a building shall be painted No. 19 Light Gray. No. 20 Medium Gray may be used as an alternate for No. 19 Light Gray where maintenance is unusually difficult.

(I) OUTDOOR MACHINERY AND EQUIPMENT

Winches, cranes and other outdoor machinery and equipment shall be painted No. 20 Medium Gray. See Section 3–13.

3–4–2, INTERIOR COLORS FOR LIGHT STATIONS

(A) CEILINGS

Ceilings are to be white throughout.

(B) TRIM

Trim including baseboards, window frames, window and door sash and doors shall be white or No. 19 Light Gray. Fixed equipment against walls such as cabinets shall be in either the wall or trim color. White is preferred for trim especially in dwellings. No. 19 Light Gray may be used as an alternate where maintenance is difficult as in generator rooms or shops. Where the No. 19 Light Gray is used for trim, window sash shall be in the wall color.

(C) SLEEPING QUARTERS

In bedrooms and sleeping quarters Principle A (No. 1 Light Green) and Principle E (No. 10 Light Blue) are recommended for hot climates or for areas having south or west exposure. Principle C (No. 25 Beach Sand), Principle D (No. 4 Soft Yellow) and Principle F (No. 6 Ivory) are recommended for cold climates, for spaces deprived of good natural light or for north and east exposure.

(D) LIVING ROOMS

Living rooms, wardrooms, day rooms and parlors are allowed a maximum number of color schemes to provide for wide latitude in functional and decorative effect. Principle A (No. 1 Light Green) and Principle E (No. 10 Light Blue) impart a cool and fresh atmosphere. These colors are best in warm or moderate climates or for south or west exposures. Principle C (No. 25 Beach Sand) and Principle D (No. 4 Soft Yellow) are warm in tone and will help compensate for low average temperatures or for north or east exposures. Principle G (No. 2 Medium Green) and Principle H (No. 8 Terra Cotta) are quiet and restful and should be restricted to casual areas in which difficult eye tasks are not performed. Principle F (No. 6 Ivory) is a general purpose color scheme suitable in all environments and where natural or artificial light is weak. Principle J (No. 9 Pearl Gray with No. 2 Medium Green end wall), Principle K (No. 9 Pearl Gray with No. 4 Soft Yellow end wall) and Principle L (No. 9 Pearl Gray with No. 8 Terra Cotta end wall), have a unique decorative effect. In each instance the colored wall should be at the end or far side of the room and never on a wall with windows.

(E) OFFICES, OPERATIONS ROOMS AND RADIO ROOMS

Offices, operations rooms and radio rooms may take a variety of color effects: Principle A (No. 1 Light Green) or Principle B (No. 9 Pearl Gray) for south and west exposures; Principle C (No. 25 Beach Sand) or Principle F (No. 6 Ivory) for north and east exposures.

(F) DINING ROOMS, MESS HALLS, AND RECREATION ROOMS

For dining rooms, mess halls and recreation rooms Principle C (No. 25 Beach Sand) is good as it provides a cheerful and appetizing effect. For a cool effect Principle A (No. 1 Light Green) may be used. Principle G (No. 2 Medium Green) and Principle E (No. 10 Light Blue) may also be applied.

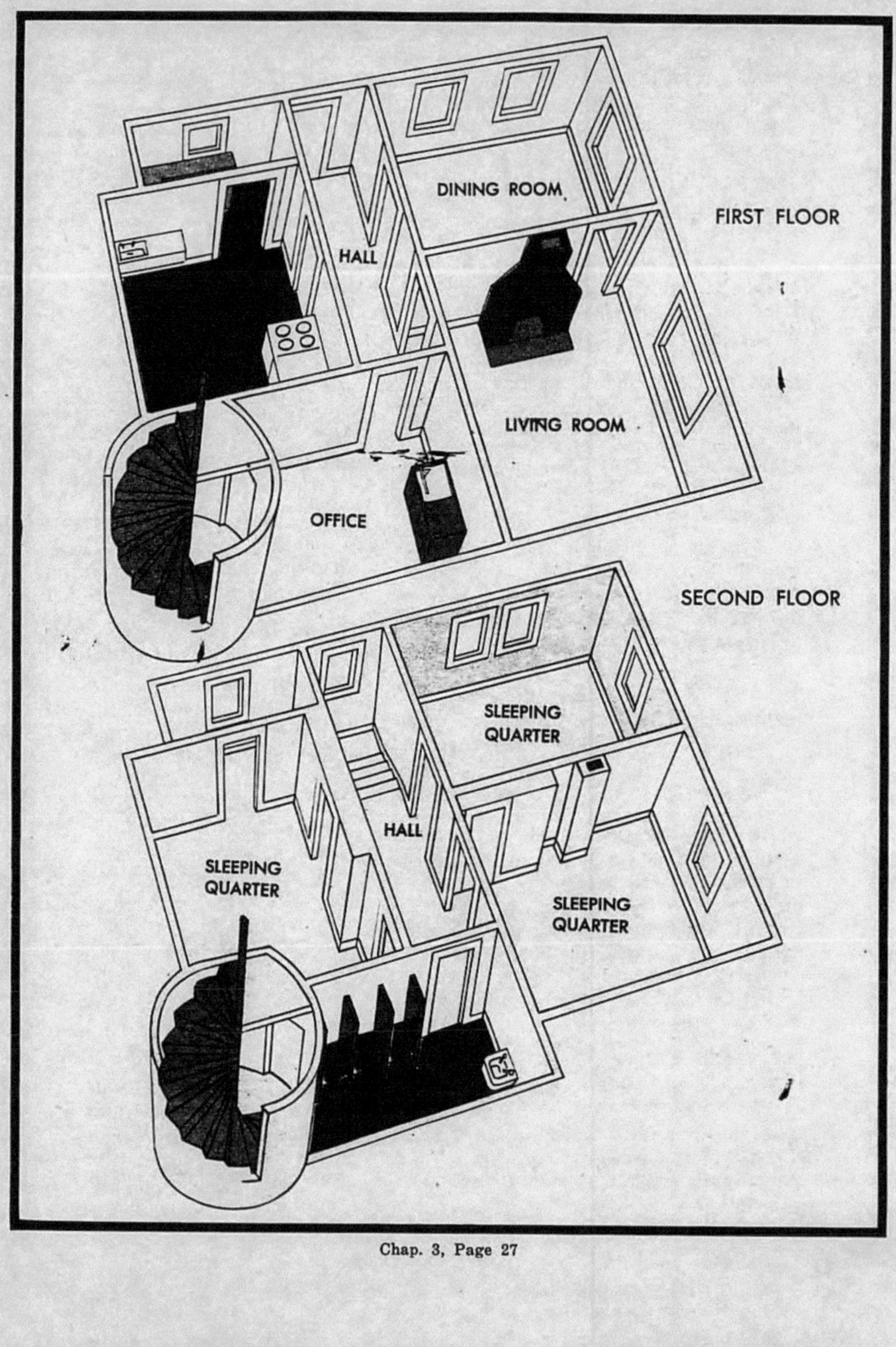
DINING ROOM
FIRST FLOOR
HALL
LIVING ROOM
OFFICE
SECOND FLOOR
SLEEPING QUARTER
HALL
SLEEPING QUARTER
SLEEPING QUARTER

(G) SANITARY SPACES

Galleys, pantries, sculleries, heads and shower rooms shall be Principle I (white). Toilet stall partitions shall be No. 19 Light Gray with No. 29 Bright Blue applied to the stall doors.

(H) UTILITY AND STORAGE SPACES

Basements, furnace rooms, battery rooms, generator rooms, emergency power rooms, laundries, miscellaneous storerooms and closets shall be painted Principle I (white).

(I) STAIRWAYS AND HALLWAYS

Stairways and hallways should be either Principle D (No. 4 Soft Yellow), Principle F (No. 6 Ivory) or Principle I (white). Soffits (under side) of stairs and landings shall be painted white. For practical reasons, stair treads, risers, stringers and balustrades may be No. 19 Light Gray.

(J) SHOPS

Small general purpose shops shall be Principle A (No. 1 Light Green) or Principle D (No. 4 Soft Yellow).

(K) GARAGES

Garages shall be Principle P (No. 4 Soft Yellow with No. 19 Light Gray dado and trim). Storage sheds shall be Principle I (white).

(L) TOWER INTERIOR

The light house tower interior, including the lens room, shall be white. The metal ladder shall be No. 19 Light Gray. Miscellaneous metal railings, metal sash, lens frames and supports shall also be No. 19 Light Gray. As an alternate No. 12 Spruce Green may be used.

(M) EQUIPMENT AND MACHINERY

Most equipment and machinery at Light Stations including desks, filing cabinets, benches, bins, racks, shelving, etc. shall be No. 30 Equipment Gray. For equipment where rougher work is involved, maintenance difficult and cleanliness not essential, such as battery racks, tool racks and bins, use No. 20 Medium Gray.

For the finishing and highlighting of machinery see Section 3–17, Machinery Colors.

(N) FLOORS

Wood floors shall be sealed and waxed. Painted floors shall normally be No. 19 Light Gray. Where maintenance is difficult, such as in shops, No. 20 Medium Gray may be used as an alternate. Painted floors of galleys, heads, furnace rooms, generator rooms and locker rooms shall be No. 26 Tile Red.

SECTION 3–5, COLORS FOR LIFEBOAT STATIONS

3–5–1, EXTERIOR COLORS FOR LIFEBOAT STATIONS

(A) WOOD, METAL AND PAINTED MASONRY BUILDINGS

The exteriors of buildings shall be painted white, including siding, eaves, gutters and downspouts (where painted), wood porch enclosures, wood porch railings and wood stair structures and balustrades. Trim shall be No. 12 Spruce Green. This includes window sash, shutters, doors, trellis under porches, roof vents, weather vanes, storm doors and windows, metal porch railings and metal stair structures and metal balustrades. Ornamental iron may be black.

In tropical regions buildings may be painted No. 11 Sun Tan to reduce glare. Where this standard is used, trim as above using white with the exception that metal porch railings and metal stair structures and metal balustrades shall be No. 12 Spruce Green.

(B) RED BRICK BUILDINGS

White shall be used as a trim color on window sash, shutters, doors, eaves, gutters and downspouts (where painted), wood columns, wood balustrades, wood ornamentation, storm doors, storm windows and trellis under porches. Metal porch railings and metal balustrades shall be No. 12 Spruce Green.

(C) UNPAINTED STONE AND MASONRY BUILDINGS

Trim with No. 12 Spruce Green in the same manner as for white buildings. It is best to avoid the painting of masonry where possible.

(D) ROOFS AND FOUNDATIONS

Roofs, where painted, shall be No. 26 Tile Red. Concrete foundations, if painted, are to be No. 19 Light Gray.

(E) MISCELLANEOUS STRUCTURES

Miscellaneous structures, lookout towers, light towers, flag towers, storm warning towers, range light towers, drill poles, lampposts, fence posts and rails, clothes poles, waste cans, etc., shall be painted white if a wooden surface and No. 12 Spruce Green if a metal surface. Flag poles are an exception and shall always be painted white.

(F) TANKS

Water tanks (metal and wood) whether on ground level or atop a structure shall be painted white on both the sides and the roof. Wood structures supporting a water tank shall be white; metal structures shall be No. 12 Spruce Green.

Large storage tanks (over 1,000 gals. capacity) containing flammable liquids are to be painted No. 12 Spruce Green and shall have the name of the contents painted in large No. 14 Brilliant Yellow letters on the exterior of the tank in a conspicuous location. Smaller containers for flammable liquids or gases shall be painted No. 14 Brilliant Yellow overall with the contents conspicuously indicated in large black letters.

(G) TRAFFIC AREAS (where painted)

Traffic areas such as porches, stair treads and risers, platforms and catwalks shall be painted No. 19 Light Gray. Railings and balustrades attached to buildings shall be painted white if of wood and No. 12 Spruce Green if of metal. No. 20 Medium Gray may be used as an alternate where maintenance is unusually difficult. This applies to launchways, metal catwalks, loading docks, piers and adjacent railings and stairways.

(H) OUTDOOR MACHINERY AND EQUIPMENT

Boat cradles, winches, cranes and other outdoor machinery and equipment shall be painted No. 20 Medium Gray. See Section 3–13.

3–5–2, INTERIOR COLORS FOR LIFEBOAT STATIONS

(A) CEILINGS

Ceilings are to be white throughout except in lookout towers.

(B) TRIM

Trim including baseboards, window frames, window and door sash and doors, shall be white or No. 19 Light Gray. Fixed equipment against walls such as cabinets shall be in either the wall or trim color. White is preferred for trim especially in dwellings. No. 19 Light Gray may be used as an alternate where maintenance is difficult as in boat houses, garages, repair shops. Where the No. 19 Light Gray is used for trim, window sash shall be in the wall color.

(C) SLEEPING QUARTERS

In bedrooms and sleeping quarters Principle A (No. 1 Light Green) and Principle E (No. 10 Light Blue) are recommended for hot climates or for areas having south or west exposures. Principle C (No. 25 Beach Sand), Principle D (No. 4 Soft Yellow) and Principle F (No. 6 Ivory) are recommended for cold climates, for spaces deprived of good natural light or for north and east exposures.

(D) LIVING ROOMS AND DAY ROOMS

Living rooms, day rooms, wardrooms and parlors are allowed a maximum number of color schemes to provide for wide latitude in functional and decorative effect. Principle A (No. 1 Light Green) and Principle E (No. 10 Light Blue) impart a cool and fresh atmosphere. These colors are best in warm or moderate climates or for south or west exposures. Principle C (No. 25 Beach Sand) and Principle D (No. 4 Soft Yellow) are warm in tone and will help compensate for low average temperatures or for north or east exposures. Principle G (No. 2 Medium Green) and Principle H (No. 8 Terra Cotta) are quiet and restful and should be restricted to casual areas in which difficult eye tasks are not performed. Principle F (No. 6 Ivory) is a general purpose color scheme suitable in all environments and where natural or artificial light is weak. Principle J (No. 9 Pearl Gray with No. 2 Medium Green end wall), Principle K (No. 9 Pearl Gray with No. 4 Soft Yellow end wall) and Principle L (No. 9 Pearl Gray with No. 8 Terra Cotta end wall), have a unique decorative effect. In each instance the colored wall should be at the end or far side of the room and never on a wall with windows.

(E) DINING ROOMS, MESS HALLS, AND RECREATION ROOMS

For dining rooms, mess halls and recreation rooms Principle C (No. 25 Beach Sand) is good as it provides a cheerful and appetizing effect. For a cool effect Principle A (No. 1 Light Green) may be used. Principle G (No. 2 Medium Green) and Principle E (No. 10 Light Blue) may also be applied.

(F) BOAT HOUSES, OFFICES, OPERATIONS ROOMS

In boat houses and garages Principle P, illustrated, shall be used (No. 4 Soft Yellow with No. 19 Light Gray trim). Offices, operations rooms, communication rooms, and radio rooms may take a variety of color effects: Principle A, illustrated (No. 1 Light Green) or Principle B, illustrated (No. 9 Pearl Gray) for south and west exposures; Principle C (No. 25 Beach Sand) or Principle F (No. 6 Ivory) for north and east exposures. In the lookout tower No. 1 Light Green (Principle A), illustrated, shall be applied to walls and overhead to reduce glare.

(G) SANITARY SPACES

Galleys, pantries, sculleries, heads and shower rooms shall be Principle I (white). Toilet stall partitions shall be No. 19 Light Gray with No. 29 Bright Blue applied to the stall doors.

(H) STAIRWAYS AND HALLWAYS

Stairways and hallways shall be either Principle D (No. 4 Soft Yellow), Principle F (No. 6 Ivory) or Principle I (white). Soffits (under side) of stairs and landings shall be painted white. For practical reasons, stair treads, risers, stringers and balustrades may be No. 19 Light Gray.

(I) UTILITY AND STORAGE SPACES

Basements, furnace rooms (illustrated), battery rooms, generator rooms, emergency power rooms, laundries, miscellaneous store-

DIESEL OIL

rooms and closets shall be painted Principle I (white), illustrated.

(J) SHOPS

Small general purpose shops shall be Principle O (No. 1 Light Green with No. 2 Medium Green trim) or Principle P (No. 4 Soft Yellow with No. 19 Light Gray trim).

(K) EQUIPMENT AND MACHINERY

Most equipment and machinery at Lifeboat Stations including desks, filing cabinets, benches, bins, racks, shelving, etc. shall be No. 30 Equipment Gray. For equipment where rougher work is involved, maintenance difficult and cleanliness not essential, such as metal working machinery, winches, boat cradles, engines, motors, tool racks and bins, use No. 20 Medium Gray, illustrated.

For the finishing and highlighting of machinery see Section 3–17.

(L) FLOORS

Wood floors shall be sealed and waxed. Painted floors shall normally be No. 19 Light Gray. Where maintenance is difficult, such as in shops, No. 20 Medium Gray may be used as an alternate. Painted floors of galleys, heads, furnace rooms, generator rooms and locker rooms shall be No. 26 Tile Red.

Operations centers and communications rooms are highly desirable in Principle A, No. 1 Light Green.

Offices may be Principle B, No. 9 Pearl Gray where a cool and non-distracting environment is wanted.

Boat house and garage interiors should be Principle P, No. 19 Light Gray dado and trim with No. 4 Soft Yellow walls.

Although most ceilings should be white, in the lookout tower No. 1 Light Green should be applied throughout.

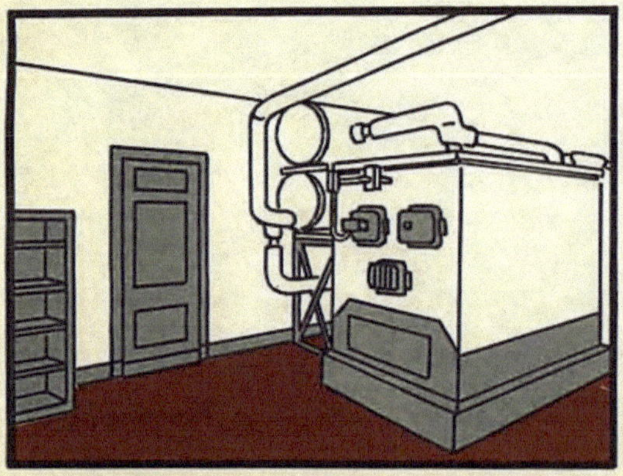

Furnace Rooms should be Principle R with white walls and No. 19 Light Gray trim. No. 26 Tile Red floors (if painted).

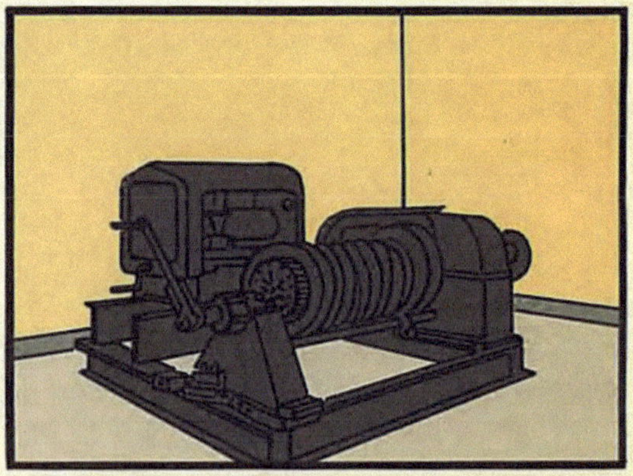

Heavily used equipment at Lifeboat Stations should be No. 20 Medium Gray to stand abuse and facilitate maintenance.

SECTION 3–6, COLORS FOR RADIO AND LORAN STATIONS

The below instructions are also applicable to Radio Direction Finder Stations, Radiobeacon Stations, and Electronics Engineering Stations

3–6–1, EXTERIOR COLORS FOR RADIO AND LORAN STATIONS

(A) WOOD, METAL AND PAINTED MASONRY BUILDINGS

The exteriors of buildings shall be painted white, including siding, eaves, gutters and downspouts (where painted), wood porch enclosures, wood porch railings and wood stair structures and balustrades. Trim shall be No. 12 Spruce Green. This includes window sash, shutters, doors, trellis under porches, roof vents, weather vanes, storm doors and windows, metal porch railings, metal stair structures and metal balustrades. Ornamental iron may be black.

In high glare regions such as the tropics or continued snow covered northern sites, all pertinent buildings at each station may be painted No. 11 Sun Tan to reduce glare. Where this standard is used, trim as above using No. 8 Terra Cotta for trim and grade beam.

~~may~~ be ... trated, to ... l is used, ... exception ... air struc- ... be No. 12

White shall be used as a trim color on window sash, shutters, doors, eaves, gutters and downspouts (where painted), wood columns, wood balustrades, wood ornamentation, storm doors, storm windows and trellis under porches. Metal porch railings and metal balustrades shall be No. 12 Spruce Green.

(C) UNPAINTED STONE AND MASONRY BUILDINGS

Trim with No. 12 Spruce Green in the same manner as for white buildings. It is best to avoid the painting of masonry where possible.

(D) ALUMINUM BUILDINGS

Aluminum buildings shall not be painted.

(E) ROOFS AND FOUNDATIONS

Roofs, where painted, shall be No. 26 Tile Red. Concrete foundations, if painted, are to be No. 19 Light Gray.

(F) MISCELLANEOUS STRUCTURES

Miscellaneous structures, towers and details such as lampposts, fence posts and rails, clothes poles, waste cans, etc. shall be painted white if a wooden surface and No. 12 Spruce Green if a metal surface. Flag poles are an exception and shall always be painted white.

In remote forward areas creosoted structures need not be painted.

(G) TANKS

Water tanks (metal) ~~and wood) whether on ground level or atop a structure~~ shall be painted white on ~~both~~ the sides and the roof. Wood structures supporting a water tank shall be white; metal structures shall be No. 12 Spruce Green. *Wood water tanks shall not be painted.*

Large storage tanks (over 1,000 gals. capacity) containing flammable liquids are to be painted No. 12 Spruce Green and shall have the name of the contents painted in large No. 14 Brilliant Yellow letters on the exterior of the tank in a conspicuous location. Smaller containers for flammable liquids or gases shall be painted No. 14 Brilliant Yellow overall with the contents conspicuously indicated in large black letters.

(H) HAZARDS TO AIR NAVIGATION

Tall structures, water tanks, towers, p... antennas, etc. frequently represent a ... to air navigation. Where this is *the* ... Aeronautics Administration ... struction Marking *and Lighting* shoul... the proper painting ... ubli- cation may ... superintend-

Large flammable liquid storage tanks in tropical semi-tropical areas may be painted aluminum to reduce heat absorption.

ent of Documents, Government Printing Office, Washington 25, D. C. Briefly, on tall towers and antenna poles alternate bands of No. 18 International Orange and white shall be applied throughout their height and the terminating bands at top and bottom shall be orange. The width of the orange bands shall be approximately one-seventh the height of the structure, and the alternate white bands shall be ~~approximately one-half~~ equal to the width of the orange bands. However, the orange bands shall never be more than 40 ft. nor less than 1½ ft. On large flat surfaces a checkerboard pattern is used. The sides of the squares or rectangles shall not measure more than 20 ft. nor less than 5 ft. (Night lighting is required for all tall structures.)

(I) TRAFFIC AREAS

Traffic areas such as porches, stair treads and risers, platforms and catwalks shall be No. 19 Light Gray. Railings and balustrades attached to buildings shall be painted white if of wood and No. 12 Spruce Green if of metal. Railings and balustrades not attached to buildings shall be painted No. 19 Light Gray. No. 20 Medium Gray may be used as an alternate for No. 19 Light Gray where maintenance is unusually difficult. This applies to catwalks, loading docks, piers, and adjacent railings and stairways.

(J) OUTDOOR MACHINERY AND EQUIPMENT

Winches, cranes and other outdoor machinery and equipment shall be painted No. 20 Medium Gray. See Section 3–13.

3–6–2, INTERIOR COLORS FOR RADIO AND LORAN STATIONS

(A) CEILINGS

Ceilings are to be white throughout except as indicated in below paragraph entitled, "Areas Requiring Low Light Levels."

(B) TRIM

Trim including baseboards, window frames, window and door sash and doors shall be white or No. 19 Light Gray. Fixed equipment against walls such as cabinets should be in either the wall or trim color. White is preferred for trim especially in dwellings. No. 19 Light Gray may be used as an alternate where maintenance is difficult as in garages and repair shops. Where the No. 19 Light Gray is used for trim, window sash should be in the wall color.

(C) SLEEPING QUARTERS

See Sleeping Quarters, Article 3–5–2(C).

(D) LIVING ROOMS AND DAY ROOMS

See Living Rooms and Day Rooms, Article 3–5–2(D).

(E) DINING ROOMS, MESS HALLS, AND RECREATION ROOMS

See Dining Rooms, Mess Halls, and Recreation Rooms, Article 3–5–2(E).

(F) OFFICES

Office areas should be Principle A (No. 1 Light Green) or Principle B (No. 9 Pearl Gray) for warm climates or for south or west exposures. Principle C (No. 25 Beach Sand), illustrated, or Principle F (No. 6 Ivory) should be used where a warm effect is wanted, as for north and east exposures. In all cases trim shall be No. 19 Light Gray.

(G) OPERATIONS AREAS

In operations rooms, broadcasting rooms, transmitting rooms, control rooms and radio rooms, Principle A (No. 1 Light Green), illustrated, shall be used. This soft, cool color will assure a comfortable and efficient seeing condition.

(H) AREAS REQUIRING LOW LIGHT LEVELS

At Radio and Loran Stations, certain areas call for low levels of illumination to aid the visibility of dials and scopes. This condition is frequently required for Loran timer rooms and supervisor's rooms. Light sources should be properly controlled and shielded. White ceilings will present no problem if illumination is well subdued. Principle G (No. 2 Medium Green), illustrated, should normally be used and the color may be carried on trim and overhead if desired. In very cold climates, Principle H (No. 8 Terra Cotta) may be applied. Principle B (No. 9 Pearl Gray,

Principle A, No. 1 Light Green is suitable for operations rooms, radio rooms and transmitting rooms.

Principle C, No. 25 Beach Sand should be used for office areas in cold climates or having north or east exposures.

Where dim light levels are maintained, as in timer rooms, Principle G, No. 2 Medium Green is desirable.

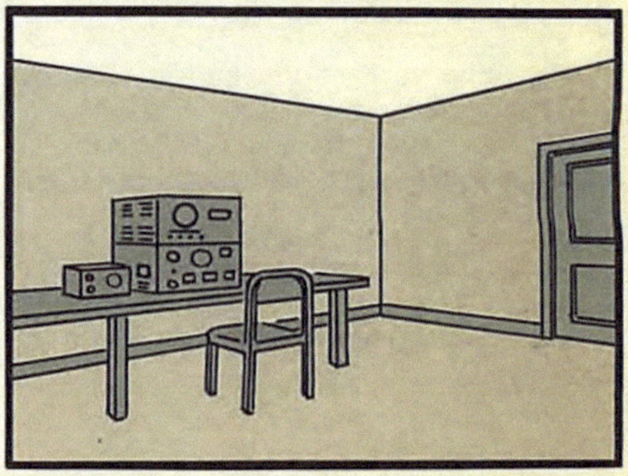

Principle B, No. 9 Pearl Gray may be used for the supervisor's room to assure an environment free of distraction.

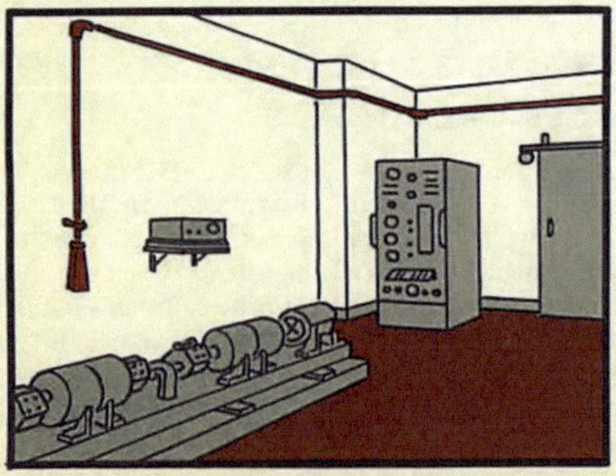

Principle I, white walls with No. 19 Light Gray trim and No. 26 Tile Red floors is best for generator rooms.

In workshops Principle P, No. 4 Soft Yellow walls with No. 19 Light Gray dado and trim may be selected.

SECTION 3–7, COLORS FOR AIR STATIONS

3–7–1, EXTERIOR COLORS FOR AIR STATIONS

(A) WOOD, METAL AND PAINTED MASONRY BUILDINGS

The exteriors of buildings shall be painted white, including siding, eaves, gutters and downspouts (where painted), wood porch enclosures, wood porch railings and wood stair structures and balustrades. Trim shall be No. 12 Spruce Green. This includes window sash, shutters, doors, trellis under porches, roof vents, weather vanes, storm doors and windows, metal porch railings, metal stair structures and metal balustrades. Ornamental iron may be black.

In tropical regions buildings may be painted No. 11 Sun Tan to reduce glare. Where this standard is used, trim as above using white with the exception that metal porch railings, metal stair structures and metal balustrades shall be No. 12 Spruce Green.

(B) RED BRICK BUILDINGS

White shall be used as a trim color on window sash, shutters, doors, eaves, gutters and downspouts (where painted), wood columns, wood balustrades, wood ornamentation, storm doors, storm windows and trellis under porches. Metal porch railings and metal balustrades shall be No. 12 Spruce Green.

(C) UNPAINTED STONE AND MASONRY BUILDINGS

Trim with No. 12 Spruce Green in the same manner as for white buildings. It is best to avoid the painting of masonry where possible.

(D) ROOFS AND FOUNDATIONS

Roofs (where painted), other than hangar roofs, shall be No. 26 Tile Red. Hangar roofs which are lettered U. S. COAST GUARD and which have directional arrows or other symbols shall be No. 19 Light Gray if painted. The lettering shall be No. 14 Brilliant Yellow with a heavy black outline. Where there is a bituminous roof, the lettering shall be No. 14 Brilliant Yellow only. For instructions on roof markings refer to Civil Aeronautics Administration Bulletin No. 12, Air Marking, obtainable from the Inquiry Branch, Civil Aeronautics Administration, Washington 25, D. C.

Concrete foundations, if painted, are to be No. 19 Light Gray.

(E) MISCELLANEOUS STRUCTURES

Miscellaneous structures, towers and details such as lampposts, fence posts and rails, clothes poles, waste cans, etc. shall be painted white if a wooden surface and No. 12 Spruce Green, if a metal surface. Flag poles are an exception and shall always be painted white.

(F) TANKS

Water tanks (metal and wood) whether on ground level or atop a structure shall be painted white on the sides and roof. Wood structures supporting a water tank shall be white; metal structures shall be No. 12 Spruce Green.

Large storage tanks (over 1,000 gals. capacity) containing flammable liquids are to be painted No. 12 Spruce Green and shall have the name of the contents painted in large No. 14 Brilliant Yellow letters on the exterior of the tank in a conspicuous location. Smaller containers for flammable liquids or gases shall be painted No. 14 Brilliant Yellow overall with the contents conspicuously indicated in large black letters.

(G) HAZARDS TO AIR NAVIGATION

Tall structures, water tanks, towers, poles, antennas, etc. should be avoided wherever possible at air stations. Where they exist Civil Aeronautics Administration publication, Obstruction Marking should be consulted for the proper painting procedures. This publication may be secured from the Superintendent of Documents, Government Printing Office, Washington 25, D. C. Briefly, on tall towers and antenna poles alternate bands of No. 18 International Orange and

U.S. COAST GUARD
N

3–7–2, INTERIOR COLORS FOR AIR STATIONS

(A) CEILINGS

Ceilings are to be white throughout.

(B) TRIM

Trim including baseboards, window frames, window and door sash, doors, stair balustrades and stringers shall be either white or No. 19 Light Gray. Fixed equipment against the walls such as cabinets shall be in either the wall or trim color. White is preferred for trim, especially in living spaces, with No. 19 Light Gray used as an alternate where maintenance is difficult as in shops. Where No. 19 Light Gray is used for trim, window sash shall be in the wall color. In either case the selected trim color shall be used uniformly throughout a building or an area in a building.

(C) BARRACKS AND SLEEPING QUARTERS

In barracks and sleeping quarters Principle A (No. 1 Light Green) and Principle E (No. 10 Light Blue) are recommended for hot climates or for areas having south or west exposures. Principle C (No. 25 Beach Sand), Principle D (No. 4 Soft Yellow) and Principle F (No. 6 Ivory) are recommended for cold climates, for spaces deprived of good natural light or for north and east exposures.

(D) WARDROOMS

To provide for wide latitude in functional and decorative effect, wardrooms, are allowed a maximum number of color schemes. Principle A (No. 1 Light Green), Principle B (No. 9 Pearl Gray) and Principle E (No. 10 Light Blue) impart a cool and fresh atmosphere. These colors are best in warm or moderate climates. Principle C (No. 25 Beach Sand) and Principle D (No. 4 Soft Yellow) are warm in tone and will help compensate for low average temperatures. Principle G (No. 2 Medium Green) and Principle H (No. 8 Terra Cotta) are quiet and restful. Principle F (No. 6 Ivory) is a general purpose color scheme suitable in all environments. Principle J (No. 9 Pearl Gray with No. 2 Medium Green end wall), Principle K (No. 9 Pearl Gray with No. 4 Soft Yellow end wall) and Principle L (No. 9 Pearl Gray with No. 8 Terra Cotta end wall) have a unique decorative effect. In each instance the colored wall should be at the end or far side of the room and never on a wall with windows.

(E) OFFICES AND OPERATIONS ROOMS

Offices and operations rooms may take a variety of color effects: Principle A (No. 1 Light Green), illustrated, or Principle B (No. 9 Pearl Gray) for south and west exposure; Principle C (No. 25 Beach Sand) or Principle F (No. 6 Ivory) for north and east exposures. In control towers, where glass is widely used, any sash, wall areas and overheads shall be No. 1 Light Green to curtail glare.

(F) RADIO ROOMS AND COMMUNICATIONS CENTERS

Principle A (No. 1 Light Green) or Principle B (No. 9 Pearl Gray) shall be used. These colors have good reflectance for seeing tasks yet insure a restful and glare-free environment.

(G) MESS HALLS, CAFETERIAS AND DINING ROOMS

For mess halls and cafeterias Principle C (No. 25 Beach Sand) may be used as it provides a cheerful and appetizing effect. For a cool effect Principle A (No. 1 Light Green) may be used. Either Principle M (No. 25 Beach Sand with No. 8 Terra Cotta dado and trim), illustrated, or Principle O (No. 1 Light Green with No. 2 Medium Green dado and trim) may be selected as alternates.

(H) RECREATION ROOMS

A dado is best for recreation rooms. Principle M (No. 25 Beach Sand with No. 8 Terra Cotta dado and trim), Principle N (No. 10 Light Blue with No. 29 Bright Blue dado and trim), Principle O (No. 1 Light Green with No. 2 Medium Green dado and trim), illustrated, or Principle A (No. 1 Light Green) may be used.

A good recreation room may be planned with Principle O, No. 1 Light Green walls and No. 2 Medium Green trim.

A green recreation room is complemented with mess in Principle M, No. 25 Beach Sand and No. 8 Terra Cotta.

Operations offices are desirable in Principle A, No. 1 Light Green, but also may use Principle B, No. 9 Pearl Gray.

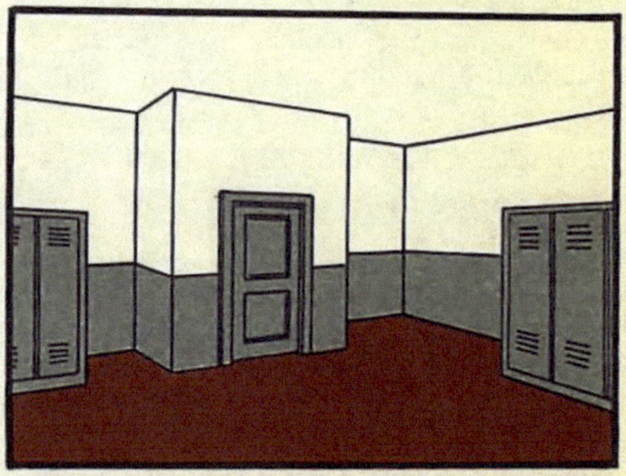

Locker rooms and laundries shall be Principle R, white with No. 19 Light Gray trim. Floors No. 26 Tile Red.

A white hangar with No. 19 Light Gray dado, trim and steel work, Principle R, will be efficient in light reflection.

Principle P, No. 4 Soft Yellow with No. 19 Light Gray dado and trim is good for hangars where climate may be cool.

(I) SANITARY SPACES

The use of Principle I (white) in galleys, pantries, sculleries, heads and shower rooms will insure cleanliness and high standard of housekeeping. In white heads, toilet stall partitions shall be No. 19 Light Gray with No. 29 Bright Blue doors. As an alternate to white, heads and showers in barracks may be painted Principle O (No. 1 Light Green with No. 2 Medium Green dado and trim). Galleys, kitchens and sanitary spaces in family quarters may be painted Principle C, or D, depending upon the color scheme adopted by the district.

(K) STAIRWAYS, CORRIDORS AND PASSAGEWAYS

Stairways, corridors and passageways shall be Principle D (No. 4 Soft Yellow), Principle F (No. 6 Ivory) or Principle I (white). Principle P (No. 4 Soft Yellow with No. 19 Light Gray dado and trim, including stair treads, risers and stringers), may be used in shop areas or where traffic is heavy. To reduce stumblings and falling hazards a 4-in. band of No. 14 Brilliant Yellow shall be applied immediately under the tread on the top and bottom risers. Where the soffits (under side) of stairs and landings are plaster or concrete they should be painted white. The undersides of metal stairs are better in No. 19 Light Gray.

(L) DISPENSARIES, INFIRMARIES AND SICK BAYS

Dispensaries, pharmacies, first aid rooms, treatment rooms and examining rooms shall be Principle I (white) for cleanliness and good light reflection. Principle C (No. 25 Beach Sand) could be used as an alternate. Sick rooms, wards, dental clinics and remainder of the hospital area shall be Principle A (No. 1 Light Green) for a cool effect.

(M) EXCHANGES

The high popularity of blue among men makes it an ideal color for use in exchange facilities. In the store, Principle N (No. 10 Light Blue with No. 29 Bright Blue dado and trim) may be used. As a special effect No. 9 Pearl Gray may be used to provide a soft neutral background for the display of various items sold. No. 29 Bright Blue is then used for trim, doors, baseboards, shelving and counters. Counter tops, for bright appearance, may be in a color or material corresponding to No. 26 Tile Red.

For soda fountains and snack bars the appetizing qualities of Principle M (No. 25 Beach Sand with No. 8 Terra Cotta dado and trim) may be used. Again, No. 29 Bright Blue may be used for counters and No. 26 Tile Red for counter tops.

(N) HANGARS

Two color schemes are illustrated for hangars. The simplest effect is to use white for upper walls and overhead, and No. 19 Light Gray for dado, trim and window sash (Principle R), illustrated. Wood roof trusses and beams should ordinarily be white. However, steel trusses, beams and girders may be No. 19 Light Gray. A more luminous and bright effect for hangars in colder climates will be found in the use of white for overhead, No. 4 Soft Yellow for upper walls, and No. 19 Light Gray for dado, trim, window sash and steel work (Principle P, illustrated). It is important to have the good light reflection provided either by white or No. 4 Soft Yellow.

(O) SHOPS

For machine shops where average tasks are performed, Principle P is recommended and illustrated. This consists of a white overhead, No. 4 Soft Yellow on upper walls, and No. 19 Light Gray on dado, trim and window sash. Concrete or wood ceiling beams should be white: No. 19 Light Gray is advised, however, for overhead steel work.

As a special color scheme for parachute shops where women may be employed, No. 25 Beach Sand may be used for walls, with No. 19 Light Gray for trim, illustrated.

In electronic shops and electrical shops where fairly difficult seeing tasks are performed and where fine tolerances are demanded, Principle O, illustrated, is desirable.

Machine shops may use Principle P, No. 4 Soft Yellow upper walls, No. 19 Light Gray dado, trim and window sash.

Electronic shops may use Principle O, No. 1 Light Green walls and No. 2 Medium Green trim and dado.

Parachute shops may have Principle C, No. 25 Beach Sand with No. 19 Light Gray trim and dado.

No. 30 Equipment Gray or No. 20 Medium Gray should be for equipment; No. 14 Brilliant Yellow for test stands.

Gasoline trucks, mules, tractors, transporters used on air fields shall be No. 14 Brilliant Yellow.

Fire trucks, rescue and crash trucks shall be No. 13 Fire Red for proper identification.

Here No. 1 Light Green is used for upper walls, with No. 2 Medium Green for dado, trim and window sash. White with No. 19 Light Gray for dado and trim (Principle R) is proposed for relatively unimportant areas such as tool cribs, storage rooms and locker rooms.

(P) EQUIPMENT AND MACHINERY

There are two gray standards for equipment: No. 30 Equipment Gray and No. 20 Medium Gray, illustrated. The lighter tone, No. 30 Equipment Gray, should be used wherever possible. While this fairly light tone of gray requires more maintenance, it does improve workmanship, since it leads to greater cleanliness and care of equipment. Portable testing apparatus, desks, stools, filing cabinets, electrical equipment and fine machine tools should be No. 30 Equipment Gray. Where rougher machinery and equipment are involved—tool racks, bins, work benches, lathes, drill presses, shapers, milling machines, tool grinders, power saws, etc.—the deeper No. 20 Medium Gray is acceptable as an alternate. (For the highlighting of machinery see Section 3–17.)

Portable equipment, such as test stands, stanchions, jacks, etc. which constitute an obstruction hazard, shall be painted No. 14 Brilliant Yellow, illustrated. Horizontal surfaces, steps and platforms, subject to wear, may be No. 20 Medium Gray. (See also Coast Guard Safety Color Code, Section 3–16.)

(Q) FLOORS

Wood floors shall be sealed and waxed. Painted floors shall normally be No. 19 Light Gray. Where maintenance is difficult, such as in shops, No. 20 Medium Gray may be used as an alternate. Painted floors of galleys, heads, furnace rooms, generator rooms and locker rooms shall be No. 26 Tile Red.

SECTION 3-8, COLORS FOR ADMINISTRATIVE FACILITIES

These instructions are applicable to the following facilities:

Headquarters	Marine Inspection Offices
Area Offices	Academy
District Offices	Training Station
Group Offices	Receiving Centers
Captain of the Port Offices	Recruiting Station

3-8-1, EXTERIOR COLORS FOR ADMINISTRATIVE FACILITIES

The maintenance of the Coast Guard's administrative facilities is generally the responsibility of either the Public Buildings Administration or private owners. Only where the responsibility for maintenance rests with the Coast Guard are the following instructions applicable.

(A) WOOD, METAL AND PAINTED MASONRY BUILDINGS

The exteriors of buildings shall be painted white, including siding, eaves, gutters and downspouts (where painted), wood porch enclosures, wood porch railings and wood stair structures and balustrades. Trim shall be No. 12 Spruce Green. This includes window sash, shutters, doors, trellis under porches, roof vents, weather vanes, storm doors and windows, metal porch railings, metal stair structures and metal balustrades. Ornamental iron may be black.

In tropical regions buildings may be painted No. 11 Sun Tan to reduce glare. Where this standard is used, trim as above using white with the exception that metal porch railings, metal stair structures and metal balustrades shall be No. 12 Spruce Green.

(B) RED BRICK BUILDINGS

White shall be used as a trim color on window sash, shutters, doors, eaves, gutters and downspouts (where painted), wood columns, wood balustrades, wood ornamentation, storm doors, storm windows and trellis under porches. Metal porch railings and metal balustrades shall be No. 12 Spruce Green.

(C) UNPAINTED STONE AND MASONRY BUILDINGS

Trim with No. 12 Spruce Green in the same manner as for white buildings. It is best to avoid the painting of masonry where possible.

(D) ROOFS AND FOUNDATIONS

Roofs, where painted, shall be No. 26 Tile Red. Concrete foundations, if painted, are to be No. 19 Light Gray.

(E) MISCELLANEOUS STRUCTURES

Miscellaneous structures, towers and details such as lampposts, fence posts and rails, clothes poles, waste cans, etc. shall be painted white if a wooden surface and No. 12 Spruce Green if a metal surface. Flag poles are an exception and shall always be painted white.

(F) TANKS

Water tanks (metal and wood) whether on ground level or atop a structure shall be painted white on both the sides and the roof. Wood structures supporting a water tank shall be white; metal structures shall be No. 12 Spruce Green.

Large storage tanks (over 1,000 gals. capacity) containing flammable liquids are to be painted No. 12 Spruce Green and shall have the name of the contents painted in large No. 14 Brilliant Yellow letters on the exterior of the tank in a conspicuous location. Smaller containers for flammable liquids or gases shall be painted No. 14

3–8–2, INTERIOR COLORS FOR ADMINISTRATIVE FACILITIES

(A) CEILINGS

Ceilings are to be white throughout with the exception of spaces where dim lighting is desirable to read illuminated dials, scopes, etc. as in CIC training rooms. In such cases extend the wall color over the ceiling.

(B) TRIM

Trim including baseboards, window frames, window and door sash, doors, stair balustrades and stringers shall be either white or No. 19 Light Gray. Fixed equipment against walls such as cabinets, shall be in either the wall or trim color. White is preferred for trim especially in living spaces, with No. 19 Light Gray used as an alternate where maintenance is difficult as in shops. Where No. 19 Light Gray is used for trim, window sash should be in the wall color. In either case the selected trim color should be used uniformly throughout a building or an area in a building.

(C) BARRACKS AND SLEEPING QUARTERS

In barracks and sleeping quarters Principle A (No. 1 Light Green) and Principle E (No. 10 Light Blue) are recommended for hot climates or for areas having south or west exposures. Principle C (No. 25 Beach Sand), Principle D (No. 4 Soft Yellow) and Principle F (No. 6 Ivory) are recommended for cold climates, for spaces deprived of good natural light or for north and east exposures.

(D) WARDROOMS, RECEPTION ROOMS AND WOMEN'S REST ROOMS

To provide for wide latitude in functional and decorative effect, wardrooms, reception rooms and women's rest rooms are allowed a maximum number of color schemes. Principle A (No. 1 Light Green), Principle B (No. 9 Pearl Gray) and Principle E (No. 10 Light Blue) impart a cool and fresh atmosphere. These colors are best in warm or moderate climates. Principle C (No. 25 Beach Sand) and Principle D (No. 4 Soft Yellow) are warm in tone and will help compensate for low average temperatures. Principle G (No. 2 Medium Green) and Principle H (No. 8 Terra Cotta) are quiet and restful. Principle F (No. 6 Ivory) is a general purpose color scheme suitable in all environments. Principle J (No. 9 Pearl Gray with No. 2 Medium Green end wall), illustrated, Principle K (No. 9 Pearl Gray with No. 4 Soft Yellow end wall) and Principle L (No. 9 Pearl Gray with No. 8 Terra Cotta end wall), illustrated, have a unique decorative effect. In each instance the colored wall should be at the end or far side of the room and never on a wall with windows.

(E) OFFICES AND OPERATIONS ROOMS

Offices and operations rooms may take a variety of color effects: Principle A (No. 1 Light Green) or Principle B (No. 9 Pearl Gray) for south and west exposures; Principle C (No. 25 Beach Sand or Principle F (No. 6 Ivory) for north and east exposures. In large general offices Principle D (No. 4 Soft Yellow) may be used satisfactorily. For offices having heavy traffic or occupancy, Principle O (No. 1 Light Green with No. 2 Medium Green dado and trim) is best.

(F) RADIO ROOMS AND COMMUNICATION CENTERS

Principle A (No. 1 Light Green) or Principle B (No. 9 Pearl Gray) shall be used. These colors have good reflectance for seeing tasks yet insure a restful and glare-free environment.

(G) MESS HALLS, CAFETERIAS AND DINING ROOMS

For mess halls and cafeterias Principle C (No. 25 Beach Sand) may be used as it provides a cheerful and appetizing effect. For a cool effect Principle A (No. 1 Light Green) may be used. Principle M (No. 25 Beach Sand with No. 8 Terra Cotta dado and trim) or Principle O (No. 1 Light Green with No. 2 Medium Green dado and trim) may be selected as alternates.

(H) RECREATION ROOMS

A dado is best for recreation rooms. Either Principle M (No. 25 Beach Sand with No. 8 Terra Cotta dado and trim), Principle N

Principle J applied to reception rooms. No. 9 Pearl Gray is on three walls, with No. 2 Medium Green over end wall.

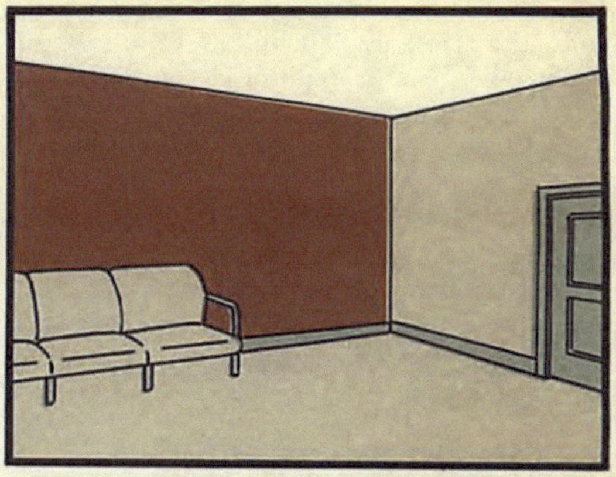

Three walls in No. 9 Pearl Gray, with end wall in No. 8 Terra Cotta, Principle L, makes an effective rest room.

Classrooms and drafting may use colored end walls. Principle J, No. 9 Pearl Gray is used with No. 2 Medium Green.

Principle K combines No. 9 Pearl Gray with No. 4 Soft Yellow on end wall. This would enliven a dark room.

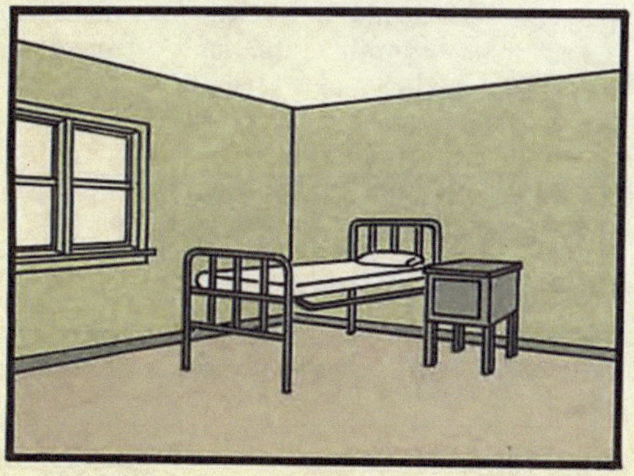

Principle A, No. 1 Light Green makes ideal hospital color. Pharmacy may be Principle C, No. 25 Beach Sand.

Principle D, No. 4 Soft Yellow with No. 19 Light Gray dado and trim is advised for passageways and stairways.

(No. 10 Light Blue with No. 29 Bright Blue dado and trim) or Principle O (No. 1 Light Green with No. 2 Medium Green dado and trim) should be used. Principle A (No. 1 Light Green) may be used as an alternate.

(I) SANITARY SPACES

~~Galleys, pantries, sculleries, heads and shower rooms shall be Principle I (white). ...let stall partitions shall be No. 19 Light ...ith No. 29 Bright Blue applied to ...rs.~~

The use of Principle I (white) in galleys,
pantries, sculleries, heads and shower
rooms will insure cleanliness and high
standards of housekeeping. In white
heads, toilet stall partitions shall be
No. 19 Light Gray with No. 29 Bright
...ue doors. As an alternate to white,
...ds and showers in barracks may be
...ted Principle O (No. 1 Light Green
No. 2 Medium Green dado and trim).
...s, kitchens and sanitary spaces in
...uarters may be painted Principle
..., depending upon the color
...pted by the district.

...CES AND

...ttery rooms, ... rooms, ...arts ...all ...ple ...).

...ways shall be el... ...ft Yellow), Principle ... Principle I (white). Pri... 4 Soft Yellow with No. 19 Lig... ...ado, trim, stair treads, risers, string... and steps), illustrated, may be used in shop areas or where traffic is heavy. To reduce stumbling and falling hazards a 4-in. band of No. 14 Brilliant Yellow shall be applied immediately under the tread on the top and bottom risers. Where the soffits (under side) of stairs and landings are plaster or concrete they should be painted white. The undersides of metal stairs are better in No. 19 Light Gray.

(L) DISPENSARIES, INFIRMARIES AND SICK BAYS

Dispensaries, pharmacies, first aid rooms, treatment rooms and examining rooms shall be Principle I (white) for cleanliness and good light reflection. Principle C (No. 25 Beach Sand) may be used as an alternate. Sick rooms, wards, dental clinic and remainder of the hospital area shall be Principle A (No. 1 Light Green) illustrated, for a cool effect.

(M) CLASS ROOMS, DRAFTING ROOMS AND LABORATORIES

For classrooms, training rooms, drafting rooms and laboratories Principle A (No. 1 Light Green), Principle B (No. 9 Pearl Gray) or Principle C (No. 25 Beach Sand) shall be used. However, where personnel are seated to face in one direction, the end wall treatment serves a wholly practical and functional purpose and may be used. Two effects are illustrated, Principle J (No. 9 Pearl Gray with No. 2 Medium Green end wall) and Principle K (No. 9 Pearl Gray with No. 4 Soft Yellow end wall). The green end wall is recommended for areas having good light; yellow may be specified where a brighter and more luminous effect is desired. Also suitable is Principle L (No. 9 Pearl Gray with No. 8 Terra Cotta end wall). In each instance the colored wall should be at the end or far side of the room and never on a wall with windows.

(N) EXCHANGES

The high popularity of blue among men makes it an ideal color for use in exchange facilities. In the store Principle N (No. 10 Light Blue with No. 29 Bright Blue dado and trim) may be used. As a special effect No. 9 Pearl Gray, illustrated, may be used to provide a soft neutral background for the display of various items sold. No. 29 Bright Blue is then used for trim, doors, baseboards, shelving and counters. Counter tops, for bright appearance, may be in a color or material corresponding to No. 26 Tile Red.

For soda fountains and snack bars the appetizing qualities of Principle M (No. 25 Beach Sand with No. 8 Terra Cotta dado and trim), illustrated, may be used. Again, No. 29 Bright Blue may be used for counters and No. 26 Tile Red for counter tops.

(O) BARBER SHOPS

Use Principle G (No. 2 Medium Green) or Principle A (No. 1 Light Green) with white trim.

(P) GYMNASIUMS

Use Principle P (No. 4 Soft Yellow with No. 19 Light Gray dado and trim), illustrated,

Exchange stores may be specially treated with No. 9 Pearl Gray on walls, No. 29 Bright Blue on trim and fixtures.

Exchange soda fountains and snack bars may use Principle M, No. 25 Beach Sand walls with No. 8 Terra Cotta trim.

Gymnasiums may use Principle P, No. 19 Light Gray for trim and dado with No. 4 Soft Yellow.

Libraries should have Principle A, No. 1 Light Green for walls with No. 19 Light Gray trim and shelving.

Auditoriums may be in Principle M, No. 25 Beach Sand and No. 8 Terra Cotta.

Chapel trim may be white or No. 19 Light Gray. Walls may be No. 10 Light Blue, No. 9 Pearl Gray, or white.

or No. 6 Ivory with No. 19 Light Gray dado and trim. Bleachers shall be No. 19 Light Gray.

(Q) LIBRARIES

Use Principle A (No. 1 Light Green), illustrated, with No. 19 Light Gray for trim and shelving.

(R) AUDITORIUMS AND THEATERS

Use Principle M (No. 25 Beach Sand with No. 8 Terra Cotta dado and trim), illustrated, or Principle O (No. 1 Light Green with No. 2 Medium Green dado and trim).

(S) CHAPELS

Chapels may use Principle E (No. 10 Light Blue), illustrated, Principle B (No. 9 Pearl Gray) or Principle I (white).

(T) SHOPS

See Bases and Depots, Section 3–9–2, (N) to (Q).

(U) EQUIPMENT AND MACHINERY

For equipment where cleanliness is important or where maintenance is not difficult such as portable testing apparatus, desks, stools, filing cabinets, electrical equipment, fine machine tools, etc. use No. 30 Equipment Gray. For equipment where rougher work is involved, maintenance is difficult and cleanliness is not essential, such as lathes, drill presses, shapers, milling machines, tool grinders, power saws, tool racks and bins, use No. 20 Medium Gray.

Portable equipment which constitutes an obstruction hazard such as test stands, stanchions and jacks should be painted No. 14 Brilliant Yellow. Horizontal surfaces, steps and platforms subject to wear may be No. 20 Medium Gray. (See Safety Color Code, Section 3–16.)

For the finishing and highlighting of machinery see Section 3–17, Machinery Colors.

(V) FLOORS

Wood floors shall be sealed and waxed. Painted floors shall normally be No. 19 Light Gray. Where maintenance is difficult such as in shops, No. 20 Medium Gray may be used as an alternate. Painted floors of galleys, heads, furnace rooms, generator rooms and locker rooms shall be No. 26 Tile Red.

SECTION 3–9, COLORS FOR BASES AND DEPOTS

These instructions are applicable to the following facilities:

Yards — Electronic Repair Shops
Moorings — Supply Centers
Supply Depots

3–9–1, EXTERIOR COLORS FOR BASES AND DEPOTS

(A) WOOD, METAL AND PAINTED MASONRY BUILDINGS

Non-industrial Type Buildings—

The exteriors of buildings shall be painted white, including siding, eaves, gutters and downspouts (where painted), wood porch enclosures, wood porch railings and wood stair structures and balustrades. Trim shall be No. 12 Spruce Green. This includes window sash, shutters, doors, trellis under porches, roof vents, weather vanes, storm doors and windows, metal porch railings, metal stair structures and metal balustrades. Ornamental iron may be black.

In tropical regions buildings may be painted No. 11 Sun Tan to reduce glare. Where this standard is used, trim as above using white with the exception that metal porch railings, metal stair structures and metal balustrades shall be No. 12 Spruce Green.

Industrial Type Buildings—

Shops, warehouses, garages, galvanized iron sheds, open steel welding sheds and other industrial type buildings shall be painted No. 19 Light Gray. The trim on such structures—doors, door frames, window sash, roof gutters and eaves—shall be No. 12 Spruce Green.

(B) RED BRICK BUILDINGS

Non-industrial Type Buildings—

White shall be used as a trim color on window sash, shutters, doors, eaves, gutters and downspouts (where painted), wood columns, wood balustrades, wood ornamentation, storm doors, storm windows and trellis under porches. Metal porch railings and metal balustrades shall be No. 12 Spruce Green.

Industrial Type Buildings—

Trim with No. 12 Spruce Green in the same manner as white is used for non-industrial type buildings.

(C) UNPAINTED STONE AND MASONRY BUILDINGS

Trim with No. 12 Spruce Green in the same manner as for white buildings. It is best to avoid the painting of masonry where possible.

(D) ROOFS AND FOUNDATIONS

Roofs, where painted, shall be No. 26 Tile Red. Concrete foundations, if painted, are to be No. 19 Light Gray.

(E) MISCELLANEOUS STRUCTURES

Miscellaneous structures, fences, lampposts and refuse containers shall be painted white if a wooden surface and No. 12 Spruce Green if a metal surface. Flag poles are an exception and shall always be painted white.

Water tanks (metal and wood) whether on ground level or atop a structure shall be painted white on both the sides and the roof. Wood structures supporting a water tank shall be white; metal structures shall be No. 12 Spruce Green.

Steel smokestacks shall be No. 19 Light Gray or, if high temperatures are involved, aluminum. Small independent fire houses containing hose carts and extinguishers shall be painted No. 13 Fire Red. This same color shall be applied to all fire hydrants, hose and sprinkler line stand pipes.

Where tall structures are a hazard to air navigation, refer to Civil Aeronautics Administration publication "Obstruction Marking," obtainable from the Superintendent of Documents, Government Printing Office, Washington 25, D. C.

(F) FLAMMABLE LIQUIDS STORAGE

Sheds, small buildings and lockers used for the storage of paints, fuels and other flammable liquids shall be painted No. 14 Brilliant Yellow and prominently identified as to contents with large black letters. Larger buildings with flammable stores, however, may be No. 19 Light Gray to avoid too conspicuous an appearance. Warnings signs using No. 14 Brilliant Yellow for lettering shall be added, however.

Large storage tanks (over 1,000 gals. capacity) containing flammable liquids are to be painted No. 12 Spruce Green and shall have the name of the contents painted in large No. 14 Brilliant Yellow letters on the exterior of the tank in a conspicuous location. Smaller containers for flammable liquids or gases shall be painted No. 14 Brilliant Yellow overall with the contents conspicuously indicated in large black letters.

(G) TRAFFIC AREAS

Porches, stair treads and risers, platforms and catwalks shall be No. 19 Light Gray. Railings and balustrades attached to buildings shall be painted white if of wood and No. 12 Spruce Green if of metal. Railings and balustrades not attached to a building shall be painted No. 19 Light Gray. No. 20 Medium Gray may be used as an alternate for No. 19 Light Gray where maintenance is unusually difficult such as on piers, floats and loading platforms. Care should be taken to add safety marking with bands of black and No. 14 Brilliant Yellow on curbing, bumpers, the edges of platforms and pits, low beams and any other obstructions which may constitute a hazard. For further details on safety markings see the Safety Color Code, Section 3–16.

(H) OUTDOOR MACHINERY AND EQUIPMENT

Winches, cranes and other outdoor machinery and equipment shall be painted No. 20 Medium Gray. Barges, drydocks, floating derricks and other floating equipment shall be No. 20 Medium Gray with superstructure No. 19 Light Gray. For further data see Shore Equipment, Section 3–13.

PAINT

3–9–2, INTERIOR COLORS FOR BASES AND DEPOTS

(A) CEILINGS

Ceilings and overheads in industrial areas shall be white. This includes wood and concrete beams, trusses and sash in monitors. Steel beams and trusses, however, may be No. 19 Light Gray. Wall colors shall extend to a line level with the bottoms of roof beams or trusses.

(B) TRIM

Trim including base boards, window frames, window and door sash, doors, stair balustrades and stringers shall be either white or No. 19 Light Gray. Fixed equipment against walls, such as cabinets, shall be in either the wall or trim color. White is preferred for trim in the living spaces while No. 19 Light Gray is preferred for industrial buildings. The trim on industrial buildings shall be used over dadoes, doors, door frames, window sash and radiators in accordance with the color principle applied.

(C) BARRACKS AND SLEEPING QUARTERS

In barracks and sleeping quarters Principle A (No. 1 Light Green) and Principle E (No. 10 Light Blue) are recommended for hot climates or for areas having south or west exposures. Principle C (No. 25 Beach Sand), Principle D (No. 4 Soft Yellow) and Principle F (No. 6 Ivory) are recommended for cold climates, for spaces deprived of good natural light or for north and east exposures.

(D) WARDROOMS, RECEPTION ROOMS, AND WOMEN'S REST ROOMS

See Article 3–8–2 (D).

(E) OFFICES

Offices may take a variety of color effects: Principle A (No. 1 Light Green) or Principle B (No. 9 Pearl Gray) for south and west exposures; Principle C (No. 25 Beach Sand) or Principle F (No. 6 Ivory) for north and east exposures. In large general offices Principle D (No. 4 Soft Yellow) may be used satisfactorily. For offices having heavy traffic or occupancy Principle O (No. 1 Light Green with No. 2 Medium Green dado and trim) is best.

(F) MESS HALLS, CAFETERIAS AND DINING ROOMS

For mess halls and cafeterias Principle C (No. 25 Beach Sand) may be used, as it provides a cheerful and appetizing effect. For a cool effect Principle A (No. 1 Light Green) may be used. Principle M (No. 25 Beach Sand with No. 8 Terra Cotta dado and trim) or Principle O (No. 1 Light Green with No. 2 Medium Green dado and trim) may be selected as alternates.

(G) RECREATION ROOMS

A dado is best for recreation rooms. Either Principle M (No. 25 Beach Sand with No. 8 Terra Cotta dado and trim), Principle N (No. 10 Light Blue with No. 29 Bright Blue dado and trim) or Principle O (No. 1 Light Green with No. 2 Medium Green dado and trim) should be used. Principle A (No. 1 Light Green) may be used as an alternate.

(H) SANITARY SPACES

~~Galleys, pantries, sculleries, heads an[illegible] shower rooms, illustrated, shall be Princi[illegible] I (white). Toilet stall partitions [illegible] No. 19 Light Gray with No. 29 [illegible] applied to the stall doors.~~

> The use of Principle I (white) in galleys, pantries, sculleries, heads and shower rooms will insure cleanliness and high standards of housekeeping. In white heads, toilet stall partitions shall be No. 19 Light Gray with No. 29 Bright Blue doors. As an alternate to white, heads and showers in barracks may be painted Principle O (No. 1 Light Green with No. 2 Medium Green dado and trim). Galleys, kitchens and sanitary spaces in family quarters may be painted Principle C. F, D, depending upon the color scheme adopted by the district.

(I) UTILITY AN[illegible]

Basements [illegible] genera[illegible] p[illegible] c[illegible] (il[illegible] ets [illegible] ciple [illegible] No. 19 [illegible] fronts s[illegible] Medium [illegible] jackets sh[illegible] be No. 20 Me[illegible] shall be black.

(J) STAIRWA[illegible]RRIDORS AND PASSA[illegible]WAYS

Stairways, corridors and passageways shall be either Principle D (No. 4 Soft Yellow), Principle F (No. 6 Ivory) or Principle I

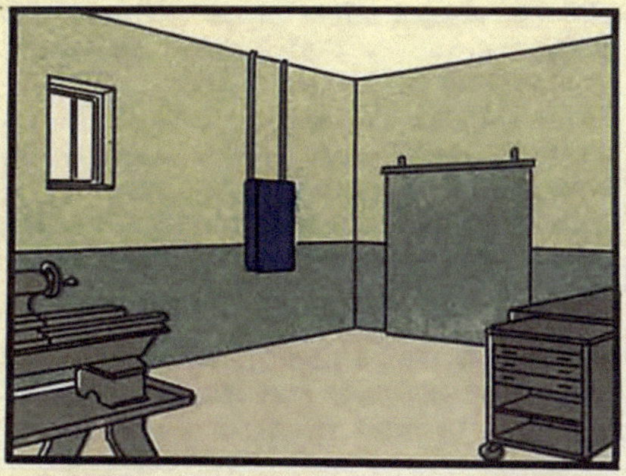

In shops and laboratories, use Principle O, No. 1 Light Green walls and No. 2 Medium Green trim and dado.

Large shops should use Principle P, No. Soft Yellow walls with No. 19 Light Gray for trim and dado.

In heat treating, Principle Q, No. 10 Light Blue walls should be combined with No. 19 Light Gray on trim.

Blacksmith, foundry and forge shops, if painted, should be Principle R, white with No. 19 Light Gray on trim.

Stock rooms, paint shops, may use Principle R, white walls with No. 19 Light Gray for dado and trim.

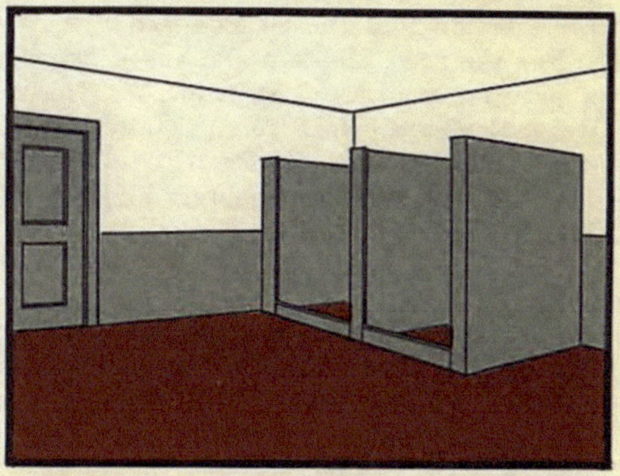

Personnel facilities, such as heads and shower rooms, should be Principle I (white). Floors No. 26 Tile Red.

(white). Principle P (No. 4 Soft Yellow with No. 19 Light Gray dado, trim, stair treads, risers, stringers and steps), should be used in shop areas or where traffic is heavy. To reduce stumbling and falling hazards a 4-in. band of No. 14 Brilliant Yellow shall be applied immediately under the tread on the top and bottom risers. Where the soffits (under side) of stairs and landings are plaster or concrete they shall be painted white. The undersides of metal stairs are better in No. 19 Light Gray.

(K) DISPENSARIES, INFIRMARIES AND SICK BAYS

Dispensaries, pharmacies, first aid rooms, treatment rooms and examining rooms should be Principle I (white) for cleanliness and good light reflection. Principle C (No. 25 Beach Sand) could be used as an alternate. Sick rooms, wards, dental clinic and remainder of the hospital area shall be Principle A (No. 1 Light Green) for a cool effect.

(L) EXCHANGES

The high popularity of blue among men makes it an ideal color for use in exchange facilities. In the store, Principle N (No. 10 Light Blue with No. 29 Bright Blue dado and trim) may be used. As a special effect No. 9 Pearl Gray may be used to provide a soft neutral background for the display of various items sold. No. 29 Bright Blue is then used for trim, doors, baseboards, shelving and counters. Counter tops, for bright appearance, may be in a color or material corresponding to No. 26 Tile Red.

For the soda fountain and snack bar the appetizing qualities of Principle M (No. 25 Beach Sand with No. 8 Terra Cotta dado and trim) may be used. Again, No. 29 Bright Blue may be used for counters and No. 26 Tile Red for counter tops.

(M) BARBER SHOPS

Use Principle G (No. 2 Medium Green) or Principle A (No. 1 Light Green) with white trim.

(N) SHOPS—FINE WORK

In shops where personnel are concentrated or where difficult tasks are performed, Principle O (No. 1 Light Green walls, No. 2 Medium Green dado and trim), illustrated, shall be used. This would include small workshops and machine shops, electrical shops, electronic repair shops, camera repair shops, laboratories, testing areas, or wherever problems of visibility are present and where the use of a soft green will reduce wall glare.

(O) SHOPS—AVERAGE WORK

Principle O (No. 1 Light Green walls, No. 2 Medium Green dado and trim), illustrated, may also be used in larger shops having fairly heavy occupancy. For very large interiors, however, Principle P shall be used (No. 4 Soft Yellow walls, No. 19 Light Gray dado and trim), illustrated. This includes large machine shops, engine repair shops, carpenter shops and mills, boat repair shops, large erection shops, rigging shops, garages.

(P) SHOPS—HEAVY WORK

In blacksmith shops, foundries, forge shops, plating shops, Principle R shall be used (white walls, No. 19 Light Gray dado and trim), illustrated. Where trouble is encountered with steel work because of fumes, aluminum may be substituted as an alternate for No. 19 Light Gray. Ventilating ducts and hoods should be either No. 19 Light Gray or aluminum.

For heat treating, Principle Q shall be applied (No. 10 Light Blue walls, No. 19 Light Gray dado and trim), illustrated. The purpose here is to introduce a cool color to offer psychological compensation for high temperatures. Heat treating devices, hoods, ducts, may be aluminum if exposure to high temperatures is involved. Otherwise, No. 19 Light Gray should be used. Tanks and equipment should be No. 20 Medium Gray.

(Q) SHOPS—SPECIAL WORK

If low levels of illumination are necessary for the viewing of instruments, dials or scopes, such as in some electronic shops, No. 2 Medium Green (Principle G), illustrated, shall be used uniformly over walls and ceilings to reduce reflections. In graphic units and photographic departments, Principle B may be used (No. 9 Pearl Gray walls with No. 19 Light Gray dado and trim), illus-

Laboratories and materials testing use Principle O, No. 1 Light Green walls, No. 2 Medium Green trim and dado.

Where dim light is required Principle G, No. 2 Medium Green may be used uniformly over walls and trim.

Photographic units may use Principle B, No. 9 Pearl Gray walls with No. 19 Light Gray trim and dado.

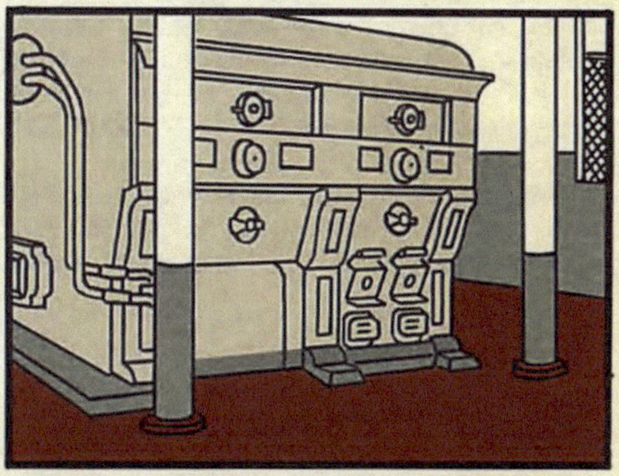

Boiler and engine rooms should be Principle R, white with No. 19 Light Gray trim and dado, No. 26 Tile Red floors.

Equipment shall be No. 30 Equipment Gray or No. 20 Medium Gray, depending on use.

No. 20 Medium Gray for piers, barges, with dockside buildings and barge superstructures No. 19 Light Gray.

trated, for the camera room. Dark rooms shall follow Principle P (No. 4 Soft Yellow walls, No. 19 Light Gray dado and trim). Principle O (No. 1 Light Green walls, No. 2 Medium Green dado and trim) may be used as an alternate. Tanks in dark rooms shall be No. 20 Medium Gray.

(R) EQUIPMENT AND MACHINERY

No. 20 Medium Gray shall be used for miscellaneous areas in industrial units, such as catwalks, ladders, metal walkways. Overhead cranes shall be No. 19 Light Gray. The exterior of the crane cab shall be No. 14 Brilliant Yellow to aid location. No. 14 Brilliant Yellow shall also be applied to crane hooks and pulleys. Use No. 19 Light Gray or No. 20 Medium Gray for heating units, blowers, ducts, conveying equipment, tanks. Smaller hoists and jib cranes shall be No. 20 Medium Gray. The ends of crane beams shall be marked with bands of No. 14 Brilliant Yellow and black.

For equipment where cleanliness is important or where maintenance is not difficult, such as portable testing apparatus, desks, stools, filing cabinets, electrical equipment, fine machine tools, etc. use No. 30 Equipment Gray. For equipment where rougher work is involved, maintenance is difficult and cleanliness is not essential, such as lathes, drill presses, shapers, milling machines, tool grinders, power saws, tool racks and bins, use No. 20 Medium Gray.

Portable equipment which constitutes an obstruction hazard such as test stands, stanchions and jacks shall be painted No. 14 Brilliant Yellow. Horizontal surfaces, steps and platforms subject to wear may be No. 20 Medium Gray. (See Safety Color Code, Section 3–16.)

For the finishing and highlighting of machinery, see Section 3–17, Machinery Colors.

(S) FLOORS

Wood floors shall be sealed and waxed. Painted floors shall normally be No. 19 Light Gray. Where maintenance is difficult such as in shops, No. 20 Medium Gray may be used as an alternate. Painted floors in galleys, heads, locker rooms, furnace rooms, emergency power rooms, shall be No. 26 Tile Red.

SECTION 3–10, MISCELLANEOUS FURNISHINGS FOR SHORE ESTABLISHEMENTS

The following general suggestions shall be considered in the use of various furnishings at Coast Guard units. Simplicity and standardization are desired. For the most part decorative features should follow those more commonly seen in office buildings than in homes. Patterned textiles by and large shall be avoided.

3–10–1, LINOLEUM AND COMPOSITION FLOORS

Where linoleum, asphalt tile or composition flooring is purchased, deep colors formerly used, such as black, dark green and dark brown, shall be avoided. Such colors waste light, show minor dirt marks, and do not invite orderly maintenance. The preferred colors—all of medium tone—are marbleized gray, marbleized green, marbleized tan and, for some purposes, marbleized red. These finishes, being fairly neutral, will harmonize with practically any wall color. The irregular pattern of marbled materials in medium rather than dark tone reduces maintenance by concealing normal foot marks and minor damage to the surface.

Where this Manual refers to No. 26 Tile Red for floors and where a linoleum or composition flooring material is to be used, the color may be plain but shall match No. 26 Tile Red as nearly as possible.

3–10–2, CARPETING

In carpeting, hard twist weaves are preferred in plain or heather (varicolored) pile. Recommended colors are medium gray, green or tan.

3–10–3, COUNTER AND DESK TOPS

Linoleum counters and desk tops are now being manufactured to approximate the tone of No. 30 Equipment Gray. This is a very excellent color for such furnishings as desks, tables, benches. Its high light reflection and clean appearance are both desirable. For food service, however, a plain blue or marbleized blue may be specified as an alternate choice for gray in mess halls; blue should not be used elsewhere.

3–10–4, UPHOLSTERING MATERIALS

There are three basic types of materials commonly used for upholstery. Woven or pile, natural-fiber materials, such as cotton and wool, are practical and economical. Because of their soft texture they are comfortable and suitable for all climates. These fabrics tend to "breathe" and will not prove slippery or sticky under normal use. To their disadvantage, they may fade or be difficult to clean. They may be damaged by water or burns.

Second are sheet materials, such as leather and vinyl. Leather is not recommended because it tends to crack and deteriorate with age, and to develop a mold in a moist climate. Vinyl upholstery materials are very reasonable in cost. They have a more or less impervious surface and are easy to clean. However, because they fail to absorb moisture they may prove slippery and may be uncomfortable in a warm climate by "sticking" to the skin or clothing.

Third are woven filament materials, such as Saran and Velon. These are made of synthetic fibers and may be high in cost. While they are easy to clean and will tend to "breathe," they are generally cold and slippery to the touch, but will wear almost indefinitely.

In selecting colors for upholstery, Coast Guard standards shall be matched, wherever possible. Gray is a good color, also maroon, deep green, deep blue and brown. Pale shades and bright colors should be avoided.

3–10–5, DRAPERY MATERIALS

Drapery materials, where purchased, shall be

simple and lacking in ornamentation or design. Natural tones, beige, pale gray, will give an interior an attractive appearance and will serve a functional purpose by avoiding distractions and unfavorable contrast with adjacent walls. Soft tones of green, tan, rose and blue are similarly good. For the most part, sharp colors such as brilliant red and yellow shall not be considered. So-called interior decoration should be set aside for a simpler and more direct approach to color that suggests efficiency rather than mere ornamentation. This will assure a suitable and appropriate setting.

SECTION 3–11, COLORS FOR VESSELS 65 FEET AND OVER IN LENGTH

The following paragraphs describe the color scheme to be employed on vessels over 65 ft. in length and on harbor tugs of any length. These instructions do not apply to barges which are covered in Section 3–12–1. An attempt has been made to specify the color for all major details. It is obviously impossible to foresee every circumstance which may arise in the future. Therefore when it is found that the instructions contained herein do not specify the color of a particular detail the general scheme shall be followed. Article 3–11–1(L) lists those items which are not to be painted.

3–11–1, EXTERIOR COLORS FOR VESSELS 65 FEET AND OVER IN LENGTH

(A) HULL

Ship hulls from the top of the antifouling or underwater area paint up to the top of the hull, bulwark or sheer line, including all fittings, shall be painted in accordance with the accompanying schedule.

On steel, iron or plastic vessels the boot-topping area shall be defined as follows: the bottom edge of the boot-topping shall coincide with the normal light operating waterline of the vessel. For vessels 150 ft. and over in length the top edge of boot-topping shall, at amidships, be above the normal full load waterline a distance equal to 1/8 the freeboard measured to the full load waterline. For vessels less than 150 ft. in length the upper edge of the boot-topping shall, at amidships, be above the normal full load waterline, a distance equal to 1/6 the freeboard measured to the full load waterline. At the bow the total width (top edge to bottom edge) of the boot-topping shall be 1.33

Class of Vessel	*Boot-topping Area*	*Above Boot-topping*
WAG	Black	White
WAGB	"	"
WAVP	"	"
WDE	"	"
WIX	"	"
WPC	"	"
WPG	"	"
WSC	"	"
WPB	"	"
WAGL	Red	Black
WAK	"	"
WARC	"	"
WAT	"	"
WYT	Black	Black
WD	"	"
WAL (except LAKE HURON)	Black	Red
LAKE HURON	Black	Black
WAGR	Painting to follow General Specifications For Building Vessels of the United States Navy, Appendix 6.	

times the total width amidships, and at the stern the total width shall be 0.66 times the total width amidships. All measurements are to be taken in a vertical plane and projected on to the hull. The upper edge of the boot-topping shall be a fair curve drawn through the above three points.

On wooden vessels the bottom edge of the boot-topping shall coincide with the *full load* waterline of the vessel. The upper edge of the boot-topping shall be identical with that described above for steel, iron and plastic vessels. Where metal sheathing is installed in way of the boot-topping area the instruction for steel vessels shall be followed.

Limited Drafting Marks are marks designating the maximum draft to which a vessel may be loaded. The marks will be applied only to those vessels specifically authorized by Headquarters to display them. The letter of authorization will locate the marks by stating the limiting draft and freeboard of the vessel as well as the frame number on which the mark will be centered. This is illustrated.

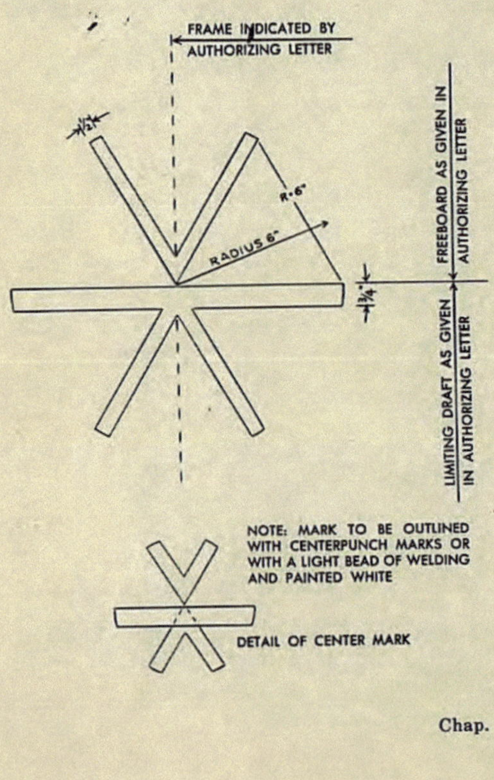

(B) SUPERSTRUCTURE

The superstructure and all attachments thereto shall be white. This includes the following:

- Deck Houses
- Aircastles
- Breakwaters
- Cabins
- Pilot Houses
- Engine Trunks
- Gun Tubs
- Gun Shields
- Bridge Wings

Canvas dodgers attached to superstructure, bulwarks or bridge wings.

Bulwarks including bulwark supports and brackets. Except that bulwark and bulwark supports forming an extension of the hull shall be painted the hull color both inboard and outboard.

Stanchions, lockers, electrical controllers, switch boxes, ventilation ducts, ladders or other details secured to the superstructure.

Ventilation ducts, electrical conduits and pipelines bracketed to the superstructure or immediately adjacent to the superstructure.

Overheads of exterior passageways and shelters and the inboard side of bulwarks and aircastles.

(C) DECKS

Wood decks shall be left unpainted. Steel decks shall be painted No. 32 Blue Gray. Where there is a coaming at the deck edge the deck paint shall be extended up the inboard surface of the coaming. Deck paint shall also be applied to bounding angles between decks and superstructure or hatch coamings. Where superstructure and hatch coamings are welded to the deck a dado shall not be used. Dadoes shall not be applied to miscellaneous deck fittings such as bitts, chocks, pedestals, stanchions, etc.

Waterways shall be painted No. 32 Blue Gray.

(D) TOPS OF CABINS, PILOT HOUSES, AND ENGINE TRUNKS

On vessels whose main decks are all wood or nearly all wood the tops of cabins, pilot houses and engine trunks shall be No. 24

BIBB CLASS CUTTER

Spar except in way of aerial identification markings. Other vessels shall have these areas painted No. 32 Blue Gray.

(E) STACKS

Stacks shall be painted No. 24 Spar with a broad black band around the top of the stack. The width of the band shall be equal to ½ the fore and aft diameter of the stack or ⅓ the height of the stack whichever dimension is the smaller. If the stack has a hood the hood shall also be painted black in addition to the above band and shall form an extension of the band.

(F) MAST AND SPARS

Masts (including ladders, crows nests, platforms, etc.), booms, cranes, davits, flagstaffs, jackstaffs, yardarms and their supports and foundations (other than rigging) shall be painted No. 24 Spar if metal, varnished if wood. Equipment secured to the masts such as radar antennas, radomes, searchlights, bull horn, etc. shall also be No. 24 Spar.

Mast areas directly in way of stack gases may be black. Such areas on masts as are painted black when the vessel is commissioned shall be kept painted black in service. The extension of these areas beyond that originally authorized, or the painting of additional areas black without specific authorization from Headquarters is prohibited.

(G) RUNNING LIGHT SCREENS

In order to contain the reflected light from running light screens within the limits allowed by Rules of the Road, light reflections from the forward part of the screen must be kept to a minimum. The geometry of light screens and the location of the light sources are such that only light emitted by specular reflection lies within the prescribed limits, whereas light emitted by diffuse reflection from the forward part of the screen exceeds the prescribed limits. Therefore, a paint providing high specular reflection and low diffuse reflection should be used. A high gloss black enamel performs this function best and shall therefore be used on running light screens.

(H) DECK MACHINERY AND FITTINGS

Deck winches, capstan, winch and capstan controllers, sounding machines, and other deck machinery shall be No. 24 Spar.

"(I) GUNS AND ORDNANCE EQUIPMENT

Guns, ordnance equipment, and their foundations and mounts, with the exception of working parts and nameplates, shall be painted black. This includes gun barrels of 5-in., 40mm guns, depth charge tracks, K-guns, roller loaders, depth charge arbor trays, rocket launchers, and interior of hedgehog projectors.

Depth charges need to be mentioned herein because of their stowage in an easily visible and exposed location. Depth charges fall in the category of ammunition. Therefore, units shall adhere to the painting and marking instructions contained in OP 2238, pertinent parts of which are quoted: "All depth charge cases are painted ocean gray, distinguishing markings are usually not applied to the case. The stenciled lettering is one inch high and is yellow in color."

Gun mounts in totally enclosed gunshields, such as the 5"/38 shall be No. 30 Equipment Gray.

Fire control equipment including gun directors shall be No. 30 Equipment Gray.

Gun Shields, including 20mm Shields, exterior or hedgehog projectors, ready service lockers and pyrotechnic lockers shall be white. In addition, pyrotechnic lockers shall have the word PYROTECHNIC stenciled on the box in conspicuous location, using the largest practicable No. 13 Fire Red letters.

Detonator boxes shall be No. 13 Fire Red overall with the word DETONATORS stenciled in white on the box in conspicuous location." (Amendment No. 6)

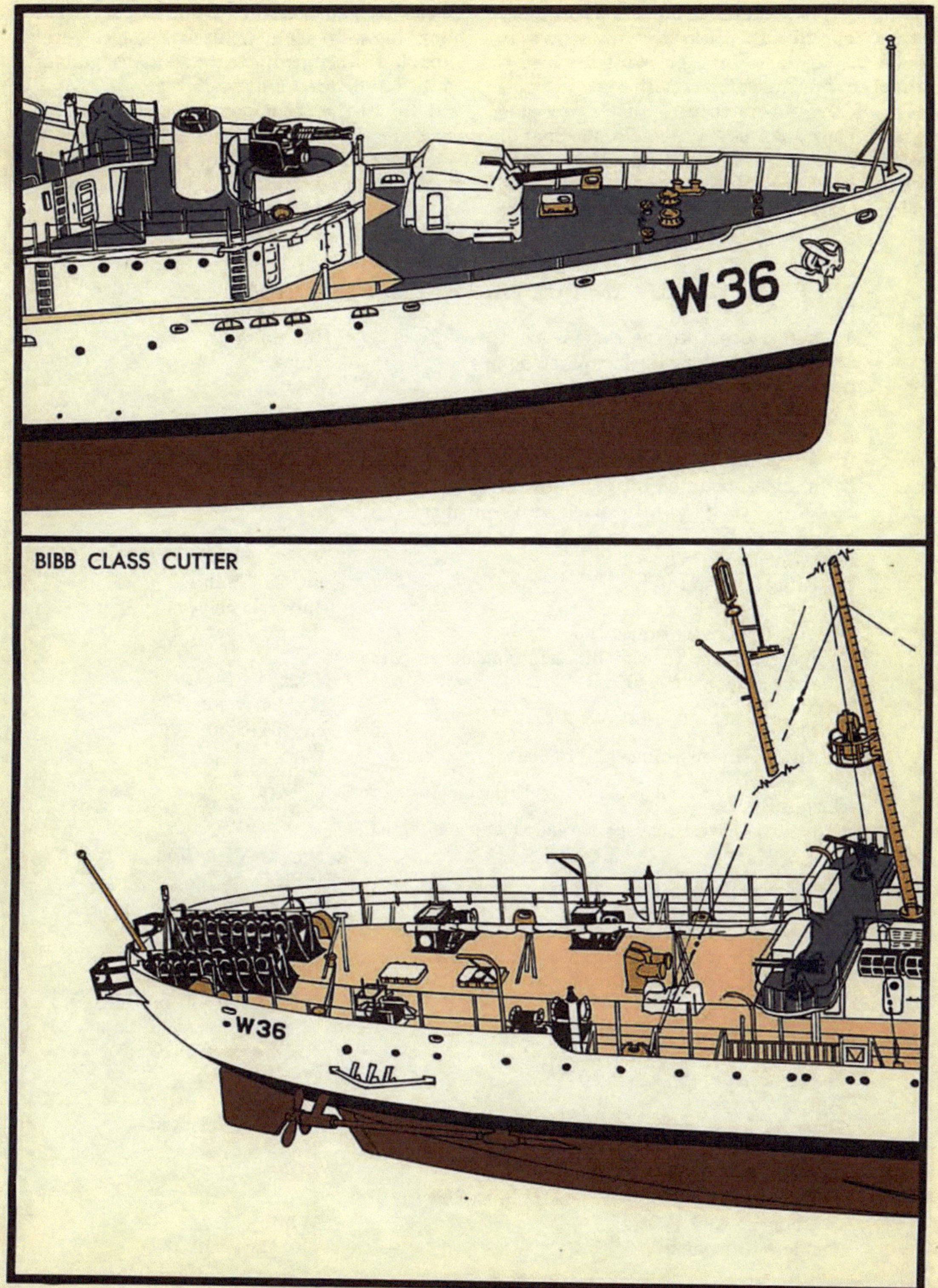
W36
BIBB CLASS CUTTER
W36

In general, miscellaneous objects whose color is not specifically prescribed elsewhere in these instructions, shall be painted white if attached or immediately adjacent to some part of the superstructure that is painted white. They shall be painted No. 24 Spar if attached or immediately adjacent to mast, spars, or other objects whose prescribed color is No. 24 Spar. Objects standing alone on the deck shall also be No. 24 Spar. Stumbling hazards such as deck padeyes, deck clips and other projections shall be painted white for better night visibility.

Refer to the accompanying list of miscellaneous objects most frequently encountered. While not all of these follow the general rules above, most do, and the general rules shall be applied to objects not listed.

MISCELLANEOUS PAINTING INSTRUCTIONS

Anchors stowed in the hawse	Hull color
Anchors stowed against superstructure	White
Anchors stowed on deck	Spar
Awning Ridgepoles	White
Awning Stanchions	White
Bitts (except those mounted on gunwales of tugs)	No. 24 Spar
Bitts, side, mounted on gunwales of tugs	Black
Blocks (except those in black areas on mast or stack)	No. 24 Spar
Boat Booms	No. 24 Spar
Boarding Ladders	No. 24 Spar
Chocks, bulwark	Bulwark color
Chocks (except bulwark type)	No. 24 Spar
Compass Stands (except binnacle and other parts required to be kept bright)	White
Davits (all types)	No. 24 Spar
Diaphones	No. 24 Spar
Direction Finder Stands and Loops	White
Deck Chests	White
Ensign Staffs	No. 24 Spar
Firemain Valves (except threaded and machined parts)	No. 13 Fire Red
Flag Lockers	White
Flag Staffs	No. 24 Spar
Gangplanks, metal (except aluminum)	No. 24 Spar
Gasoline Drums	No. 14 Brilliant Yellow with GASOLINE in black letters
Gasoline Stowage Racks	No. 24 Spar
Hatch Coamings	No. 24 Spar
Hatch Covers	White
Hose Racks, Saddles and Reels (for fire hose)	No. 13 Fire Red
Jack Staffs	No. 24 Spar
K-Guns	Black
Ladders (except those on masts) including stringers and rails	White
Ladders (on masts)	No. 24 Spar

MISCELLANEOUS PAINTING INSTRUCTIONS

Liferails (pipe or canvas covered)	White
Life Rafts, Life Rings, Life Buoys and Life Floats	No. 18 International Orange
Pelorus Pedestals	White
Radio Direction Finder Stands and Loops	White
Radar Antenna	No. 24 Spar
Ready Service Lockers	White
Searchlights (except those on masts)	White
Searchlights (on masts)	No. 24 Spar
Tiller, Spare, stowed on deck	No. 24 Spar
Tiller, Spare, stowed against bulkhead	White
Rigging (where painted)	Black
Vegetable Lockers (on deck)	White
Ventilators (all types including gooseneck pipe vents not attached or immediately adjacent to superstructure)	No. 24 Spar
Ventilators attached or immediately adjacent to the superstructure	White
Wheelhouse Visors (underside)	No. 2 Medium Green

(L) ITEMS NOT TO BE PAINTED

In general paint shall not be applied to surfaces where it will cause interference with the functions for which the surfaces were designed, nor shall it be applied to those surfaces which are traditionally kept bright. Refer to the accompanying list of typical items not to be painted and the method of treatment for each.

ITEMS NOT TO BE PAINTED

Accommodation ladders, wood	Varnish
Applicator, nozzles	Polish
Boat booms, wood	Varnish
Boatswains' chairs	Oiled
Brightwork	Polish and coat with clear plastic or lacquer
Canvas covers (removable)	No coating
Chain, galvanized	No coating
Deck treads, non-skid	No coating
Deck, wood	No coating
Dogs	No coating
Fire hose nozzles	Polish
Gangplanks, wood	Oiled or varnished
Gaskets, rubber for water-tight doors	Pulverized graphite
Glass	No coating
Gratings, wood	Oiled
Grease cups	No coating
Gypsy head whelps	No coating
Hose	No coating,
Insulators	No coating

ITEMS NOT TO BE PAINTED

Knife edges on watertight doors and hatches	Sandpapered
Ladders, pilot	Oiled
Leather coverings	Oiled with preservative
Masts and spars, wood	Varnish
Name plates	No coating, polish if brass
Oars	No coating
Oil cups	No coating
Oil holes	No coating
Railing, wood	Varnish
Release mechanisms	Greased
Rigging, running	Greased
Rigging, standing	Apply preservative
Searchlight, shutters	No coating
Stages	Oiled
Strongbacks, wood	Varnish
Working or machined parts of valves, machinery, blocks, guns or other equipment	Greased or no coating
Zincs	No coating

(M) DAMAGE CONTROL AND SAFETY MARKINGS

Hose racks, liquid foam containers, spanner wrenches, fire main valves, casualty power terminals, damage control shores and similar equipment used for damage control purposes shall be painted No. 13 Fire Red. Damage control equipment stored in lockers shall be identified by a suitable label painted on the locker door in No. 13 Fire Red. Covers of all sound power jack boxes and switch boxes shall also be No. 13 Fire Red.

Gasoline and other flammable liquid containers shall be painted No. 14 Brilliant Yellow with the name of the contents conspicuously indicated in large black letters.

Stumbling hazards such as deck padeyes, deck clips and other small projections from the deck shall be painted white for better night visibility.

In painting the above markings do not paint those surfaces listed in paragraph (L) above.

(N) COMPRESSED GAS CYLINDER IDENTIFICATION

Vessels shall follow the color scheme set forth in Section 3–18, Compressed Gas Cylinder Identification.

(O) MARKINGS ON FLOATABLE EQUIPMENT

Ring Type Life Buoys. The vessel's name, or if not named, the vessel's number, shall be placed on the top semicircle of the ring. On the bottom semicircle the legend U. S. COAST GUARD shall appear. Black letters 2 in. high shall be used.

Life Rafts. Markings shall always be placed on the longer legs of the raft. The vessel's name or, if not name, the vessel's number, shall be placed on one leg and the legend U. S. COAST GUARD shall be placed on the other leg. The markings shall be so placed as to be readily seen, and when read in the normal manner of reading the vessel's name or number shall be read first.

Lifejackets. The vessel's name or, if not named, the vessel's number shall be placed across the middle of the back of the lifejacket in letters ¾ in. in height.

Wood Articles. Deck chests, boat boxes, oars or other wood articles which may be washed overboard shall have the vessel's name preceded by USCGC, or designating number preceded by CG, burned into the article in ½ in. letters so as to leave a clear impression. Burning pencils item No. 3201 Class 37 of the Catalog of Navy Material may be used for this purpose.

(P) ANCHOR CHAIN MARKING

Anchor chain, with the exception of the below markings, is to be painted black. Light vessels shall paint only that part of the chain which is normally out of water when anchored on station. The detachable link between shots, excluding the detachable link for the 5-fathom connecting shot, shall be painted as follows starting from the anchor end and working inboard:

15-fathom detachable link, No. 13 Fire Red.

30-fathom detachable link, white.

45-fathom detachable link, No. 29 Bright Blue.

60-fathom detachable link, No. 13 Fire Red.

The same manner of marking detachable links shall be continued on to the end of the chain with the exception of the detachable links in the last and next to last shot.

Additional markings shall be as follows:

The first link on each side of the 15-fathom detachable link shall be painted white. Each of the white links shall be marked by one turn of wire around the stud.

The first two links on each side of the 30-fathom detachable link shall be painted white. The second link at each side of the detachable link shall be marked by two turns of wire around the stud.

The first three links on each side of the 45-fathom detachable link shall be painted white. The third link at each side of the detachable link shall be marked by three turns of wire around the stud.

The same manner of marking shall be continued on to the end of the chain with the exception of the last and next to the shot.

All of the links in the next to the last inboard shot shall be painted No. 14 Brilliant Yellow, and all of the links in the last shot inboard shall be painted No. 13 Fire Red.

(Q) WAR SERVICE INSIGNIA

Display of war service insignia is optional. War service insignia when displayed shall consist of replicas of the area campaign medal ribbons and Navy Unit Commendation Ribbon together with operation and engagement stars to which the ship would be entitled under the same rules and regulations as prescribed for individuals. No painted replicas of the Presidential Unit Citation ribbon shall be worn. The following ribbons only are to be displayed:

(a) Navy Unit Commendation.
(b) American Area.
(c) European-African-Middle East Area.
(d) Asiatic-Pacific Area.

Replicas of Navy Unit Commendation Ribbons and area campaign medal ribbons shall be painted by ship's force. The sizes and locations designated in the accompanying table shall be adhered to for the ship types listed and shall be used as a general rule for ship types not listed. Where symmetrical arrangements permit, the painted replicas will be displayed on both sides of the vessel in corresponding locations, port and starboard. See tabulation on page 3–76.

Painted replicas of area campaign medal ribbons shall be arranged in the same order as ribbons worn by personnel, with a maximum of three replicas in a horizontal line. Ships entitled to display the four replicas named will show the Navy Unit Commendation Ribbon over the row of three area campaign ribbons.

The diameter of the operation and engagement stars shall be ⅜ the width of the ribbon, and the stars shall be painted on the replicas with the point down.

(R) IDENTIFICATION MARKINGS ON HULL

Named Vessels—

Named vessels, other than lightships, shall have their visual call numbers on each bow.

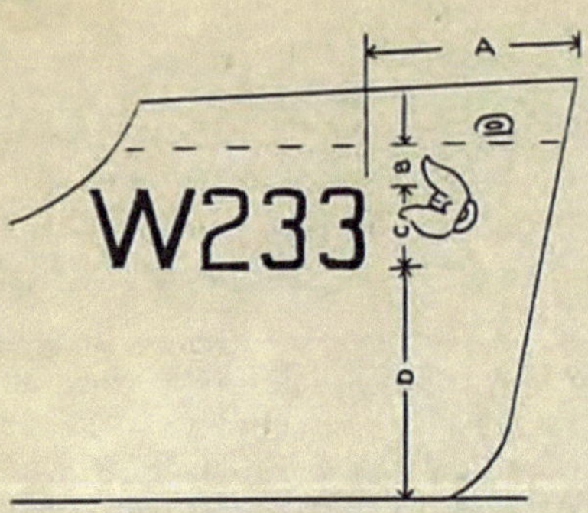

With Bulwark, Without Boot-topping

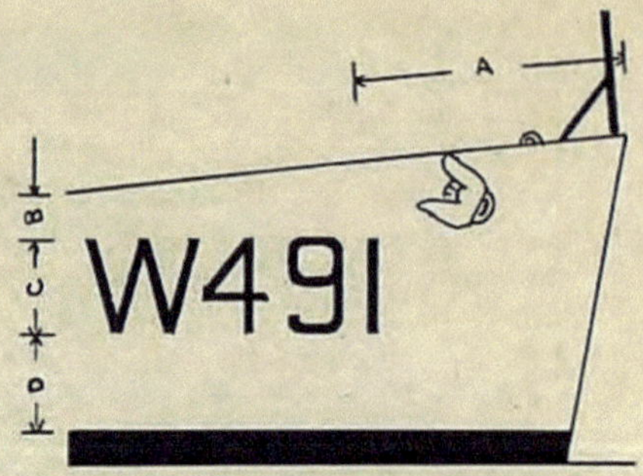

Without Bulwarks, With Boot-topping

Vessel Length Overall	A	B	C	D
65 ft. and over up to 100 ft.	5 in.	1 in.	2 ft.	*
100 ft. and over up to 150 ft.	8 in	1 ft.-6 in.	3 ft.	*
150 ft. and over up to 200 ft.	10 ft.	2 ft.	4 ft.	*
200 ft. and over up to 300 ft.	12 ft.	2 ft.-6 in	5 ft.	*
300 ft. and over	18 ft.	3 ft.	6 ft.	*

*D must be equal to or greater than B.

A = Distance between foremost part of foremost character to stem at forecastle deck.

B = Distance between top of aftermost lettering and hull at deck edge or in the case of vessels with bulwarks, the deck edge projected through the hull.

C = Height of letters.

D = Distance between bottom edge of lettering and boot-topping or waterline where boot-topping is not applied.

Note: Distance A is measured in a fore and aft plane. The remainder of the distances are measured along the hull.

The size and location of the numerals shall be in accordance with the illustrations and table. Letters of the type described in Section 3–22, Letters and Numerals, shall be used. Black letters shall be used on white hulls and white letters shall be used on black hulls. It should be noticed that where application of the rules results in distance B being greater than D, the vertical location of the lettering shall be adjusted upward so that B and D are equal. If such readjustment results in B and D being less than ¼C, then C shall be decreased until B and D are approximately ¼C. Where, by application of the above rules the number lies in way of the anchor, air port or other interferences which would substantially reduce the legibility of the numeral, its position shall be adjusted the minimum amount necessary to be clear of all such interferences.

Named vessels, other than lightships, shall have their names painted across the stern or transom in 12-in. letters of the type specified in Section 3–22, Letters and Numerals. Black letters shall be used on white hulls and white letters on black hulls. The name shall be located 12 in. from the top of the hull. If in the case of vessels of extremely low freeboard this results in the name being closer to the boot-topping or waterline than to the top of the hull, it shall be centered between the top of the hull and the top of the boot-topping or the waterline if there be no boot-topping. Where a bulwark is fitted the name shall be placed on the bulwark so as to be equidistant from the top and bottom. The

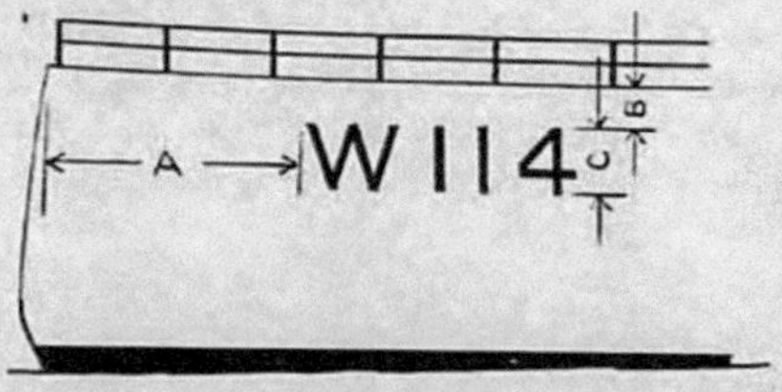

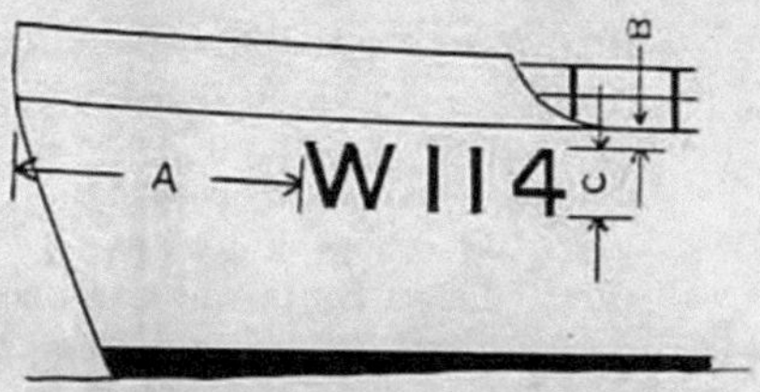

Vessel	A	B	C
165 ft. Cutter	6 ft.	1 ft.	2 ft.
269 ft. Cutter	20 ft.	2 ft.	3 ft.
290 ft. Cutter	28 ft.	2 ft.	3 ft.
327 ft. Cutter	13 ft.	2 ft.	3 ft.
250 ft. & 255 ft. Cutters	10 ft.	2 ft.	3 ft.
WAVPs	33 ft.	2 ft.	3 ft.
WDEs	6 ft.	2 ft.	3 ft.

Dimension A is measured in a fore and aft plane. All other measurements are along the hull. Dimension B is measured from main deck downward.

name shall be centered on the centerline of the vessel, or where this is impractical it shall be painted on either quarter located as far aft as practicable. Vessels authorized to display name boards are excepted from these rules.

In addition to the above markings, WAGBs, WAVPs, WDEs, WPGs and WPCs shall have their visual call numbers on each quarter in black letters. The location and size of the numbers shall be as shown in the illustration and table.

Numbered Vessels—

Numbered vessels shall have their designating number preceded by CG (such as CG 83400) on each bow. Buoy boats shall add the suffix D to the designating number—as an example CG 80004-D. The size and location of numerals shall be in accordance with the rules stated for visual call numbers on the bows of named vessels. Black numerals shall be used on white hulls and white numerals shall be used on black hulls.

Numbered vessels shall have their designating number as used on the bow painted across the stern or transom in 12-in. letters of the type specified in Section 3–22, Letters and Numerals. Black letters shall be used on white hulls and white letters shall be used on black hulls. The number shall be located 12 in. from the top of the hull. If in the case of vessels of extremely low freeboard this results in the name being closer to the boot-topping or waterline than to the top of the hull it shall be centered between the top of the hull and the top of the boot-topping or the waterline if there be no boot-topping. Where a bulwark is fitted, the number shall be placed on the bulwark so as to be equidistant from the top and bottom. The number shall be centered on the centerline of the vessel, or where this is impractical it shall be painted on either quarter located as far aft as practicable.

Light Vessels—

Light vessels shall have the name of the station to which assigned or the word RELIEF as applicable, painted in white on both sides of the hull in the largest letters permitted by the freeboard (about 6 ft.-6 in.). The style and spacing of the letters shall be as specified in Section 3–22, Letters and Numerals.

(S) AERIAL IDENTIFICATION

All radio equipped vessels and boats shall have their international radio call letters in

No. 18 International Orange against a black background area on the largest unobstructed and horizontal clear panel available, such as the superstructure deck, cabin trunk top, engine trunk top, housetop or compass platform. Those boats not equipped with radio and not assigned to floating units or shore stations shall have their designating numerals in this same relative position as far as practicable. The lettering shall be placed athwartships with the tops of the characters toward the vessel's bow and shall be as large as can be placed in the space available with a minimum of 3 stroke widths between the end characters and the edge of the black background area. The fore and aft dimension of the black area shall extend beyond the top and bottom of the letters to a minimum distance of ½ the letter height. The style and spacing of the letters shall be as specified in Section 3–22, Letters and Numerals.

(T) DRAFT MARKS

Draft figures on metal hulls shall be made by running a bead of weld around the outline of draft figures projected on the hull. The figures shall be so located that the bottom of the figure is tangent with the draft it represents measured from the bottom of the keel line extended. Draft figures shall conform to those shown in Section 3–22, Letters and Numerals, and shall be used only on vessels having a full load draft in excess of 4 ft.

(U) CAMOUFLAGE INSTRUCTIONS

Camouflage of Coast Guard vessels, except when otherwise directed by competent authority, will consist of painting all horizontal surfaces to a gray color approximately Federal Color No. 1615 (Deck Gray Navy Formula 20). Vertical surfaces above boot-topping shall be painted to a color approximating haze gray color Federal Color 2635 (Haze Gray Formula 5H).

Coast Guard vessels will not normally carry camouflage paints. The following class of vessels, WAVP, WPG, WDE, WAGB, WAG, WAGL, WPC, WSC, WPB are directed to carry tinting material so that stocks of white paint may be converted to gray for application of camouflage. Amount of tinting material shall not exceed 1 gal. for each 20 gals. of white paint carried on board except that a minimum of 1 gal. is authorized for all classes. One gal. of Deck Gray Navy Formula 20 and Haze Gray Navy Formula 5H is authorized for matching.

PAINTED REPLICAS OF CAMPAIGN RIBBONS

Type Vessel	*Insignia Size*	
BIBB Class	6 in. wide x 18 in. long	Outboard end of Bridge Bulwark 12 in. below top
WIND Class	Same	For'd of Life Buoy on Bridge Bulwark 12 in. below top.
OWASCO Class	Same	Outboard end of Bridge Bulwark 6 in. below top.
CACTUS Class	Same	Same
IRONWOOD Class	Same	Same
IRIS Class	Same	Same
NAUGATUCK Class	4 in. wide x 12 in. long	Centered on Superstructure Bulwark between Bridge and Break, frame 20-23, 6 in. below top.
165 ft. "A" Class	Same	On end of Bridge wing 12 in. below top.
165 ft. "B" Class	Same	Side of Chart House 2 ft. below top.

PAINTED REPLICAS OF CAMPAIGN RIBBONS

125 ft. Class	Same	Side of Chart House 2 ft. below top.
83 ft. Cutters	Same	On Bridge Wings center on Machine Gun Bulge 6 in. below top.

3–11–2, INTERIOR COLORS FOR VESSELS 65 FEET AND OVER IN LENGTH

(A) COMPARTMENT COLORS—SYSTEM A

Color System A is applicable to all vessels 65 ft. and over in length and harbor tugs unless approval for the use of Color System B has been specifically authorized.

Where the overhead color is different from the bulkhead color the overhead color shall be extended down to the level of the bottom of the overhead beams.

Ducts, pipes and cables are to be painted the same color as the overhead or bulkhead along which they run.

Door frames for both watertight and joiner doors shall be No. 30 Equipment Gray.

Stanchions adjacent to bulkheads shall be painted the bulkhead color. All other stanchions shall be No. 30 Equipment Gray. Stanchions shall have the same height and color dado as is supplied to the bulkheads.

Deck paints may be extended up adjoining bulkheads to form a dado 4 to 8 in. in height. Dado shall be brought up to the bottom of the doors. Kick pipes shall be painted the dado color. Installed deck coverings shall not be painted but at time of installation shall be procured in colors which most nearly match the prescribed deck colors. Deck coverings which are dark in color or which clash with prescribed color principles are not to be used.

COMPARTMENT COLORS—SYSTEM A

Color Principle		
Principle No. 1	Bulkheads, No. 1 Light Green Overheads, White Deck, No. 31 Deck Green	Cabins Officers' Staterooms Wardrooms Sick Bays (Berthing Area Only) Weathermen's Berthing Passageways in Officers' Country
Principle No. 2	Bulkheads, No. 1 Light Green Overheads, White Deck, No. 19 Light Gray	Radio Room Offices (Ship's, Engineer's, First Lt.'s, Aerological, etc.) Barber Shops
Principle No. 3	Bulkheads, White Overheads, White Deck, No. 19 Light Gray	Crew's and CPO Quarters Balloon Shelters Store Rooms Magazines IC Rooms Passageways not in Wardroom Country, Sanitary or Machinery Spaces

COMPARTMENT COLORS—SYSTEM A

Principle No. 4	Bulkheads, White Overheads, White Deck, No. 26 Tile Red	Galleys Sculleries Pantries Heads Laundries Miscellaneous Sanitary Spaces Engine Rooms Boiler Rooms Steering Engine Room Anchor Windlass Room; Motor Rooms Emergency Generator Room Passageways in Machinery or Sanitary Spaces Shops
Principle No. 5	Bulkheads, No. 1 Light Green Overheads, No. 1 Light Green Deck, No. 19 Light Gray	Pilot House Chart House
Principle No. 6	Bulkheads, No. 2 Medium Green Overheads, No. 2 Medium Green Deck, No. 19 Light Gray	CIC Sonar Rooms Radar Rooms

(B) COMPARTMENT COLORS — SYSTEM B

Color System B consists of the six color principles used in Color System A plus four additional color principles. This system with its greater color variety is designed for use on larger vessels or vessels remaining at sea for extended periods. Use of this color system on WAVP class cutters is optional. Specific authority from Commandant (O) is required if it is desired to use this color system on other cutters.

Where the overhead color is different from the bulkhead color the overhead color shall be extended down to the level of the bottom of the overhead beams.

Ducts, pipes and cables are to be painted the same color as the overhead or bulkhead along which they run.

Door frames for both watertight and joiner doors shall be No. 30 Equipment Gray.

Stanchions adjacent to bulkheads shall be painted the bulkhead color and all other stanchions shall be No. 30 Equipment Gray. Stanchions shall have the same height and color dado as is applied to the bulkheads.

Deck paints may be extended up adjoining bulkheads to form a dado 4 to 8 in. in height. Dado shall be brought up to the bottom of the doors. Kick pipes shall be painted the dado color. Installed deck coverings shall not be painted but at time of installation shall be procured in colors which most nearly match the prescribed deck colors. Deck coverings which are dark in color or which clash with the prescribed color principle are not to be used.

COMPARTMENT COLORS—SYSTEM B

Color Principle		
Principle No. 1	Bulkheads, No. 1 Light Green Overheads, White Deck, No. 31 Deck Green	Cabins Officers' Staterooms Wardrooms

Principle 1, white overhead, No. 1 Light Green bulkhead, No. 31 Deck Green for wardrooms, officers' staterooms.

Principle 2, No. 1 Light Green bulkheads, No. 19 Light Gray deck is applied to offices, radio rooms, radio workshops.

Principle 7, crew's berthing, white overhead, No. 1 Light Green bulkheads, No. 26 Tile Red. Gray equipment.

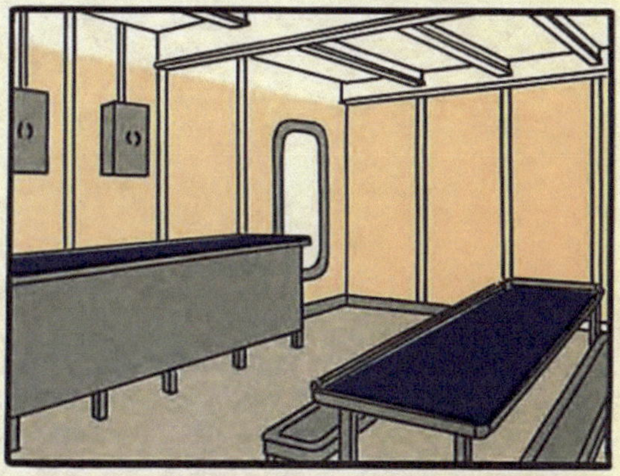

Principle 8 for crew's mess and recreation uses No. 25 Beach Sand for bright and cheerful variety.

The dispensary in No. 25 Beach Sand, with the sick bay in No. 1 Light Green, offers pleasing harmony.

Principle 10 in passageways uses No. 4 Soft Yellow with No. 30 Equipment Gray trim for a sunny effect.

COMPARTMENT COLORS—SYSTEM B

Principle No. 1	Bulkheads, No. 1 Light Green Overheads, White Deck, No. 31 Deck Green	Sick Bays (Berthing Area Only) Weathermen's Berthing Warrant Officers' Berthing Offices (Ship's, Engineer's, First Lt's., Aerological, etc.) Radio Rooms
Principle No. 2	Bulkheads, No. 1 Light Green Overheads, White Deck, No. 19 Light Gray	Radio and Electronic Workshops Barber Shops
Principle No. 3	Bulkheads, White Overheads, White Deck, No. 19 Light Gray	IC Rooms Storerooms and Magazines Passageways and Lobbies in storeroom areas
Principle No. 4	Bulkheads, White Overheads, White Deck, No. 26 Tile Red	Galleys Sculleries Pantries Heads Laundries Steering Engine Rooms Anchor Windlass Rooms Emergency Generator Room Boiler Rooms Engine Rooms Motor Rooms
Principle No. 5	Bulkheads, No. 1 Light Green Overheads, No. 1 Light Green Deck, No. 19 Light Gray	Pilot House Chart House
Principle No. 6	Bulkheads, No. 2 Medium Green Overheads, No. 2 Medium Green Deck, No. 19 Light Gray	CIC Sonar Rooms Radar Rooms
Principle No. 7	Bulkheads, No. 1 Light Green Overheads, White Deck, Where no deck covering is used paint No. 26 Tile Red. Where deck covering is used, paint margin area and dado No. 31 Deck Green.	CPO Berthing Spaces Crew Berthing Spaces
Principle No. 8	Bulkheads, No. 25 Beach Sand Overheads, White Deck, No. 19 Light Gray	CPO Messing Spaces Crew's Messing Spaces Crew's Recreation Dispensaries
Principle No. 9	Bulkheads, No. 4 Soft Yellow Overheads, White Deck, No. 19 Light Gray	Shops (Machine, Electric, Carpenter, etc., except Radio and Electronic Workshops) Post Offices Retail Stores Balloon Shelters

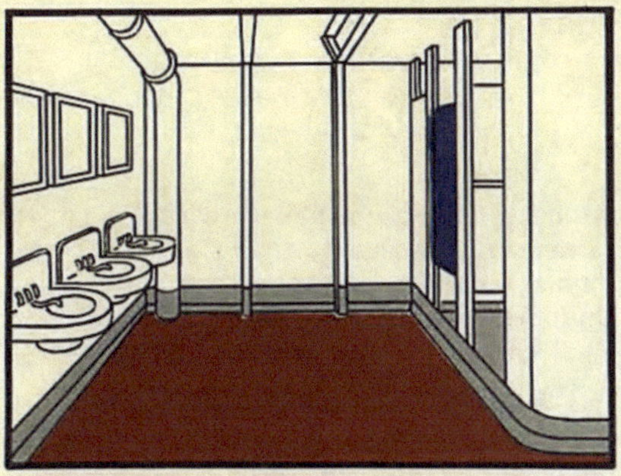

Principle 4 in heads uses white with No. 26 Tile Red for deck to encourage cleanliness and good maintenance.

In the Pilot House, No. 1 Light Green (Principle 5) is used for bulkheads and overhead to curtail glare.

Radio rooms take Principle 2, which combines white overhead, No. 1 Light Green bulkhead and gray deck.

For CIC, to aid dark adaptation, No. 2 Medium Green (Principle 6) is used for good visibility.

Principle 9 in shops combines No. 4 Soft Yellow with No. 19 Light Gray deck to relieve monotony.

Engine room (Principle 4) uses white with No. 26 Tile Red deck and No. 30 Equipment Gray on machinery.

COMPARTMENT COLORS—SYSTEM B

Principle No. 10	Bulkheads, No. 4 Soft Yellow Overheads, White Deck, No. 26 Tile Red	Passageways in living and working areas Hobby Rooms

(C) ~~(B)~~ BILGES, VOIDS AND COFFERDAMS

Bilges shall be the red color of Quick Drying Red Lead Primer prescribed for bilge painting.

(D) ~~(C)~~ DOORS, HATCH COVERS, AND ACCESS PLATES

Watertight doors and raised hatch covers shall be white. Flush deck hatches shall be the color of the deck. Joiner doors shall be No. 30 Equipment Gray. Access plates, other than fuel oil or diesel oil tank tops, shall be the color of the overhead, bulkhead or deck through which they provide access. Tank tops for fuel oil or diesel oil tanks shall be No. 13 Fire Red.

(E) ~~(D)~~ FURNITURE, FIXTURES AND EQUIPMENT

Painted surfaces of furniture, fixtures and [illegible]ment such as decks, tables, chairs, [illegible] lockers, cabinets, switch boxes, [illegible] connection boxes, con[illegible] etc., shall be No. 30 [illegible]

Gray linoleum closely matching No. 30 Equipment Gray is recommended for desk tops. Linoleum or plastic laminates in gray, blue or green colors and solid, linen or marbelized patterns may be used for mess tables, benches and counter tops.

[illegible]INE

[illegible] engine and mo[illegible]chinery and machine[illegible] No. 30 Equipment Gray. [illegible] on machine tools, except those [illegible] to be left bright, shall be No. 22 Hig[illegible] Light Buff. Switch boxes and starting boxes shall be No. 29 Bright Blue. Dangerous surfaces likely to cut, burn, shock or thrash operators shall be No. 18 International Orange. Machined working surfaces such as cutting edges, threads and sliding parts shall be left unpainted. For additional information on this subject refer to Section 3–17, Machinery Colors.

G ~~(F)~~ DAMAGE CONTROLS AND SAFETY MARKINGS

Hose racks, liquid foam containers, spanner wrenches, fire main valves, casualty power terminals, submersible pump outlets, battle lanterns, ~~damage control shores~~ and similar equipment used for damage control purposes shall be painted No. 13 Fire Red.

H ~~(G)~~ SAFETY COLOR MARKINGS

Gasoline and other flammable liquid containers shall be painted No. 14 Brilliant Yellow with the name of the contents conspicuously indicated in large black letters.

Stumbling, striking, and falling hazards shall be painted No. 14 Brilliant Yellow. Since interiors of vessels contain large numbers of such hazards discretion must be used in applying No. 14 Brilliant Yellow to prevent overusage of this intense color. In general these colors shall be applied only to hazards where experience indicates its application would serve a useful purpose.

(I) ~~(H)~~ COMPRESSED GAS CYLINDERS

Vessels shall follow the requirements set forth in Section 3–18, Compressed Gas Cylinder Identification.

(J) ~~(I)~~ HEATED SURFACES

Surfaces whose operating temperature is above 300° F. shall be painted aluminum with Heat-Resisting Paint.

(K) ~~(J)~~ LADDERS AND COMPANIONWAYS

Metal ladders and companionways shall be No. 30 Equipment Gray. This includes stringers, clips, piperails, pipe stanchions, tread support rungs and screens. Stainless steel or composition treads and uncovered galvanized chain handrails shall not be painted.

Wood ladders and companionways shall be varnished. Stainless steel, composition or galvanized parts shall not be painted. Noncorrosion resistant parts shall be No. 30 Equipment Gray.

(L) ~~(K)~~ SURFACES NOT TO BE PAINTED

Those surfaces listed under article 3–11–1 (M) ~~(L)~~ shall not be painted.

SECTION 3–12, COLORS FOR VESSELS LESS THAN 65 FEET IN LENGTH

The following paragraphs describe the color scheme to be employed in vessels less than 65 ft. in length and on barges of any length. These instructions do not apply to harbor tugs which are covered by Section 3–11. Due to the great variety of small boat types it is impractical to write painting instructions for every item. When situations arise which are not specifically provided for by the following instructions the general color scheme shall be followed.

3–12–1, EXTERIOR COLORS FOR VESSELS LESS THAN 65 FEET IN LENGTH

(A) HULL

Vessels of this class (shipboard boats, motor lifeboats, motor cargo boats, picket boats, utility boats, etc.) with the below exceptions shall have white hulls above the boot-topping area.

On steel, iron or plastic vessels the boot-topping area shall be defined as follows: the bottom edge of the boot-topping shall coincide with the light load waterline of the vessel. The top edge of the boot-topping shall, at amidships, be above the full load waterline a distance equal to ¼ the free board measured to the full load waterline. At the bow the height of the top edge of the boot-topping above the full load waterline shall be equal to twice that amidships. At the stern the height of the top edge of the boot-topping above the full load waterline shall be equal to that amidships. All measurements are to be taken in a vertical plane and projected on to the hull.

Where rubbing rails or spray rails located along the quarter of the vessel interfere with or lie close to the top edge of the boot-topping, the height of the boot-topping shall be adjusted the minimum amount to be either completely above or completely below the rubbing or spray rails and shall normally be parallel to these rails. In all cases the top edge of the boot-topping shall be a fair and optically pleasing curve from stem to stern.

No change in the top edge of the boot-topping is contemplated for 40-ft. Utility Boats previously painted in accordance with Painting Instruction Memorandum 2–52.

On wooden vessels the bottom edge of the boot-topping shall coincide with the *full load* waterline of the vessel. The upper edge of the boot-topping shall be identical with that for steel, iron and plastic vessels. Where metal sheathing is installed in way of the boot-topping area the instructions for steel vessels shall be followed.

Exceptions to the above rule are as follows:

(a) Barges, Buoy Boats, Fireboats and LCMs shall have black hulls above the upper edge of the antifouling paint. The upper edge of the antifouling paint shall coincide with the full load waterline.

(b) Flood Relief Punts shall have No. 18 International Orange hulls, including the underwater area.

(c) DUKWs shall have white hulls including the underwater area except that wheels and wheel wells shall be black.

(B) SUPERSTRUCTURE

Cabins, pilot houses, engine trunks extending appreciably above the sheer line, hatches and other superstructure shall be painted white on the vertical surfaces. The horizontal surfaces shall be No. 32 Blue Gray if steel and No. 24 Spar if of wood or canvas, with the exception of utility boats which shall be No. 32 Blue Gray. Where there is no distinct demarcation between vertical and horizontal surfaces, such as in the canopies of gigs and the cabins of the 36' motor lifeboat, the white color shall be used over the entire canopy or cabin. The entire superstructure of fireboats shall be No. 13 Fire Red.

tion of work boats such as buoy boats, motor

cargo boats and utility boats which shall have No. 32 Blue Gray decks. LCMs, fireboats and barges shall have black decks.

(D) INSIDE OF OPEN HULLS,
COCKPITS AND CARGO SPACES

The inside of open hulls and the cockpits and cargo spaces of decked hulls shall be white down to the side seats or risers, if any, or down to the cockpit deck, grating or floor boards if there be no side seats or risers. No. 24 Spar shall be applied below the white including seats, thwarts, inside of hull, cockpit decks, gratings, floor boards or open bilges. Engine trunks not extending appreciably above the sheer line shall be No. 24 Spar overall. This color scheme shall apply to all boats except as noted below.

40-ft. Utility Boats shall have No. 32 Blue Gray cockpit decks.

LCMs, Barges and Buoy Boats shall have the entire interior of their cargo spaces black.

DUKWs and motor Cargo Boats shall have the entire interior of their cockpit and cargo space painted No. 24 Spar.

Flood Relief Punts shall have the entire interior, including hull, thwarts, floor boards, and bilges No. 18 International Orange.

(F) GUNS AND ORDNANCE
EQUIPMENT

Guns, ordnance equipment and their foundations and mounts with the exception of working parts shall be painted black. Ready service lockers and pyrotechnic lockers shall be white. Pyrotechnic lockers shall in addition have the word PYROTECHNICS stencilled on the box in a conspicuous location using the largest practicable No. 13 Fire Red letters.

Detonator boxes shall be No. 13 Fire Red overall with the word DETONATORS stencilled in white on the box in a conspicuous location.

(G) RUNNING LIGHT SCREENS

Running light screens shall be painted a high gloss black enamel.

(H) MISCELLANEOUS DETAILS
AND FITTINGS

In general miscellaneous metal fittings and equipment shall be painted white if attached or immediately adjacent to some part of the superstructure that is painted white and shall otherwise be painted No. 24 Spar. Wood fittings shall normally be varnished.

LCMs, fireboats and barges are exceptions to rule and shall have all miscellaneous objects painted black. (See Art. 3-11-1(K))

(I) ITEMS NOT TO BE PAINTED

In general paint shall not be applied to surfaces where it will cause interference with the functions for which the surfaces were designed nor shall it be applied to those surfaces which are traditionally kept bright. Refer to the accompanying typical items not to be painted and the method of treatment for each.

ITEMS NOT TO BE PAINTED

Anchor chains, galvanized	No coating
Anchors, galvanized	No coating
Boatswains' chairs	Oiled
Brightwork	Polish and coat with clear plastic or lacquer
Canvas covers (removable)	No coating
Chain, galvanized	No coating
Deck treads, non-skid	No coating
Deck, wood	No coating
Dogs	Grease moving parts
Fire hose nozzles	Polish

52 FT. BUOY BOAT

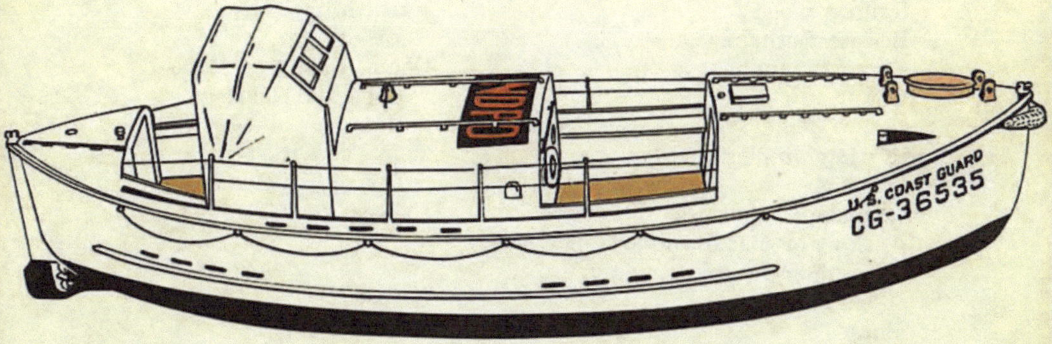

36 FT. MOTOR LIFE BOAT

40 FT. UTILITY BOAT

ITEMS NOT TO BE PAINTED

Gaskets, rubber for water-tight doors	Pulverized graphite
Glass	No coating
Gratings, wood	Oiled
Grease cups	No coating
Gypsy head whelps	No coating
Handrails and stanchions cold drawn finished	Polish
Hose	No coating
Insulators	No coating
Knife edges on watertight doors and hatches	Sandpapered
Ladders, wood	Oiled
Leather coverings	Oiled with preservative
Masts and spars, wood	Varnish
Name plates	No coating, polish or lacquer, if brass
Oars	No coating
Oil cups	No coating
Oil holes	No coating
Railing, wood	Varnish
Release mechanisms	Greased
Rigging, running	Chain, Gear and Wire Rope Lubricating Grease
Rigging, standing	Thin Film Rust Preventive Grade I
Searchlight, shutters	No coating
Working or machined parts of valves, machinery, blocks, guns or other equipment	Greased
Zincs	No coating

(J) DAMAGE CONTROL AND SAFETY MARKINGS

Hose racks, spanner wrenches, fire-main valves, fire axes and similar fire fighting equipment shall be No. 13 Fire Red. Damage control equipment stored in lockers shall be identified by a suitable label painted on the locker door in No. 13 Fire Red. Covers of all sound power jack boxes and switch boxes shall also be No. 13 Fire Red.

Gasoline and other flammable liquid containers shall be painted No. 14 Brilliant Yellow with the name of the contents conspicuously indicated in large black letters.

Stumbling hazards such as deck padeyes, deck clips and other small projections from the deck shall be painted white for better night visibility.

In painting the above markings do not paint those surfaces listed in paragraph (I) above.

(K) COMPRESSED GAS CYLINDER IDENTIFICATION

The color schemes set forth in Section 3–18 shall be followed.

(L) MARKINGS ON FLOATABLE EQUIPMENT

Ring Type Life Buoys: the vessel's name, or if not named, the vessel's number, shall be placed on the top semicircle of the ring. On the bottom semicircle the legend U. S.

38 FT. PICKET BOAT

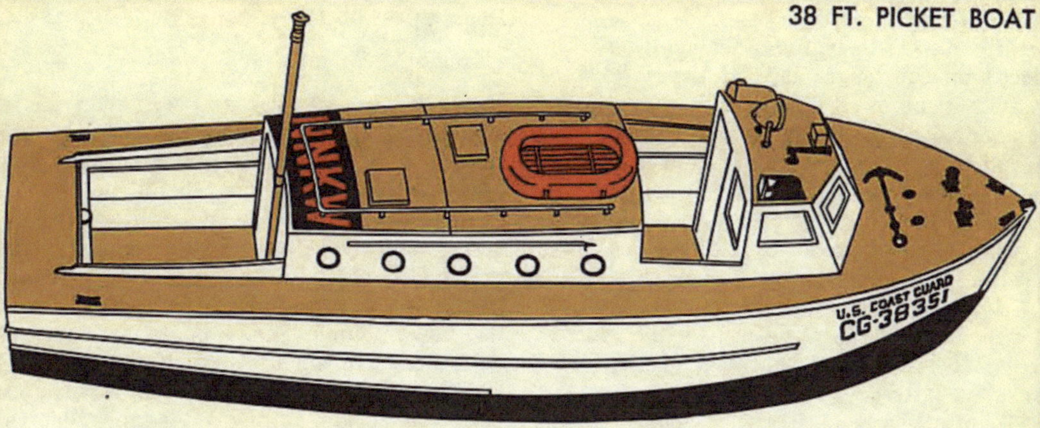

30 FT. RESCUE BOAT

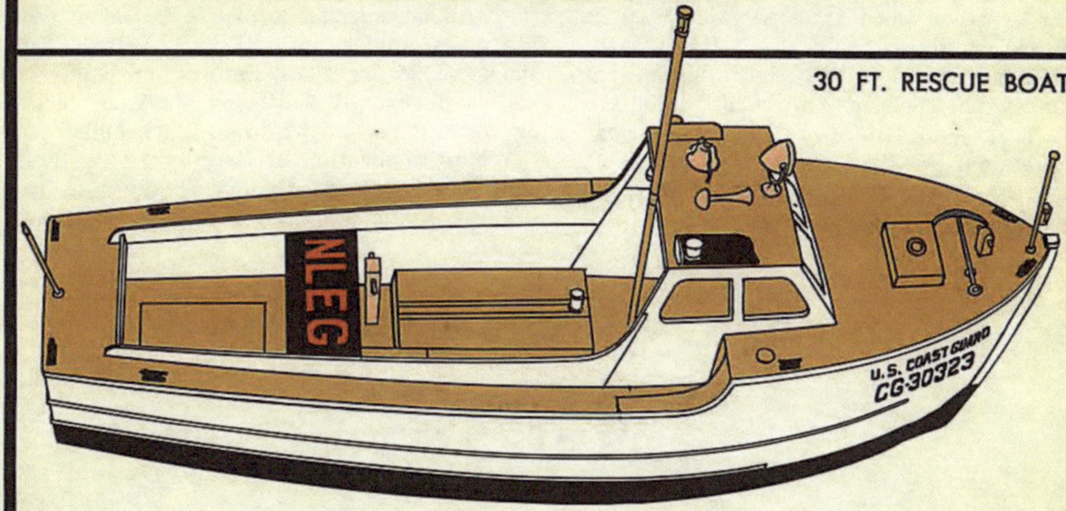

26 FT. MOTOR LAUNCH

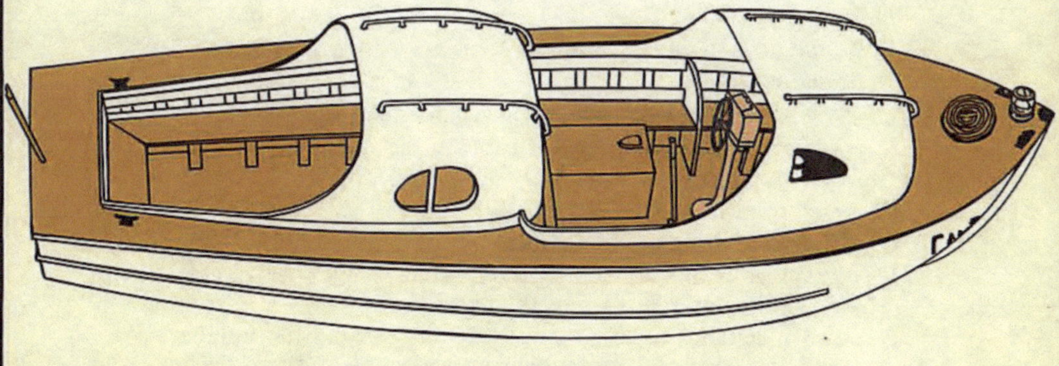

COAST GUARD shall appear. Black letters 2 in. high shall be used.

Life Rafts: markings shall always be placed on the longer legs of the raft. The vessel's name or, if not named, the vessel's number, shall be placed on one leg and the legend U. S. COAST GUARD shall be placed on the other leg. The markings shall be so placed as to be readily seen and when read in the normal manner of reading the vessel's name or number shall be read first.

Lifejackets: the vessel's name or, if not named, the vessel's number shall be placed across the middle of the back of the lifejacket in letters ¾ in. in height.

Wood articles: deck chests, boat boxes, oars or other wood articles which may be washed overboard shall have the vessel's name preceded by USCGC or designating number preceded by CG burned into the article in ½-in. letters so as to leave a clear impression. Burning pencils item No. 3201 Class 37 of the Catalog of Navy Material may be used for this purpose.

(M) IDENTIFICATIONS MARKINGS

Boats 30 to 65 ft. In Length

(Excluding Tugs)—

Boats 30 ft. and over and less than 65 ft. (excluding tugs) shall be identified by the legend U. S. COAST GUARD on each bow centered over the boat's official number including the prefix CG. In addition, the official number of the boat, including the prefix CG, shall be placed on the stern in letters of the same height as the numbers in the bow (double ended boats are not required to have the stern marking). Size and location of the letters shall be in accordance with the accompanying table and illustration. All measurements are to be taken on the outer surface of the hull. Letters as shown in Section 3–22, Letters and Numerals, shall be used. Lettering shall be black on white hulls and white on black hulls.

Where application of the above rules results in E being less than A, the location of the legend and designating number shall be

Overall Length	A	B	C	D	E
30 ft. and over to 40 ft.	4 in.	2 in.	6 in.	18 in.	*
40 ft. and over to 50 ft.	6 in.	3 in.	9 in.	24 in.	*
54 ft. and over to 65 ft.	8 in.	4 in.	12 in.	30 in.	*

*E must be equal to or greater than A.

A = Distance from lower edge of rubbing rail, gunwale, or sheer line to top edge of legend.

B = Height of legend.

C = Height of designating number.

D = Distance from foremost edge of foremost number to stem along the outer surface of the hull.

E = Least distance between lower edge of designating number and top edge of boot-topping or waterline if there be no boot-topping.

26 FT. MONOMOY SURF BOAT

24 FT. MOTOR CARGO BOAT

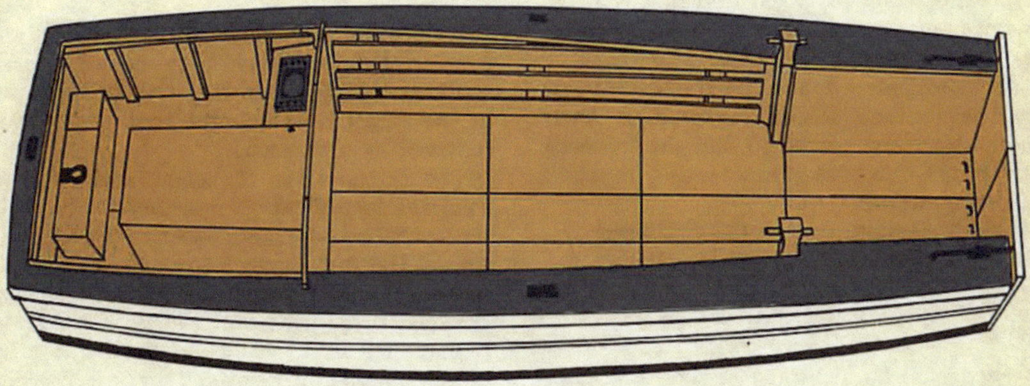

DUKW

adjusted so that A equals E and the remaining dimensions reduced proportionately to achieve a suitable appearance.

The legend and designating number may be adjusted both vertically and horizontally and letter size reduced as necessary to provide suitable clearance from all hull projections exceeding 1 in.

Boats Under 30 ft. In Length—

Boats less than 30 ft. in length, (excluding barges, DUKWs, ~~and~~ ship's boats, and mobile boarding unit boats) shall be identified by the legend U. S. COAST GUARD in 3-in. high letters. Lettering shall be black on white hulls and white on black hulls. Location of the lettering shall be as illustrated. If the lettering so located lies across a plank seam the lettering shall be raised or lowered as necessary so as to avoid the seam. All measurements shall be taken along the hull. For style of lettering, see Section 3–22, Letters and Numerals.

As an exception to the rule the below boats shall have their legend centered on their sheer plank and located so that the foremost part of the foremost character is 4 in. abaft the ~~stern.~~ stem.

20-ft Dinghy 16-ft. Dinghy
19-ft. Dory 13-ft. Dinghy
10-ft. Dinghy

Barges—

Barges shall be identified in the same manner as boats 30 ft. to 65 ft. in length except that the prefix CGB shall be substituted for the prefix CG. Barges obtained from the Army or Navy on a loan basis shall retain their Navy or Army identification symbols unless otherwise directed by the Commandant. White letters shall be used on black hulls.

DUKWs—

These amphibians are now classified as small boats and will be administered as such including identification markings and records.

Identification markings on DUKWS shall consist of the legend UNITED STATES COAST GUARD placed on each side in black letters. This legend shall be centered in the top panel which is outlined at the top and bottom respectively by the sheer half-round and chafing rail next below and shall extend fore and aft for a distance of aproximately 24 feet.

The small boat number CG---- shall be centered directly on the next panel below in figures of the same size and style as the legend UNITED STATES COAST GUARD.

On the vertical surface at the stern the small boat number shall be applied to the two outboard panels on either side of the winch opening, reading from left to right across the stern. The figures shall be five inches high. Attach boat number plate in accordamce with current "Operations Instructions". The marking USCG shall also be applied on the bow, one letter in each of the four panels, the size and style of letters being identical to those on the stern. For style of lettering see Section 3-22, Letters and Numerals

Boats Assigned To Named Vessels—

Identification markings on boats 20 ft. and over in length assigned to named vessels shall consist of an abbreviation of the name of the vessel to which the boat is assigned followed by a numeral.

The abbreviation of vessels' names to be used for identification markings shall, with the exception of those listed herewith, consist of the first three letters of the vessel's name. The first letter in the abbreviation shall be 6 in. high and the remaining two letters shall be 4 in. high. Following is a list of authorized abbreviations for vessels which do not fall under the general rule. Capital and lower case letters indicate 6-in. and 4-in. letters respectively.

The boat numbers which follow the abbreviations shall be 6 in. high and are determined by the location of the boat on the vessel. Starboard boats are assigned odd numbers starting forward and working aft. In cases where boats are stowed one above the other, the upper boat shall have the lower number. Numbers on the bows shall be omitted when only one boat is carried by a vessel.

Markings shall be located on both bows in accordance with the accompanying illustrations.

All measurements shall be taken along the surface of the hull. The spacing between the

ABBREVIATIONS FOR VESSEL IDENTIFICATION

Name	*Abbreviation*	*Name*	*Abbreviation*
ALERT	Alr	IRIS	Irs
APALACHEE	Apr	IRONWOOD	Irw
BERING STRAIT	B St	IROQUOIS	Irq
BLACKHAW	Blh	MACKINAW	Maw
BLACKROCK	Blr	OJIBWA	Ojb
BLACKTHORN	Blt	PAPAW	Paw
BLUEBELL	Blb	PLANETREE	Plt
CARTIGAN	Ctn	PRIMROSE	Prm
CASTLE ROCK	C Ro	ROCKAWAY	Rok
CLEMATIS	Clm	SEDGE	Seg
CHERRY	Chr	SWEETBRIER	Swb
CHINOOK	Chn	SWEETGUM	Swg
COLUMBINE	Clb	TAMPA	Tpa
COOK INLET	C In	VERBENA	Veb
COOS BAY	C Ba	WHITE ALDER	W Al
EVERGREEN	Evr	WHITE BUSH	W Bu
EWING	Ewn	WHITE HEATHER	W He
FERN	Frn	WHITE HOLLY	W Ho
FIREBRUSH	Fbu	WHITE LUPINE	W Lu
FORSYTHIA	Frs	WHITE PINE	W Pi
FREDERICK LEE	F Le	WHITE SAGE	W Sa
GENERAL GREENE	G Gr	WHITE SUMAC	W Su
GOLDENROD	Gld	WINNEBAGO	Wib
HALF MOON	H Mo	WOODBINE	Wob
HICKORY	Hik	WOODBRUSH	Wor

letters and figures shall be such that they shall be legible at a reasonable distance and present a good appearance.

Letters and figures used for these markings shall be Navy standard stock beveled boat bow letters and beveled boat figures, both with a 1-in. bevel. Refer to Catalog of Navy Material, Class 12, Item Nos. 4341 to 4445 and 4661 to 4679.

When applying these markings to clinker built boats, the outside surface of the hull shall be built up to a smooth surface where necessary and the markings shall be parallel to the bottom edge of the plank on which they are secured instead of parallel to the bottom edge of the guard rail.

Identification markings on boats less than 20 ft. assigned to named vessels shall be the same as boats 20 ft. and over assigned to named vessels except that the markings shall be painted on both bows in 3-in. letters, all capitals. For style of lettering, see Section 3–22, Letters and Numerals.

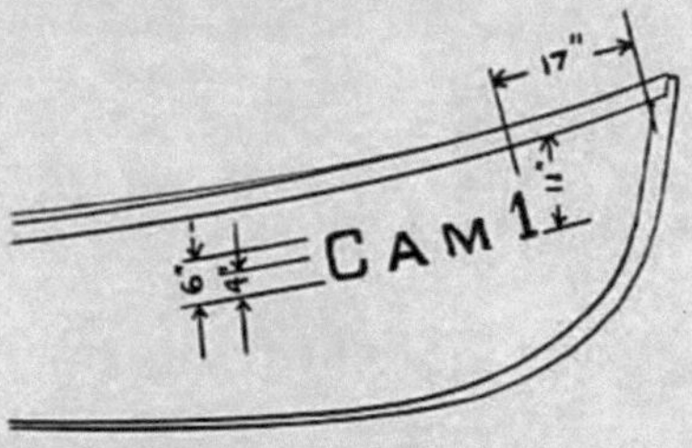

26 ft. MONOMOY SURFBOAT

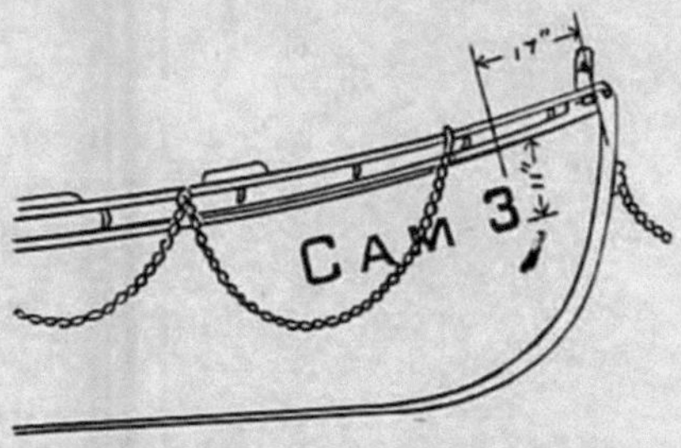

25 ft.-10 in. MOTOR S.B. SURFBOAT

Boats Assigned To
Numbered Floating Units—

Boats assigned to numbered floating units including lightships and barges shall be identified by placing the designating number of the unit, such as CG 63005, on both bows in black 3-in. block letters. For style of lettering, see Section 3–22, Letters and Numerals.

Special Purpose Identification—

In addition to the identification markings above, the following types of vessels (and boats carried by those vessels) which are used for special purposes shall have additional markings as follows:

Fireboats and Firefighting Barges: a dash and the designator F shall follow the last digit of the boat or barge number. As an example, CG 40341-F.

Buoy Boats (WDs): a dash and the designator D shall follow the last digit of the boat number. As an example, CG 80004-D.

3–12–2, INTERIOR COLORS FOR VESSELS LESS THAN 65 FEET IN LENGTH

The interior colors for vessels less than 65 ft. in length shall consist of one color scheme only as described below.

(A) OVERHEADS AND BULKHEADS

Overheads and bulkheads including all pipes, ducts and cables running along the overheads or bulkheads shall be white.

Door frames for both watertight and steel joiner doors shall be No. 30 Equipment Gray. Wood door frames shall be varnished.

Stanchions adjacent to bulkheads shall be painted the bulkhead color. All other stanchions shall be No. 30 Equipment Gray.

(B) DECKS

Decks shall be No. 19 Light Gray. A dado of the deck color may be extended up adjoining bulkheads for a distance of 4 to 8 in. The dado shall be brought up to the bottom of the doors. Kick pipes shall also be the deck color. Installed deck coverings shall not be painted but shall be procured in a color matching as nearly as possible the No. 19 Light Gray color.

(C) BILGES AND VOIDS

Bilges of wooden boats are not to be painted: The green color of the wood preservative is to be maintained by retreating with preservative as necessary.

On metal vessels the natural color of the anticorrosive paint prescribed for the type of metal used shall be considered the proper bilge color.

(D) DOORS, HATCH COVERS AND ACCESS PLATES

Doors shall be No. 30 Equipment Gray. Raised hatch covers shall be white. Flush deck hatches shall be the deck color. Access plates, other than fuel oil or diesel oil tank tops, shall be the color of the surrounding bulkhead or deck.

Tank tops for fuel oil or diesel oil tanks shall be No. 13 Fire Red.

(E) FURNITURE, FIXTURES AND EQUIPMENT

Painted surfaces of furniture, fixtures and equipment such as desks, tables, chairs, dressers, lockers, cabinets, switch boxes, lighting fixtures, connection boxes, controllers, switchboards, etc., shall be No. 30 Equipment Gray.

(F) MACHINERY AND MACHINE TOOLS

Engines, (including main propulsion engine and motors), motors, generators, machinery and machine tools shall be painted No. 30 Equipment Gray. Working areas on machine tools, except those surfaces to be left bright, shall be No. 22 High-Light Buff. Switch boxes and starting boxes shall be No. 29 Bright Blue. Dangerous surfaces likely to cut, burn, shock or thrash operators shall be No. 18 International Orange. Machined working surfaces such as cutting edges, threads and sliding parts shall be left unpainted. For additional information on this subject refer to Section 3–17, Machinery Colors.

(G) FIREFIGHTING EQUIPMENT

All equipment used primarily for firefighting shall be painted No. 13 Fire Red.

Stumbling, striking and falling hazards

shall be painted No. 14 Brilliant Yellow. Since interiors of vessels contain large numbers of such hazards discretion must be used in applying No. 14 Brilliant Yellow to prevent overusage of this intense color. In general this color shall be applied only to hazards where experience indicates its application would serve a useful purpose.

(H) HEATED SURFACES

Surfaces whose operating temperature is above 300° F. shall be painted aluminum with Heat-Resistant Paint.

(I) LADDERS AND COMPANIONWAYS

Metal ladders and companionways shall be No. 30 Equipment Gray. This includes stringers, clips, piperails, pipe stanchions, tread supports, rungs and screens. Stainless steel or composition treads and uncovered galvanized chain handrails shall not be painted.

Wood ladders and companionways shall be varnished. Stainless steel, composition or galvanized parts shall not be painted. Non-corrosion resistant parts shall be No. 30 Equipment Gray.

(J) SURFACES NOT TO BE PAINTED

Those surfaces listed under paragraph 3–11–1(L) shall not be painted.

SECTION 3–13, COLORS FOR SHORE EQUIPMENT

Recommended colors for outdoor and indoor equipment are given throughout this Manual. To supplement these recommendations, here is further data regarding miscellaneous equipment found particularly at large Bases.

3–13–1, CRANES AND SHIPWAYS

Tall dry dock cranes, Gantry cranes, crawler cranes, hammerhead cranes and steel shipways shall be painted No. 12 Spruce Green to tie in with the same standard recommended for tall structures such as towers located at Coast Guard units. For purposes of safety, however, the ends of crane beams shall be painted No. 14 Brilliant Yellow (over a space at least 6 ft. from the top). Yellows shall also be used on crane pulleys and hooks. Bands of No. 14 Brilliant Yellow and black shall be applied conspicuously at ground level on guards over wheels so that movement of cranes will be fully noted by personnel.

3–13–2, DRYDOCKS

Floating dry docks shall be painted black, including railings and ladders. Any unguarded platform or dock edges, and any stumbling or obstruction hazards, shall be finished in No. 14 Brilliant Yellow. Large valves controlling pump water flooding and discharge shall have the following symbolism applied: No. 13 Fire Red for the flooding valve; No. 16 Bright Green for the discharge valve; No. 14 Brilliant Yellow for the emergency gate which connects tanks.

3–13–3, FLOATING EQUIPMENT

Floating derricks, dredges and barges shall be painted No. 20 Medium Gray. (As traditional aboard ship, wood spars may be painted No. 24 Spar.) The sheds thereon may be No. 19 Light Gray with No. 12 Spruce Green Trim or they may be white with green trim as specified for Coast Guard buildings and dwellings.

3–13–4, MOBILE LAND EQUIPMENT

Mobile equipment, such as railroad locomotives, railroad cars, railway cranes, automotive and power cranes, road rollers, shall be No. 20 Medium Gray. Safety markings in No. 14 Brilliant Yellow (or black and yellow) shall be applied to crane ends, pulleys, hooks, bumper plates, handrailings and grips. Buoy carriers shall be painted No. 14 Brilliant Yellow to reduce hazards. For further information on vehicles see Section 3–14, Vehicle Colors.

3–13–5, DOCK EQUIPMENT

Miscellaneous dock equipment shall be No. 20 Medium Gray, such as winches, engines, pumps, compressors. Welding generator and compressor units (illustrated) shall be No. 20 Medium Gray. However, electric monitors and control boxes, including the control handles, shall be No. 29 Bright Blue to agree with the Safety Color Code.

Trash scoops shall be No. 20 Medium Gray. Tanks for water shall be white. Other tanks—fuel, compressed air, etc.—shall be No. 12 Spruce Green and shall be lettered as to contents with No. 14 Brilliant Yellow. Portable or small sized flammable liquid tanks and containers shall be painted No. 14 Brilliant Yellow throughout, with contents indicated with lettering in black.

Bollards and cleats on docks shall be black. Scaffolding used in ship repair may be No. 20 Medium Gray.

Cranes and shipway structures shall be No. 12 Spruce Green with safety marking in No. 14 Brilliant Yellow.

Floating derricks, ~~barges~~, dredges shall be No. 20 Medium Gray. Sheds may be white or No. 19 Light Gray.

No. 20 Medium Gray shall be used for locomotives, railway cars, railway and automotive cranes and winches.

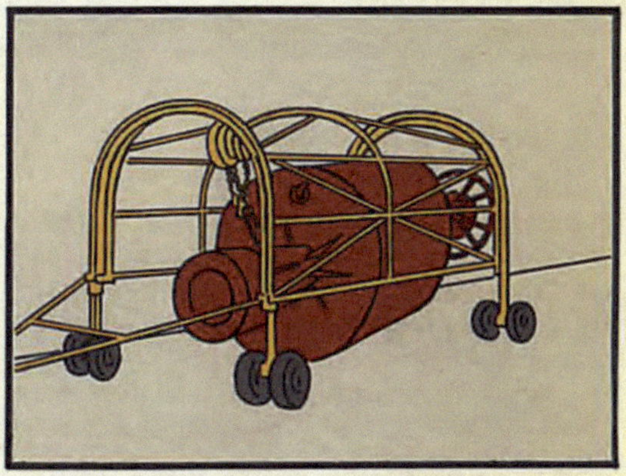

Buoy carriers shall be painted No. 14 Brilliant Yellow so framework will be conspicuous when handling buoys.

Engines and compressors shall be No. 20 Medium Gray. Electric monitors and control boxes No. 29 Bright Blue.

Trash scoops shall be No. 20 Medium Gray. Large tanks shall be No. 12 Spruce Green, with yellow lettering.

SECTION 3–14, VEHICLE COLORS

3–14–1, EXTERIOR COLORS

Vehicles in the Coast Guard shall be painted one of four colors: No. 19 Light Gray, No. 14 Brilliant Yellow, No. 13 Fire Red, or black. When new vehicles are purchased they should be procured in the prescribed color provided this can be done without undue delay in delivery or increase in cost. When this is not possible vehicles shall be procured in the manufacturer's color which most closely matches the prescribed color.

New vehicles, or old vehicles whose paint is in good condition, shall not be repainted to conform to these color requirements. All vehicles shall be painted the prescribed color when routine repainting is accomplished.

(A) VEHICLES PAINTED NO. 19 LIGHT GRAY

Automobiles, ambulances, trucks, trailers, carryalls, large truck-tractors, station wagons, jeeps and beach carts shall be painted No. 19 Light Gray. Certain automobiles used in investigative work and specifically authorized by the Commandant (O) to display state vehicle plates in lieu of Government license plates shall be black.

(B) VEHICLES PAINTED NO. 14 BRILLIANT YELLOW

Scooters, snow plows, snowmobiles, toboggans, tractors, airport mules, fork-lift trucks, bulldozers and buoy carriers shall be No. 14 Brilliant Yellow.

(C) VEHICLES PAINTED NO. 13 FIRE RED

Fire trucks, crash and rescue trucks or other vehicles used for these purposes shall be painted No. 13 Fire Red.

(D) VEHICLE IDENTIFICATION

Gray Vehicles—

Vehicles painted No. 19 Light Gray shall have identification markings on the center of both front door panels or in an equivalent position relative to the driver's seat if there is no door. The identification shall consist of a 6-in. x 6-in. U. S. Government Shield Decalcomania, the inscription UNITED STATES COAST GUARD and the vehicle license number. The lettering shall be in No. 29 Bright Blue and shall be 2 in. high. The arrangement of the identification markings shall be similar to that illustrated. U. S. Government Shield decalcomanias are obtainable from the Commandant (CHS) at no cost. See Section 3–22, Letters and Numerals, for illustration of lettering to be used.

Yellow Vehicles—

Vehicles painted No. 14 Brilliant Yellow shall bear identification markings similar to those painted No. 19 Light Gray. If these vehicles are used primarily on aircraft landing areas, identification shall be in accordance with current directive of Bureau of Aeronautics with appropriate alterations in wording for adoption to Coast Guard use. On those vehicles where space limitations prevent the use of the full inscription it may be abbreviated to U. S. C. G.

Red Vehicles—

Vehicles painted No. 13 Fire Red shall have identification markings on the center of both front door panels or in an equivalent position relative to the driver's seat if there is no door. The identification shall consist of a 6-in. x 6-in. U. S. Government Shield Decalcomania, the inscription U. S. Coast Guard (Name of Unit) and the location of the vehicle. The lettering shall be in gold leaf and shall be 2 in. high. The arrangement of the identification markings shall be similar to that illustrated with the decalcomania over the lettering and the order of lettering as shown. U. S. Government Shield decalcomanias are obtainable from the Commandant (CHS) at no cost.

Ambulances—

Ambulances shall in addition to the above identification for gray vehicles carry a 6-in. red cross on a white circular background 9

Common vehicles, sedans, trucks, shall be No. 19 Light Gray and shall display insignia.

Buses shall be No. 19 Light Gray. Safety markings in black and No. 14 Brilliant Yellow shall be added.

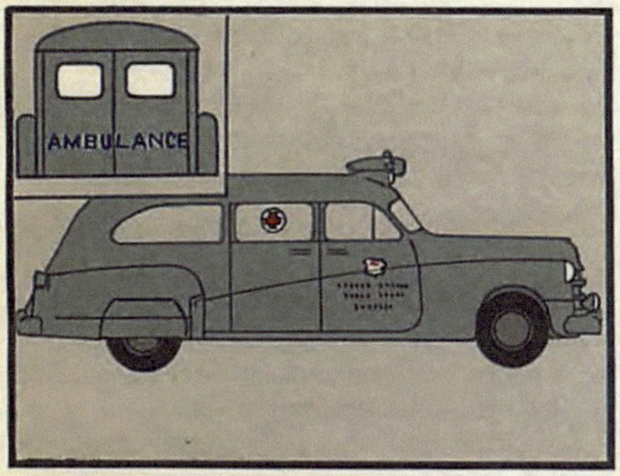

Ambulances shall carry insignia and red cross. Body of vehicle shall be No. 19 Light Gray.

Communication trucks shall have call letters on roof, using No. 18 International Orange on black.

No. 14 Brilliant Yellow is to be applied to bulldozers, materials handling and earth moving equipment.

No. 13 Fire Red is standard for fire trucks and for rescue and crash trucks on airfields.

in. in diameter centered on the forward windows of the litter compartment on each side of the ambulance. The word AMBULANCE in No. 29 Bright Blue letters 3 in. high shall be centered on the rear door panel as illustrated.

Communications Vehicles—

Communication trucks shall, in addition to the above identification for gray vehicles, have air identification markings as follows. An area approximately 3 ft. wide and 11 ft. long on the truck top shall be painted black to provide a background for the truck's radio call letters. The call letters in No. 18 International Orange shall be painted on the black background area in letters 18 in. high by 16 in. wide with a stroke width of 4 in. Refer to Section 3–22, Letters and Numerals for type lettering to be employed.

Communication trucks shall have ground identification markings as follows. A Coast Guard decalcomania 11 in. high by 11½ in. wide shall be placed on both side panels of the truck body. The words COMMUNICATION TRUCK shall appear over the decalcomania, and the district and district location shall appear below the decalcomania. All lettering shall be in No. 29 Bright Blue letters 3 in. high. The arrangement shall be as illustrated.

Communication Auxiliary Trucks shall have air identification and ground identification markings similar to Communication Trucks with the markings reduced to scale as necessary. They shall use a Coast Guard decalcomania 8 in. high by 8 in. wide and the lettering on the side panels of the trucks shall be 2 in. in height.

Coast Guard decalcomanias are obtainable from Coast Guard Supply Center, Jersey City, N. J.

Tank Trucks—

Tank trucks, in addition to the identification markings required of all gray vehicles, shall have the word FLAMMABLE displayed on each side and on the rear of the tank body in No. 14 Brilliant Yellow letters not less than 3 in. high.

Tank trucks used solely on airfields shall be painted No. 14 Brilliant Yellow overall with identification markings in Black.

(E) SAFETY MARKINGS

Buses and other appropriate vehicles shall have alternate bands of No. 14 Brilliant Yellow and black applied to front and rear body panels or bumpers. See Safety Color Code, Section 3–16, for further information on this subject.

3–14–2, INTERIOR COLORS

Vehicle (except bus and ambulance) interiors, where painted, shall use the same color as applied to the exterior of the body.

Bus and ambulance interiors shall be painted according to the following scheme.

Area	*Color*
Overhead	No. 6 Ivory
Window Section	No. 1 Light Green
Lower Bulkheads and Seat Backs	No. 2 Medium Green

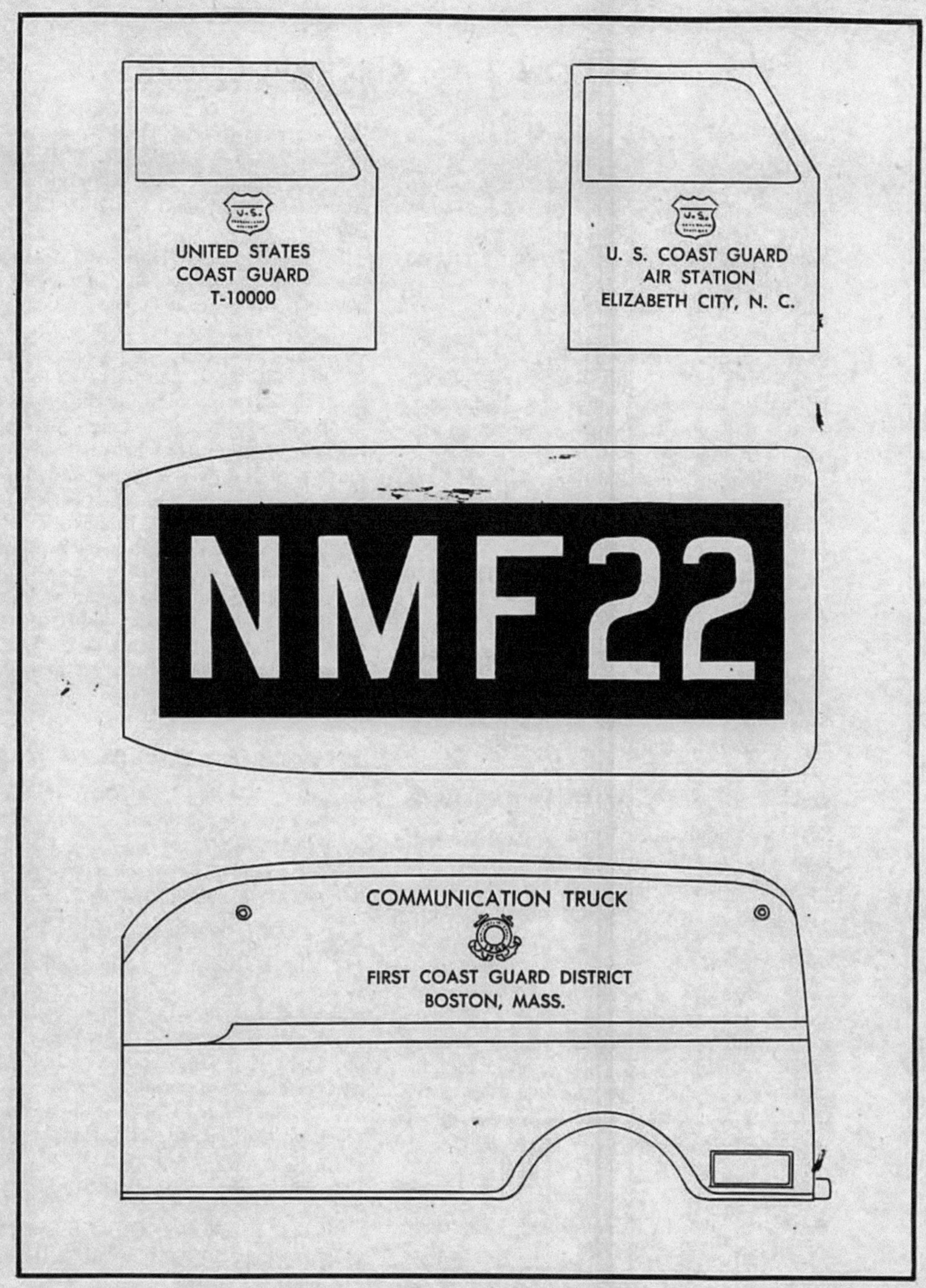
UNITED STATES
COAST GUARD
T-10000
U. S. COAST GUARD
AIR STATION
ELIZABETH CITY, N. C.
NMF22
COMMUNICATION TRUCK
FIRST COAST GUARD DISTRICT
BOSTON, MASS.

SECTION 3–15, AIRCRAFT COLORS

Coast Guard aircraft shall be painted in accordance with the current edition of below references as modified or supplemented by the instructions in following paragraphs.

(a) Navy Aeronautical Specification SR-2 (Insignia Markings and Color of Naval Aircraft).

(b) Navy Aeronautical Specification SR-70 (Application of Protective Coatings to Fabric Surfaces of Aircraft).

(c) Military Specification MIL–F–5055–1 (Finishes; Aircraft, Application and Treatment of, For Producing Aerodynamically Smooth Surfaces).

(d) Military Specification MIL–F–7179 (Finishing and Coatings; General Specifications for Protection of Aircraft and Aircraft Parts).

(e) Military Specification MIL–I–6140 (Insignia; National Aircraft).

(f) Military Specification MIL–I–6142 (Identification of Escape Hatches; Design Requirements for).

(g) Military Specification MIL–I–6143 (Insignia and Markings for Search and Rescue Aircraft; Design Requirements for).

3–15–1, GENERAL EXTERIOR PAINTING

The exterior finish of Coast Guard aircraft generally agrees with Navy Aeronautical Specification SR–2 and shall be as follows:

(A) FIXED WING

Landplanes—

Landplanes are generally unpainted, except that all surfaces other than aluminum or its alloys shall be given an aluminum pigmented finish. Landplanes may be painted aluminized lacquer finish when the condition of the aluminum surfaces indicates signs of minor corrosion or to improve its appearance.

Seaplanes—

Seaplanes, including amphibians, shall have aluminized lacquer finish.

(B) ROTARY WING

Rotary wing aircraft shall be given a No. 14 Brilliant Yellow lacquer finish. Rotor blades shall be finished with aluminum lacquer. Rotor blade tips shall be painted identifying colors.

NOTE: On landplanes and seaplanes, to provide a non-reflecting surface for pilot comfort, the section of the hull or fuselage forward of the windshield including the cowl ring shall be painted a non-specular (dull) black. This black painted area will extend from the extreme edges of the windshield or cabin glass foward to the nose section, tapering as necessary to conform to the general pattern of the nose. Low or midwing multi-engined aircraft which have the engine nacelles practically in line with the pilot's position shall have an inboard section of the inboard nacelles painted a non-specular (dull) black. This area shall extend from the wing surface to a line approximately 15° past the vertical center line of the nacelle from the nose of the intersection of nacelle and wing.

3–15–2, GENERAL EXTERIOR MARKING

Exterior markings of aircraft shall conform to the following general requirements. Specific instructions for marking the different aircraft models based on these general requirements are given below.

(A) FIXED WING AIRCRAFT

(1) The National Insignia shall be applied to the top side of the left wing, to the under side of the right wing and on both sides of the fuselage (or vertical fin) in a conspicuous place. The single point of the star shall be forward or uppermost.

(2) The letters USCG shall be applied in black on the top side of the right wing and the under side of the left wing. The lettering shall read from inboard toward the wing tip.

(3) U. S. COAST GUARD shall be applied in black on both sides of the fuselage (or hull).

(4) The Coast Guard Insignia shall be applied to each side of the fuselage (or hull) in the vicinity of the pilot's compartment. Decalcomanias for this purpose may be obtained by requisition from the Aircraft Supply and Repair Base, Elizabeth City, N. C. Stock number for 12 in. is L82–ARSB–PS–1224–D1. Stock number for 18 in. is L82–ARSB–PS–1224–D.

(5) The Coast Guard model and serial number shall be applied in black on both sides of the vertical fin.

(6) The partial serial number of the aircraft shall be applied in black to both sides of the fuselage (or hull) utilizing the last four digits only, except where only three have been assigned, in which case the whole serial number shall be used. In addition, this partial serial number shall be applied to both sides at the nose or bow of the aircraft.

(B) ROTARY WING AIRCRAFT

(1) The National Insignia shall be applied to both sides of the helicopter.

(2) The letters USCG shall be applied in black to the under side of the body or tail cone with the top of the letters to the port side.

(3) U. S. COAST GUARD shall be applied in black on both sides of the fuselage or tail cone.

(4) The Coast Guard Insignia shall be applied on both sides of the fuselage. (See 3–15–2(A), (4) above for decalcomania ordering information.)

(5) The model and serial number shall be applied in black on each side of the fuselage or tail cone.

(6) A partial serial number shall be applied in black to both sides and on top of the fuselage, utilizing the last four digits only, except where only three have been assigned in which case the whole serial number shall be used.

(7) Helicopters with a tail rotor shall have applied to the tail cone forward of the disc of the rotor, a red warning sign consisting of the words DANGER KEEP AWAY and a red unfeathered arrow. The shaft of the arrow shall be centered as specified. The word DANGER shall be spaced above the shaft, while the words KEEP AWAY shall be similarly located below the shaft.

(C) SEARCH AND RESCUE MARKINGS

The below search and rescue markings shall be applied to all helicopters and the following types of fixed wing aircraft:

All PB–1G Aircraft
All PBM–5G Aircraft
All PBY–5AG Aircraft
All JRF–5G Aircraft
All UF–1G Aircraft
All P4Y–2G Aircraft

(1) Upper and lower surfaces of both wing tips shall be painted No. 14 Brilliant Yellow from the wing tip inboard approximately 7 per cent of the span (excluding span increased by wing tip float). This area shall be extended inboard by a black border of 6 in.

(2) A No. 14 Brilliant Yellow band shall encircle the body forward of the horizontal stabilizer. This band shall be 18 in. in width on JRF aircraft and 36 in. in width on others; and be inclosed by two 3-in. or 6-in. black borders respectively. The aft edge of the band, including the border, shall be approximately 9 and 18 in. respectively, forward of the leading edge of the horizontal stabilizer.

3–15–3, PAINTING AND MARKING DETAILS FOR PB–1G AIRCRAFT (Refer to AN–01–20EG–2)

(A) GENERAL, MARKINGS

1. (a) The National Insignia applied to the wings shall have a 60-in. star diameter. The insignia shall parallel the wing rear spar and its center shall be located at a point 10 in. forward of the wing rear spar at wing station 24. (Fig. 2 of reference.)

1. (b) The National Insignia shall be applied to both sides of the vertical stabilizer and shall have a 30-in. star diameter. The center shall be located on the front spar at station 5 of the vertical stabilizer. The insignia shall be perpendicular to the stabilizer spars. (Fig. 2 of reference.)

2. USCG shall be applied to both wings in letters 72 in. high. The letters shall be cen-

tered between wing stations 10 and 28. They shall be applied so that their bottoms parallel the aileron cut-out, and tangent to the flap hinge line extended. (Fig. 2 of reference.)

3. U. S. COAST GUARD shall be applied on both sides of the fuselage in letters 12 in. high. The letters shall be centered between fuselage stations 6 and 61. The tops of the lettering shall be on the fuselage center line. (Fig. 2 of reference.)

4. A large Coast Guard insignia shall be applied on both sides of the fuselage with its center located at the intersection of fuselage station 2E and stiffener K. The vertical center line of the insignia shall be along fuselage station 2E. (Refer to Fig. 5G, TO–01–20E–3.)

5. (a) The Model (PB–1G) shall be applied in letters 4 in. high on both sides of the vertical stabilizer. The numbers shall be centered between vertical stabilizer front and rear spars with the bottoms of the numerals on station 9. (Fig. 2 of reference.)

5. (b) The serial number shall be applied in letters 4 in. high on both sides of the vertical stabilizer, 4 in. below the model designation and centered fore and aft therewith. (Fig. 2 of reference.)

6. (a) The partial serial number (last 4 digits) shall be applied to both sides of the fuselage in letters 12 in. high. The number shall be centered between fuselage stations 6A and 6D. The bottoms shall be on a line parallel to and 15 in. above the fuselage center line. (Fig. 2 of reference.)

6. (b) The partial serial number (last 4 digits) shall be applied to both sides of the fuselage in numerals 6 in. high. The number shall be centered on fuselage station 1A. The tops of the numbers shall be on the fuselage center line. (Fig. 2 of reference.)

(B) SEARCH AND RESCUE MARKINGS

1. The inboard edge of the 6-in. black border shall be along wing station 31 and the remainder of the tip shall be painted No. 14 Brilliant Yellow (Fig. 2 of reference.)

2. The after edge of the rear black border shall be located 7 in. aft of fuselage station 7 and shall be parallel thereto. (Fig. 2 of reference and Fig. 13 of TO–01–20E–3.)

3–15–4, PAINTING AND MARKING DETAILS FOR PBM–5G AIRCRAFT (Refer to AN–01–35EE–2)

(A) GENERAL, MARKINGS

1. (a) The star diameter of the wing National Insignia shall be 60 in. The insignia shall parallel the wing center line and centered at the intersection of the wing center line and wing station 457.125. (Fig. 6 of reference.)

1. (b) The 60-in. diameter Star National Insignia shall be applied to both sides of the hull. The insignia shall be centered fore and aft at bulkhead number 5, and vertically with respect to the anchor hatch at frame No. 4. (Fig. 6 of reference and Fig. 79 of AN–01–35E–3A.)

2. USCG shall be applied to the wings in letters 60 in. high. The letters shall be centered between wing stations 307.310 and 532.310. The tops of the letters shall be parallel to and 20 in. aft of the wing leading edge. (Fig. 6 of reference.)

3. U. S. COAST GUARD shall be applied on both sides of the fuselage in letters 12 in. high. The tops of the letters shall be along the side frames between hull stations 601 and 716. The forward letter shall begin on hull station 579. (Refer to Figs. 79 and 80 of AN–01–35E–3A.)

4. A large Coast Guard insignia shall be applied on both sides of the hull centered on station 91 and a line 12 in. below and parallel to the bottom of the pilots' sliding windows. (Fig. 6 of reference.)

5. (a) The Model (PBM–5G) shall be applied in letters 4 in. high on the outboard sides of both vertical stabilizers. The lettering shall be centered at the median of stabilizer station 106.75. (Fig. 7 of reference.)

5. (b) The serial number shall be applied in letters 4 in. high, located on the stabilizers 4 in. below the model designation and centered vertically therewith. (Fig. 7 of reference.)

6. (a) The partial serial number (last 4 digits) shall be applied to the hull on both sides in letters 12 in. high. The center of this number shall be located on hull station 513. The bottoms of the numerals shall be along

PB-1G AIRCRAFT

the top frames of the portholes. (Fig. 79 of reference.)

6. (b) The partial serial number (last 4 digits) shall be applied to both sides of the hull in letters 6 in. high. The center of this number shall be on frame 26⅞ and the bottom of the letters shall be on a line extended from the bottom of the door coaming of the port anchor hatch. (Fig. 79 of reference.)

(B) SEARCH AND RESCUE

1. The inboard edge of the 6-in. black border shall be along wing station 615.875 and the remainder of the tip shall be painted No. 14 Brilliant Yellow. The wing tip float and struts shall be painted No. 14 Brilliant Yellow. (Fig. 6 of reference.)

2. The rear black border shall be located between fuselage frame 36 and station 795½. (Fig. 6 of reference.)

3–15–5, PAINTING AND MARKING DETAILS FOR PBY-5AG AIRCRAFT (Refer to AN–01–5MA–2)

(A) GENERAL, MARKINGS

1. (a) The National Insignia applied to the wings shall have a 60-in. star diameter. The insignia shall be parallel and tangent to the wing rear spar and its center shall be located on wing station 16. (Fig. 2 of reference.)

1. (b) The National Insignia shall be applied to both sides of the hull with a 30-in. star diameter. The insignia shall be centered fore and aft between hull stations 193 and 268 and vertically 25 in. below the deck line. The insignia shall parallel the deck line. (Refer to Figs. 4–2 and 4–3 of AN–01–5M–3.)

2. USCG shall be applied to both wings in letters 50 in. high. The letters shall be centered between wing stations 12 and 19. The letters shall parallel, and their bottoms shall be on the rear spar center line. (Fig. 2 of reference.)

3. U.S. COAST GUARD shall be applied on both sides of the hull in letters 6 in. high, centered between hull stations 7 and 8, with the tops of the letters 10 in. below, and parallel with the deck line. (Fig. 4–2 of AN–01–5M–3.)

4. A large Coast Guard insignia shall be applied on both sides of the hull with its center at hull belt frame 1.33 and 18 in. below the deck line. (Fig. 4–2 of AN–01–5M–3.)

5. (a) The Model (PBY–5AG) shall be applied in letters 4 in. high on both sides of the vertical stabilizer. The tops of the letters shall be on the center line extended of the upper rudder hinge bearings, and they shall be centered fore and aft on the stabilizer. (Fig. 2 of reference.)

5. (b) The serial number shall be applied in letters 4 in. high on both sides of the vertical stabilizer, 4 in. below the model designation and centered vertically therewith. (Fig. 2 of reference.)

6. (a) The partial serial number (last 4 digits) shall be applied to both sides of the hull in numerals 6 in. high. The numerals shall be centered between hull stations 346 and 368. The tops of the numerals shall be 10 in. below, and parallel with the deck line. (Fig. 2 of reference.)

6. (b) The partial serial number (last 4 digits) shall be applied to both sides of the hull in numerals 6 in. high. The numerals shall be centered between hull stations 25 and 49. The tops of the numerals shall be 3 in. below and parallel to the bottom line of the bow turret. (Fig. 2 of reference.)

(B) SEARCH AND RESCUE

1. The inboard edge of the wing 6-in. black border shall be along the line 15 in. inboard and parallel to wing station 21. The remainder of the wing tip and the float shall be painted No. 14 Brilliant Yellow. (Fig. 2 of reference.)

2. The after edge of the hull 6-in. rear black border shall be parallel to hull station 9 and 4 in. forward thereof. (Fig. 2 of reference.)

3–15–6, PAINTING AND MARKING DETAILS FOR R4D–5 AIRCRAFT (Refer to AN–01–40NC–2)

1. (a) The National Insignia applied to both wings shall have a 60-in. star diameter. The insignia shall be parallel to the trailing edge of the wing and tangent to the aileron cut-out and centered at wing station 271.0. (Fig. 5 of reference.)

UF–2G AIRPLANE

For dimensioned drawing, see Coast Guard UF Airplane Bulletin No. 112.

R5D–3 AIRPLANE

For dimensioned drawing, see Coast Guard R5D Airplane Bulletin No. 23.

SC–130B AIRPLANE

For dimensioned drawing, see Coast Guard SC–130B Airplane Bulletin No. 1.

RM–1Z AIRPLANE

For dimensioned drawing, see Coast Guard RM Airplane Bulletin No. 18.

C–123B AIRPLANE

For dimensioned drawing, see Coast Guard C–123 Airplane Bulletin No. 2

This type being phased out

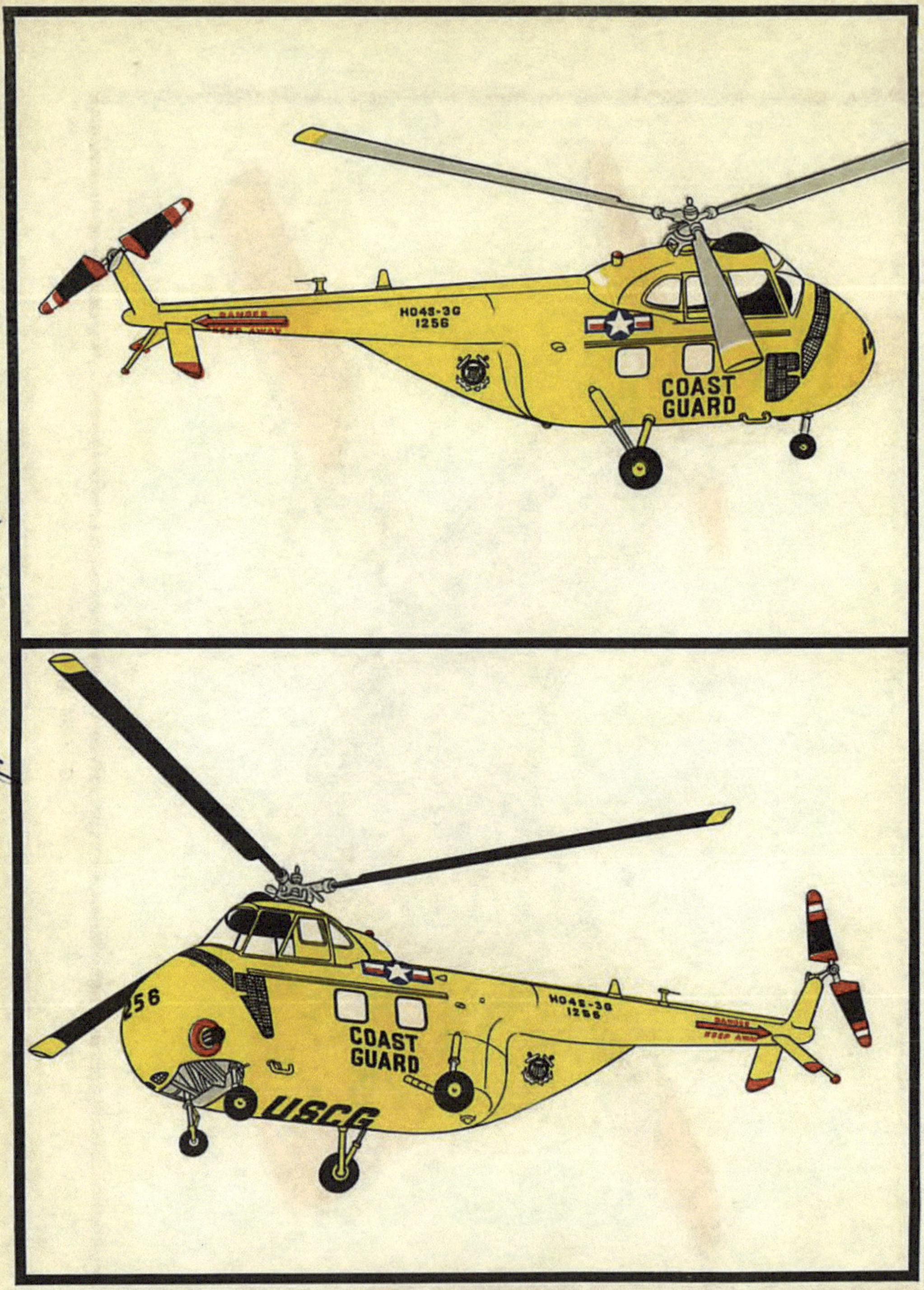

HO4S–3G HELICOPTER

For dimensioned drawing, see Coast Guard HO4S Helicopter Bulletin No. 193.

1. (b) The National Insignia applied to both sides of the fuselage shall have a 60-in. star diameter. The insignia shall be centered on fuselage station 559.0 at the horizontal stabilizer chord line extended, which it shall parallel. (Fig. 5 of reference.)

2. USCG shall be applied to both wings in letters 50 in. high. The letters shall be centered between wing stations 89.5 and 286.4 and be centered with, and parallel to the wing center spar. (Fig. 5 of reference.)

3. U.S. COAST GUARD shall be applied to both sides of the fuselage in letters 12 in. high. The letters shall be centered between fuselage stations 258.5 and 429.5. The bottoms of the letters shall be 6 in. above and parallel to the line through the side window top frames. (Fig. 5 of reference.)

4. A large Coast Guard insignia shall be applied on both sides of the fuselage with its center located on fuselage station 40, 20 in. below longitudinal No. 9 (Fig. 5 of reference.)

5. (a) The Model (R4D–5) shall be applied in letters 4 in. high on both sides of the vertical stabilizer. These letters shall be centered on vertical stabilizer station 97 and between the de-icer rivnuts and the rudder cut-out. (Fig. 6 of AN–01–40NC–3.)

5. (b) The serial number shall be applied in letters 4 in. high on both sides of the vertical stabilizer, 4 in. below the model designation and centered fore and aft therewith.

6. (a) The partial serial number (last 4 digits) shall be applied to both sides of the fuselage in letters 12 in. high. The numerals shall be centered between fuselage stations 372.5 and 429.5 with the bottom of the numerals on a line 6 in. above and parallel to the floor line. (Fig. 5 of reference.)

6. (b) The partial serial number (last 4 digits) shall be applied to both sides of the fuselage in letters 6 in. high. The numeral shall be centered between fuselage stations 18 and 40. The bottom of the numeral shall be along the floor line at station 40. (Fig. 5 of reference.)

3–15–7, PAINTING AND MARKING DETAILS FOR R5D–3/4 AIRCRAFT (Refer to AN–01–40NM–2)

1. (a) The National Insignia applied to both wings shall have a 60-in. star diameter. The insignia shall be parallel to the center spar and centered therewith at wing station 537. (Fig. 7 of reference and Fig. 14 of AN–01–40NM–3.)

1. (b) The National Insignia applied to both sides of the fuselage shall have a 60-in. star diameter. The insignia shall be centered on fuselage station 742, 26 in. above the fuselage reference plane and parallel thereto. (Fig. 6 of reference.)

2. USCG shall be applied to both wings in letters 50 in. high. The letters shall be centered on the wing center spar and between wing stations 457 and 644. (Fig. 7 of reference.)

3. U.S. COAST GUARD shall be applied to both sides of the fuselage in letters 12 in. high. It shall be centered between fuselage stations 501 and 663. The bottoms of the letters shall be 6 in. above and parallel with the top frame line of the cabin windows. (Fig. 6 of reference.)

4. A large Coast Guard insignia shall be applied on both sides of the fuselage with its center located on fuselage station 64, 30 in. above the fuselage reference plane. (Fig. 6 of reference and Fig. 3, Sheet 1, of AN–01–40NM–3.)

5. (a) The Model (R5D–3/4) shall be applied in letters 4 in. high on both sides of the vertical stabilizer. The bottom of the letters shall be on vertical stabilizer station 90. The most forward letter shall have its forward bottom corner at the junction of the nose section of the stabilizer. (Fig. 6 of reference.)

5. (b) The serial number shall be applied in letters 4 in. high on both sides of the vertical stabilizer 4 in. below the model designation and centered fore and aft therewith. (Fig. 6 of reference.)

6. (a) The partial serial number (last 4 digits) shall be applied to both sides of the fuselage in letters 12 in. high. The numerals shall be centered between fuselage stations 828 and 881. The tops shall be on the fuse-

lage reference plane. (Fig. 6 of reference.)

6. (b) The partial number (last 4 digits) shall be applied to both sides of the fuselage in letters 6 in. high. The numerals shall be centered between fuselage stations 8 and 30. The bottoms of the numerals shall be on the fuselage reference plane. (Fig. 6 of reference.)

3–15–8, PAINTING AND MARKING DETAILS FOR R5O–4 AND R5O–5 AIRCRAFT

(Refer to AN–01–75C–3)

1. (a) The National Insignia applied to both wings shall have a 30-in. star diameter. The bar shall parallel the main beam, the center shall be on wing station 289, and the circle shall be tangent at the aileron cut-out. (Fig. 75 of reference.)

1. (b) The National Insignia applied to both sides of the fuselage shall have a 30-in. star diameter, the insignia shall be centered at fuselage station 494 and 23 in. above the fuselage reference line. (Fig. 36 of reference.)

2. USCG shall be applied to both wings in letters 40 in. high, the letters shall be centered between wing stations 165 and 315. The bottoms of the letters shall be parallel and tangent to the aileron cut-out. (Fig. 75 of reference.)

3. U.S. COAST GUARD shall be applied to both sides of the fuselage in letters 12 in. high. It shall be centered between fuselage stations 212¼ and 375¼. The bottoms of the letters shall be parallel to the fuselage reference line and 34 in. above. (Fig. 36 of reference.)

4. A large Coast Guard insignia shall be applied on both sides of the fuselage with its center located at the intersection of fuselage station 92¾, and 15 in. above the fuselage reference line. (Fig. 36 of reference.)

5. (a) The Model (R5O–4 for CG Nos. 12447, and 12453; R5O–5 for CG No. 12456) shall be applied in letters 4 in. high on the outboard side of each vertical stabilizer. The letters shall be centered on fin station 31.125 midway between the rudder hinge line and the de-icer boot rivnut line. (Fig. 32 of reference.)

5. (b) The serial number shall be applied in letters 4 in. high on the outboard side of each vertical stabilizer 4 in. below the model designation and centered fore and aft therewith.

6. (a) The partial serial number (last 4 digits) shall be applied to both sides of the fuselage in letters 12 in. high. The numerals shall be centered between fuselage stations 325$\frac{7}{16}$ and 370$\frac{7}{16}$. The tops of the numerals shall be on the fuselage reference line. (Fig. 36 of reference.)

6. (b) The partial serial number (last 4 digits) shall be applied to both sides of the fuselage in letters 6 in. high. The numerals shall be centered between fuselage stations 14½ and 36½ with the tops of the numerals on a line 9 in. below the fuselage reference line. (Fig. 36 of reference.)

3–15–9, PAINTING AND MARKING DETAILS FOR JRB–4 AIRCRAFT

(Refer to AN–01–90CD–2)

1. (a) The National Insignia applied to both wings shall have a 40-in. star diameter. The insignia center shall be on wing rib 7 with the circle tangent to the aileron cut-out and the bar perpendicular to the wing ribs. (Fig. 7 of reference.)

1. (b) The National Insignia applied to both sides of the fuselage shall have a 20-in. star diameter. The insignia shall have its center on fuselage bulkhead 12, 17 in. above the bottom longeron. The bar shall parallel the fuselage stringers. (Fig. 7 of reference and Fig. 28 of AN–01–90CD–3.)

2. USCG shall be applied to both wings in letters 24 in. high. The letters shall be centered between wing ribs 3 and 8 of the leading and trailing edges, parallel to the wing center line. (Fig. 7 of reference.)

3. U.S. COAST GUARD shall be applied to both sides of the fuselage in letters 6 in. high. The letters shall be centered between fuselage bulkheads 7 and 10. They shall have their tops on a line 6 in. below and parallel to the cabin windows bottom frame line. (Fig. 7 of reference.)

4. A small Coast Guard insignia shall be applied on both sides of the fuselage with its center located at the intersection of fuselage bulkhead 3 and stringer 4. (Fig. 7 of reference and Fig. 25 of AN–01–90CD–3.)

5. (a) The Model (JRB-4) shall be applied in letters 2 in. high on the outboard side of both vertical stabilizers. The letters shall be centered between the rudder hinge line and the leading edge of the stabilizer, 18 in. above and parallel with the chord line of the horizontal stabilizer.

5. (b) The serial number shall be applied in letters 2 in. high on the outboard side of both vertical stabilizers, 2 in. below and centered fore and aft with the model designation. (Fig. 7 of reference.)

6. (a) The partial serial number (last 4 digits) shall be applied to both sides of the fuselage in numerals 4 in. high. It shall be centered between fuselage bulkheads 10 and 11. The tops of the numerals shall be on a line parallel to and 3 in. below the bottom line extended, of the U.S. COAST GUARD specified in paragraph 3 above. (Fig. 7 of reference.)

6. (b) The partial serial number (last 4 digits) shall be applied to both sides of the fuselage in numerals 4 in. high. It shall be centered between fuselage bulkheads 1 and 2. The tops shall be on the rivet line of stringer 20. (Fig. 7 of reference and Fig. 25 of AN-01-90CD-3.)

3-15-10, PAINTING AND MARKING DETAILS FOR JRF-5G AIRCRAFT (Refer to NAVAER-01-85VF-2)

(A) GENERAL, MARKINGS

1. (a) The National Insignia applied to the wings shall have a 40-in. star diameter. The center of the insignia shall be on wing station 191. The insignia shall parallel the 25 per cent line of the wing and be tangent to the aileron cut-out. (Fig. 5 of reference.)

1. (b) The National Insignia applied to both sides of the hull shall have a 30-in. star diameter. The insignia shall be centered on hull station 17 and the hull reference line, which it shall parallel. (Fig. 5 of reference.)

2. USCG shall be applied to both wings in letters 32 in. high. The letters shall be centered between wing stations 105 and 226. The forward edges of the letters shall be on the 25 per cent line of the wing.

3. U.S. COAST GUARD shall be applied on both sides of the hull in letters 6 in. high. The letters shall be centered between hull stations 22 and 29. The letters shall be centered on the hull reference line. (Fig. 5 of reference.)

4. A large Coast Guard insignia shall be applied to both sides of the hull centered at station 8, 3 in. above the hull reference line. (Fig. 5 of reference.)

5. (a) The Model (JRF-5G) shall be applied in letters 2 in. high on both sides of the vertical stabilizer. The bottoms of the letters shall be on vertical stabilizer station $90\frac{3}{16}$ and centered between the de-icer rivnuts and the rudder hinge line. (Fig. 5 of reference.)

5. (b) The serial number shall be applied in letters 2 in. high on both sides of the vertical stabilizer, 2 in. below the model designation and centered fore and aft therewith. (Fig. 5 of reference.)

6. (a) The partial serial number (last 4 digits or 3 digits as applicable) shall be applied to both sides of the hull in letters 6 in. high. The numerals shall be centered between hull stations 27 and 29. The bottoms of the numerals shall be 24 in. above the hull reference line. (Fig. 5 of reference.)

6. (b) The partial serial number (last 4 digits or 3 digits as applicable) shall be applied to both sides of the hull in letters 4 in. high. The numerals shall be centered on hull station 2. The tops of the numerals shall be along the hull reference line. (Fig. 5 of reference.)

(B) SEARCH AND RESCUE MARKINGS

1. The inboard edge of the 6-in. black border shall be along wing station 247 and the remainder of the tip shall be painted No. 14 Brilliant Yellow. In addition, wing tip float and struts shall be No. 14 Brilliant Yellow.

2. The forward edge of the after black border shall be located at hull station 32 and shall be parallel thereto. (Fig. 5 of reference.)

3-15-11, PAINTING AND MARKING DETAILS FOR OY-1 AIRCRAFT (Refer to AN-01-50DB-3)

1. (a) The National Insignia applied to both wings shall have a 30-in. star diameter,

the insignia shall be centered between the front and rear spar and on wing station 126.69. (Fig. 10, sheet 7, of reference.)

1. (b) The National Insignia applied to both sides of the fuselage shall have a 20-in. star diameter. The insignia shall be centered on fuselage station 133.25 between the upper and lower longerons and parallel with the thrust center line. (Fig. 10, sheet 4, of reference.)

2. USCG shall be applied to both wings in letters 24 in. high, between wing stations 43.25 and 133.25 and centered between the front and rear spars. (Fig. 10, sheet 7, of reference.)

3. U.S. COAST GUARD shall be applied to both sides of the fuselage in letters 4 in. high, parallel with the line of thrust midway between the lower longeron and the first stringer above and between fuselage stations 77.00 and 133.25. (Fig. 10, sheet 4, of reference.)

4. A small Coast Guard insignia shall be applied on both sides of the fuselage with its center on fuselage station 8.0 and 4 in. below the thrust center line. (Fig. 10, sheet 4, of reference.)

5. (a) The Model (OY–1) shall be applied in letters 4 in. high on both sides of the vertical stabilizer. The bottoms of the letters shall be 26 in. below the vertical stabilizer top cut-off. The letters shall be centered fore and aft directly under the leading edge of the rudder overhang. (Fig. 10, sheet 2, of reference.)

5. (b) The serial number shall be applied in numerals 6 in. high on both sides of the vertical stabilizer, 4 in. below the model designation and centered fore and aft therewith.

6. (a) Not applicable.

6. (b) The partial serial number (last 4 digits) shall be applied to both sides of the engine cowling in numerals 4 in. high. The extreme after edge of the numerals shall be 8 in. forward of fuselage station O. The numerals shall have their tops on the thrust center line. (Fig. Page IV of reference.)

3–15–12, PAINTING AND MARKING DETAILS FOR HO3S–1G HELICOPTER (Refer to AN–01–23OHD–2)

1. (a) Not applicable.

1. (b) The National Insignia applied to both sides of the tail cone shall have a 20-in. star diameter. The insignia shall be centered at the intersection of tail cone station 75 and the tail cone center line. The bar shall parallel the tail cone center line. (Fig. 1–1 of reference.)

2. USCG shall be applied in letters 16 in. high to the underside of the fuselage with tops of the letters to the port side. The letters shall be between tail cone stations 21 and 81 and centered on the fuselage center line. (Figs. 1–1 and 3–1 of reference.)

3. U.S. COAST GUARD shall be applied to both sides of the fuselage in letters 10 in. high. The forward letter shall begin at fuselage station 131. The bottoms of the letters shall be 2 in. above WLO. (Fig. 1–1 of reference.)

4. A small Coast Guard insignia shall be applied to both sides of the fuselage, the port insignia shall be centered on the bottom panel of the sliding door assembly and the starboard insignia shall be in a corresponding place on the starboard side of the fuselage. (Fig. 1–1 of reference.)

5. (a) The Model (HO3S–1G) shall be applied in letters 4 in. high to both sides of the tail cone, the letters shall be centered fore and aft on tail cone station 141¾ with the bottoms of the letters 2 in. above the tail cone center line. (Fig. 1–1 of reference.)

5. (b) The serial number shall be applied in numerals 4 in. high on both sides of the tail cone, centered fore and aft with the model designation and with the numeral tops 2 in. below the tail cone center line.

6. (a) The partial serial number (last 4 or 3 digits as applicable) shall be applied in numerals 10 in. high on the top of the fuselage forward of main rotor. The tops of the numerals shall be to starboard. The forward numeral shall begin at fuselage station 61.

6. (b) The partial (last 4 or 3 digits as applicable) serial number shall be applied in 10 in. numerals to both sides of the fuselage. The numerals shall be centered on fuselage station 49. Their tops shall be on a line 24 in. below WLO. (Fig. 1–1 of reference.)

7. A red unfeathered arrow with shaft width of 4 in. shall be centered on the tail

cone center line with the arrow point on tail cone station 214. The shaft shall terminate at fuselage station 178. Two inches above and centered fore and aft therewith, shall be the word DANGER in red 4-in. letters. Two inches below and centered fore and aft with the arrow shall appear the words KEEP AWAY in 4-in. letters. (Fig. 1–1 of reference.)

3–15–13, PAINTING AND MARKING DETAILS FOR P4Y–2G AIRCRAFT (Refer to AN–01–5EN–2)

(A) GENERAL, EXTERIOR

1. (a) The National Insignia applied to the wings shall have a 40-in. star diameter. The insignia shall be parallel to the wing center line and its center shall be located on wing station 21. The insignia shall be tangent to the wing rear spar center line. (Fig. 4 of reference.)

1. (b) The National Insignia applied to both sides of the fuselage shall have a 40-in. star diameter. The insignia shall be centered at the intersection of fuselage station 3.4 and W.L. 44$\frac{7}{16}$, and shall parallel W.L. 44$\frac{7}{16}$. (Fig. 3 of reference.)

2. USCG shall be applied to both wings in letters 40 in. high. The letters shall be centered between wing stations 18 and 25. The letters shall parallel and be centered on the wing center line. (Fig. 4 of reference.)

3. U.S. COAST GUARD shall be applied to both sides of the fuselage in letters 12 in. high. The letters shall be centered between fuselage stations 6.1 and 7.6. The bottom of the letters shall be along the center line of the upper longeron between stations 6.0 and 7.6. (Fig. 3 of reference.)

4. A large Coast Guard insignia shall be applied to both sides of the fuselage with its center on fuselage station 1.2, 14 in. above W.L. 57$\frac{3}{8}$. (Fig. 3 of reference.)

5. (a) The Model (P4Y–2G) shall be applied in letters 4 in. high on both sides of the vertical stabilizer. The letters shall be centered fore and aft between the rudder cut-out and the rivet line of the vertical fin leading edge, with the bottoms along the rivet line of the stabilizer rib at the middle rudder hinge. (Fig. 55 of AN–01–5EN–3.)

5. (b) The serial number shall be applied on both sides of the vertical stabilizer in letters 4 in. high, 4 in. below the model designation and centered fore and aft therewith. (Fig. 55 of AN–01–5EN–3.)

6. (a) The partial serial number (last 4 digits) shall be applied in 12 in. letters on both sides of the fuselage. The numerals shall be centered between fuselage station 7.4 and 7.6. The numerals shall have their tops parallel to and on W.L. 57$\frac{3}{8}$. (Fig. 3 of reference.)

6. (b) The partial serial number (last 4 digits) shall be applied in 6-in. letters at the bow on both sides. The numerals shall be centered between fuselage stations 0.2 and 1.0 (bottom). The tops of the numbers shall be on W.L. 44$\frac{7}{16}$ extended. (Fig. 3 of reference.)

7. SEARCH AND RESCUE MARKINGS:

(a) The upper and lower surfaces of both wing tips shall be painted No. 14 Brilliant Yellow from the wing tip inboard approximately 7 per cent of the space. This area shall be extended inboard by a 6-in. black border.

(b) A No. 14 Brilliant Yellow band shall encircle the body forward of the horizontal stabilizer. This band shall be 36 in. in width and be inclosed by two 6-in. black borders. The aft edge of the band, including the black border, shall be 18 in. forward of the leading edge of the horizontal stabilizer.

3–15–14, PAINTING AND MARKING DETAILS FOR HTL–1 HELICOPTER (Refer to AN–01–110HAA–2)

1. (a) Not applicable.

1. (b) The National Insignia applied to both sides of the tail cone shall have a 20-in. star diameter. The star circle shall be tangent and the bar shall parallel the top longeron of the tail boom. The star shall be centered 83 in. aft of the main rotor center line. (Fig. 1–3 of reference.)

2. Not applicable.

3. U.S. COAST GUARD shall be applied to both sides of the tail boom in letters 4 in. high. The forward letter shall begin at a point 136 in. aft of the main rotor center line.

The tops of the letters shall be 6 in. below the tail boom upper longeron. (Fig. 1–3 of reference.)

4. A small Coast Guard insignia shall be applied to both sides of the fuselage on the upper engine compartment cowling, centered between the rotor center line and the cabin enclosure after edge and tangent at the juncture of the upper engine compartment and the side engine compartment cowling (or screen). (Fig. 1–3 of reference.)

5. (a) The Model designation (HTL–1) shall be applied in letters 4 in. high on both sides of the tail boom. The model designation shall be centered above and parallel to the U. S. COAST GUARD (Par. 3) and the tops shall be tangent to the upper tail boom longeron. (Fig. 1–3 of reference.)

5. (b) The serial number shall be applied in numerals 4 in. high on both sides of the tail boom. The numbers shall be centered with and their tops 2 in. below the bottom of the U. S. COAST GUARD (Par. 3). (Fig. 1–3 of reference.)

6. (a) Not applicable.

6. (b) Not applicable.

7. A red unfeathered arrow with shaft width of 2 in. shall be centered vertically between the upper and lower longeron on both sides of the tail boom. The length shall be 30 in. and the point of the arrow shall terminate at a point 210.75 inches aft of station zero. The word DANGER shall be located on the tail boom fairing in red 2-in. letters tangent at the top with the upper longeron, and centered fore and aft with the arrow. The words KEEP AWAY shall be applied in red 2-in. letters, centered fore and aft and 2 in. below the red arrow. (Fig. 1–3 of reference.)

3–15–15, PAINTING AND MARKING DETAILS FOR UF–1G AIRCRAFT (Refer to AN–01–85AB–2 and Grumman Aircraft Engin'g Dwg. 98049)

(A) GENERAL, MARKINGS

1. (a) The National Insignia applied to the wings shall have a 50-in. star diameter. The top surface insignia shall be centered on wing station 330. The lower surface insignia shall be centered on wing station 277.5.

1. (b) The National Insignia applied to both sides of the hull shall have a 30-in. star diameter centered at hull station 187.5 and a line parallel to but 24 in. below the bottom of the pilots' inclosure.

2. USCG shall be applied to both wings in letters 45 in. high. The top surface lettering shall be centered on wing station 330. The lower surface lettering shall be centered on wing station 319.0.

3. U. S. COAST GUARD shall be applied on both sides of the hull in letters 12 in. high. The letters shall be centered between hull stations 349 and 465 and applied so that their bottoms are 6 in. above the top line of the window frames.

4. A large Coast Guard insignia shall be applied on both sides of the hull with its center on hull station 132, 24 in. below the bottom of the pilots' inclosure.

5. (a) The Model (UF–1G) shall be applied in letters 4 in. high on both sides of the vertical stabilizer. The bottom shall be centered fore and aft between the leading edge and the hinge line of the rudder and on stabilizer station 266½.

5. (b) The serial number shall be applied in numerals 4 in. high on both sides of the vertical stabilizer, 4 in. below the model designation.

6. (a) The serial numbers shall be applied to both sides of the hull in numerals 12 in. high. The numbers shall be centered on hull station 491, 3 in. above the main entrance door bottom frame.

6. (b) The serial number shall be applied to both sides of the hull in numbers 6 in. high. The numbers shall be centered between hull stations 32 and 62. The bottom of the numbers shall be 6 in. above the chine at station 47 and parallel with the waterline.

(B) SEARCH AND RESCUE MARKINGS

1. The outboard edge of the 6-in. black border shall be along wing station 411 and the remainder of the tip, including float and struts shall be No. 14 Brilliant Yellow.

2. The forward edge of the hull 6-in. rear black border shall parallel the hull stations and be 36 in. forward of the horizontal stabilizer leading edge.

3-15-16, PAINTING AND MARKING DETAILS FOR HO4S-1G AND HO4S-2G HELICOPTERS (Refer to Sikorsky Aircraft Division Dwg. 14-01-1020)

1. The National Insignia applied to both sides of the fuselage shall have a 20-in. star diameter. The star shall be centered 86 in. above the static ground line and 32 in. forward of the tail cone quick disconnect point. The bar shall parallel the static ground line.

2. USCG shall be applied in letters 24 in. high to the underside of the fuselage with the tops of the letters to the port side. The letters shall be centered between the wheels.

3. U. S. COAST GUARD shall be applied to both sides of the fuselage in letters 10 in. high. The forward letter shall begin at the cargo compartment forward bulkhead. The bottoms of the letters shall be 32 in. above the static ground line.

4. A small Coast Guard insignia shall be applied to both sides of the fuselage. The insignia shall be centered on the cabin access door and in approximately the same location on opposite side.

5. (a) The Model (HO4S-1G or HO4S-2G) shall be applied in letters 4 in. high on both sides of the tail cone, the letters shall parallel the static ground line and be centered fore and aft on the tail cone, 4 in. above its center line.

5. (b) The serial number shall be applied in 4-in. numerals, centered vertically but with a 4-in. separation under the model designation.

6. (a) The serial number (4 digits) shall be applied in numerals 10 in. high on top of the fuselage, forward of the rotor, and centered in the available space aft of the windshield. The tops of the numerals shall be forward.

6. (b) The serial number shall be applied in 10-in. numerals on the engine quick access doors. The bottom of the numerals shall parallel and be 60 in. above the static ground line. Two numerals shall be applied on each door.

7. A red unfeathered arrow with 4-in. shaft width shall be centered on the tail cone with the arrow point 45 in. forward of the vertical through the tail rotor gear box. The arrow shaft shall be 48 in. long. Two inches above and centered fore and aft therewith shall appear the word DANGER in red 4-in. letters. Two inches below the arrow and centered fore and aft therewith shall appear the words KEEP AWAY in red 4-in. letters.

3-15-17, PAINTING AND MARKING DETAILS FOR HO5S-1G HELICOPTER (Refer to Sikorsky Aircraft Division Dwg. 11-01-1513)

1. (a) Not applicable.

1. (b) The National Insignia applied to both sides of the tail cone shall have a 20-in. star diameter. The star circle shall be tangent to, and the bar shall parallel the top outboard longeron of the tail boom. The star shall be centered 24 in. aft of the tail boom fuselage intersection.

2. USCG shall be applied in letters 16 in. high to the underside of the fuselage with tops of the letters to port. The letters shall be centered fore and aft about a point 30 in. forward of the rotor center line and with the fuselage center line.

3. U. S. COAST GUARD shall be applied to both sides of the fuselage in letters 4 in. high. The forward letter shall begin at a point 2 in. aft of the main entrance door forward frame. The tops of the letters shall be 2 in. below the door window bottom frame and parallel thereto.

4. A small Coast Guard Insignia shall be applied to both sides of the fuselage, tangent at the top with the bottom frame of the rear side windows and centered fore and aft therewith.

5. (a) The Model designation (HO5S-1G) shall be applied in letters 4 in. high. The forward letter shall begin 66 in. aft of the fuselage tail boom junction. The letters shall be 2 in. above the tail boom center line and parallel to the static ground line.

5. (b) The serial number shall be applied in 4-in. numerals on both sides of the tail cone, 4 in. below and centered fore and aft with the model designation.

6. (a) Not applicable.

6. (b) Not applicable.

7. A red unfeathered arrow with shaft width of 2 in. shall be centered on the tail boom center line with the arrow point terminating at the junction of the angle gear box fairing with the tail boom. The shaft length shall be 23 in. The word DANGER shall be located on the tail boom in red 2-in. letters, 2 in. above the top of the arrow shaft and centered fore and aft therewith. The words KEEP AWAY shall be applied in red 2-in. letters 2 in. below the arrow shaft and centered therewith.

3–15–18, PAINTING AND MARKING DETAILS FOR AIRBORNE LIFEBOATS

Airborne lifeboats shall have their exterior surfaces and other surfaces exposed to aerial observers, except walk-ways, finished in No. 14 Brilliant Yellow.

U. S. COAST GUARD shall be applied on each side at the bow in 4-in. letters. The Air Station to which the boat is assigned; such as CGAS PORT ANGELES, WASH. shall be applied at the stern in 2-in. letters. The lettering shall conform with Aeronautical Specification SR–2 and shall be black.

SECTION 3–16, SAFETY COLOR CODE

The Coast Guard Safety Color Code contained herein follows that of the American Standards Association, National Safety Council and U. S. Navy Safety Color Codes, modified as necessary for application to the Coast Guard. Briefly, red signifies damage control and fire protection equipment; yellow indicates striking and falling hazards and also flammable liquid and gas storage; orange indicates cutting, crushing and shocking hazards; green is used for first aid facilities and equipment; blue for identification of controllers and switch boxes. The Safety Color Code does not apply to vessels.

3–16–1, USE OF RED

No. 13 Fire Red is standard for the identification of damage control and fire protection equipment. Its use for any other purposes, such as to mark points of danger or flammable liquid containers shall be discontinued.

No. 13 Fire Red shall be used for fire sheds, fire hydrants, fire alarm boxes, outside hose connections to standpipes and sprinkler lines. On the interiors of buildings it shall be applied to fire carts, hose connections and fire main valves. In painting fire main valves, the valve body, bonnet and handle shall be painted. Valve stems and threads on the valve body shall not be painted. Sprinkler heads and fusible links shall not be painted.

Areas of No. 13 Fire Red shall be painted behind fire fighting equipment on walls and columns. In order to facilitate location from a distance, 12-in. bands or squares of red shall be applied directly above the equipment and about 12 ft. from the floor. Where extinguishers of different types are used, these shall be painted red (with the exception of brass or copper containers) and appropriately lettered in black or white: CO_2, FOAM, SODA ACID, DRY CHEMICAL.

Although tradition has established red as a color for containers of gasoline, naphtha, kerosene, alcohol and solvents, this use of the color is obviously wrong because of confusion with fire protection devices. Such containers shall be refinished in No. 14 Brilliant Yellow and lettered in black to indicate contents.

3–16–2, USE OF YELLOW

No. 14 Brilliant Yellow is the accepted standard to mark strike-against, stumbling and falling hazards. Where suitable, alternate bands of black and No. 14 Brilliant Yellow shall be used. Thus yellow (or black and yellow) shall be applied to guard railings, curbing, the edges of platforms and pits, crane beams, pulleys and hooks, portable ladders, skids, hand trucks. Large overhead crane beams in shops, however, may be No. 19 Light Gray, with No. 14 Brilliant Yellow applied to the crane cab and to pulleys and hooks. Yellow is also desirable for materials handling equipment, small trucks, obstructions, dead-ends, barricades, bumper plates on vehicles, the ends of tall jib, wall or automotive cranes.

On stairways a 4-in. band of No. 14 Brilliant Yellow shall be applied immediately under the tread on top and bottom risers.

(As noted in Article 3–7–1 (I), No. 14 Brilliant Yellow is applicable to bulldozers, tractors, mules, ground maintenance and control equipment on airfields. No. 13 Fire Red is applicable to crash and rescue trucks.)

3–16–3, USE OF ORANGE

No. 18 International Orange is standard for hazards which are likely to cut, crush, burn or shock personnel. It shall be applied on or near the dangerous parts, gears, shears, rollers of saws, planers, brakes, rolling and crushing devices, forming presses, punch presses, riveting machines. It is desirable to use orange on the under side of guards over belts or gears in order to signal against carelessness by exposing a gaudy color to the eye. The exteriors of such guards shall be gray.

No. 18 International Orange shall be used

Fire sheds, fire protection equipment, hydrants, hose connections, standpipes shall be No. 13 Fire Red.

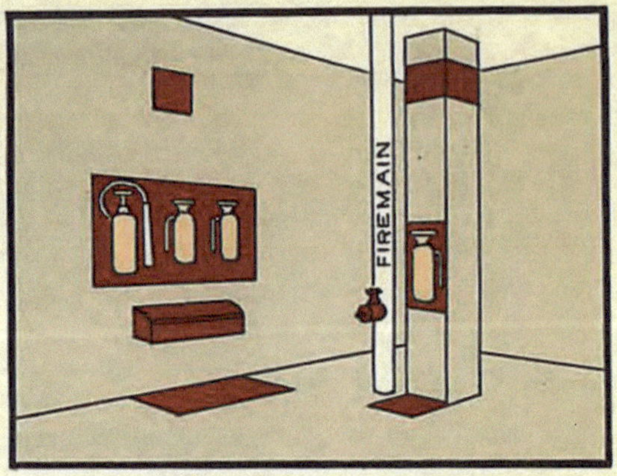

Areas of red shall be placed on walls and columns to aid the quick location of fire extinguishers.

No. 14 Brilliant Yellow shall be used for obstructions, guard railings, curbing, projections and such hazards.

Yellow is standard for small materials handling equipment, trucks, skids, ladders, shop mules, scooters.

Moving parts, projections, obstructions shall be clearly marked in yellow and black for ideal safety.

Yellow and black shall be applied to crane pulleys and hooks, the bumpers of trucks and to any projecting hazards.

for guards around hot pipes and exposed electrical wires and connections. Overhead electric wires or rails for conveying and hoisting equipment often present a hazard which should be conspicuously marked on adjacent beams, covers or supports. The inside of switch box doors and covers shall be painted No. 18 International Orange to reveal a vivid color and encourage the proper closing of panel doors at all times.

3–16–4, USE OF BLUE

This color shall be used in industrial areas for switch and fuse boxes, control panels, off-and-on control boxes on machinery. In personnel facilities and offices, however, electrical control panels may be in the wall color. In all instances, the *inside* areas of such boxes should be painted No. 18 International Orange.

Blue is also applicable to electrical control mechanisms, welding gear, the control boxes of hoists, winches, cranes. Used as a background for a sign, with the lettering OUT OF SERVICE, good safety practice makes its use mandatory in marking equipment cut down for repair: elevators, boilers, kilns, ovens, tanks, pits, scaffolding. The blue sign shall, wherever possible, be placed over controls and is not to be removed except by the person who has put it in place.

3–16–5, USE OF GREEN

No. 16 Bright Green shall be used for the identification of first aid equipment. It shall be painted on first aid and medicine cabinets, stretcher boxes, cabinets for gas masks, safety showers and for all signs relating to first aid and safety. To facilitate the location of safety devices, 12-in. green crosses shall be painted about 12 ft. from the floor on walls and columns to be clearly seen from a distance.

3–16–6, AISLE MARKS

It is good practice to use aisle marks and lines to indicate traffic lines, parking locations and storage areas. White shall be used where floors are dark, and black where floors are light. Lines and marks around hazards or along pit and platform edges shall be No. 14 Brilliant Yellow. This practice will help to avoid accidents.

It is important to understand that safety colors are for functional purposes only, not for decoration. Particularly in industrial areas, their use shall hold practical meaning and be seen by personnel solely where the distraction of color will provide a safeguard.

For the application of safety colors to machinery, see Article 3–17–4.

3-16-7, CRANE CONTROLS AND HOOKS
For standardization, crane control handles, corresponding hooks and lower blocks shall be painted according to the following color scheme, modified to individual installations:

RED - Port Vang
GREEN - Starboard Vang
WHITE - Topping Lift
YELLOW - Main Purchase
ORANGE - Relief Purchase

No. 18 International Orange shall be applied on or near the hazardous parts of shears, knives, rollers, gears.

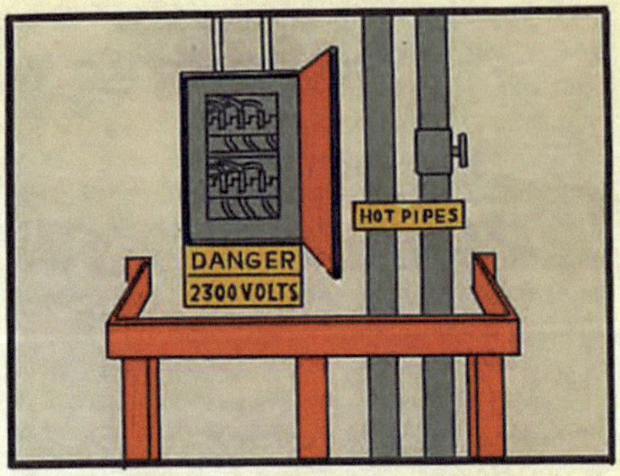

Orange should mark hot pipes, exposed electrical connections, and all cutting, crushing or shocking dangers.

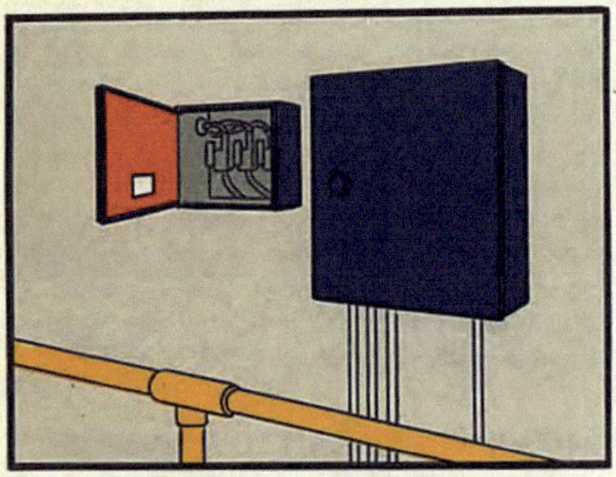

Shop switch boxes and control panels should be No. 29 Bright Blue, with No. 18 International Orange inside doors.

Out of service signs, in blue, shall be used for equipment, boilers, elevators, shut ~~cut~~ down for repair.

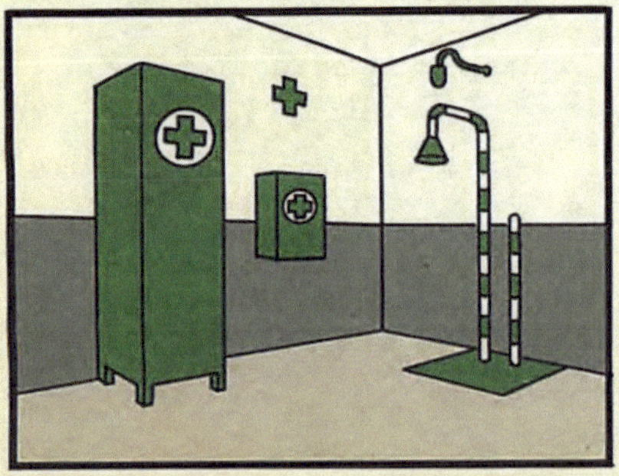

No. 16 Bright Green is standard to mark first aid equipment and devices, safety showers, stretcher boxes.

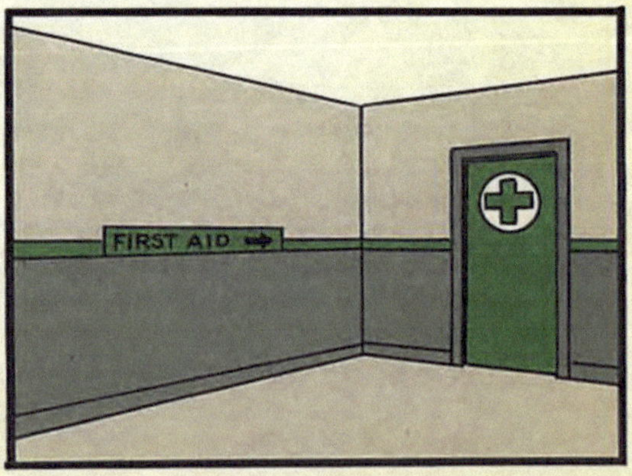

Signs relating to safety, doors to first aid departments, shall be painted No. 16 Bright Green.

SECTION 3–17, MACHINERY COLORS

The painting and finishing of Coast Guard machinery should follow simple principles. Two grays are recommended: No. 30 Equipment Gray and No. 20 Medium Gray. Gray makes an ideal surface. It is easily cleaned and touched up. It is fair in the reflection of light and therefore helps to encourage good machine care and maintenance. Having a neutral tone that resembles the color of metal, it will present a good appearance where the finish itself may be worn away. Better than any decorative color, gray does not look unsightly where it may be chipped or soiled.

3–17–1, USE OF NO. 30 EQUIPMENT GRAY

No. 30 Equipment Gray, being a light tone of gray, is the preferred finish for precision tools and machinery. Its use is recommended where high standards are expected and where a light rather than medium or deep finish will contribute to superior workmanship. It is particularly appropriate for fine tool work, the machinery and equipment in precision machine shops, electronic shops, testing laboratories, emergency power plants and the like.

3–17–2, USE OF NO. 20 MEDIUM GRAY

No. 20 Medium Gray, being a deeper and therefore more practical color than No. 30 Equipment Gray, may be accepted for more or less universal use in machine shops, carpenter shops, engine repair shops, erecting shops, boat repair shops, heat treating and plating shops, foundries, forge shops, blacksmith shops, power houses, engine rooms and compressor rooms. No. 20 Medium Gray may also be considered for engine blocks, frames and supports. Good practice would prescribe its use as an area under machinery on wood and concrete floors to seal the surface, prevent oil absorption and make cleaning easy.

3–17–3, USE OF NO. 22 HIGH-LIGHT BUFF

It is permissible to paint machinery uniformly in No. 30 Equipment Gray or No. 20 Medium Gray. However, there is great advantage in adding No. 22 High-Light Buff to concentrate attention at important points and to reflect more light. Illustrations herewith show the use of the buff color on typical machinery. Normally the buff should be restricted to heads, chucks and surfaces at vital points and in the immediate field of view. It should not be used for decorative purposes, on machined steel surfaces not subject to painting, or on remote parts of a machine which are not significant in its operation. It is usually unnecessary on wheels or levers which a worker may operate while his attention is directed elsewhere.

3–17–4, USE OF SAFETY MARKINGS

No. 29 Bright Blue is prescribed for switch boxes and controls to aid identification and yet be free of distraction. No. 18 International Orange is to be used where cutting and crushing hazards are involved. It should be painted on the guards over saws, along the edges of knives, shears and rollers. This color should also be applied with great care so that its meaning to personnel will be significant. A guard that completely encloses a hazard should, of course, be the same gray as the body of the machine. No. 14 Brilliant Yellow, or black and yellow striping, may be applied to projections, curbings and parts which constitute a strike-against or stumbling hazard. Also see Safety Color Code, Section 3–16.

ENGINE LATHE

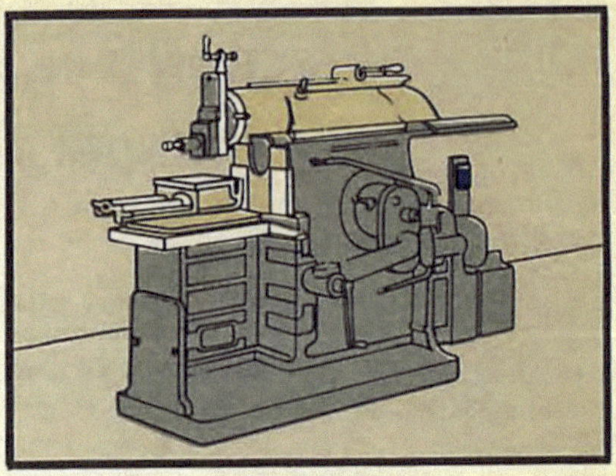

SHAPER

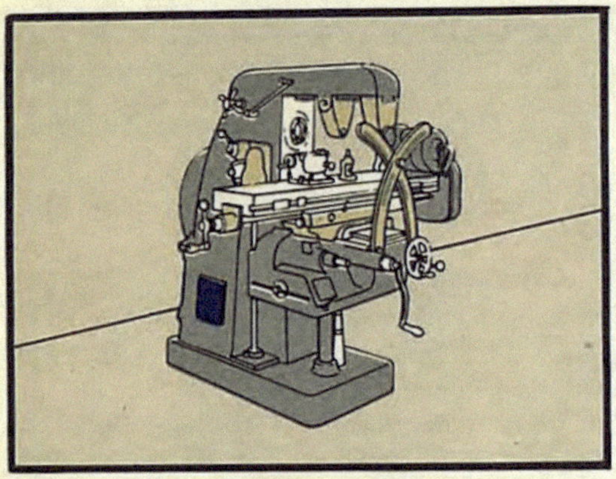

MILLING MACHINE

POWER SAW

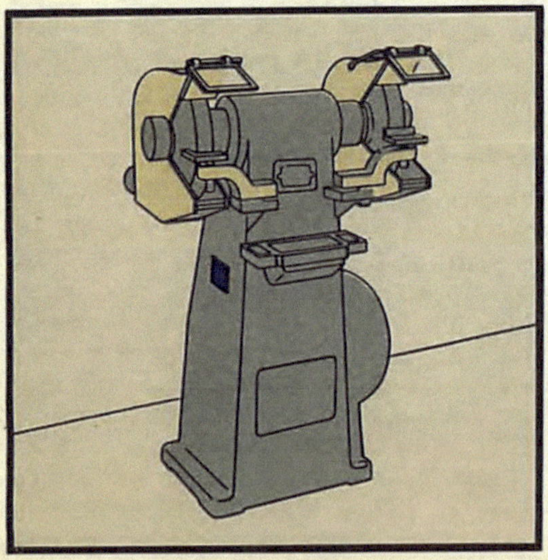

TOOL GRINDER

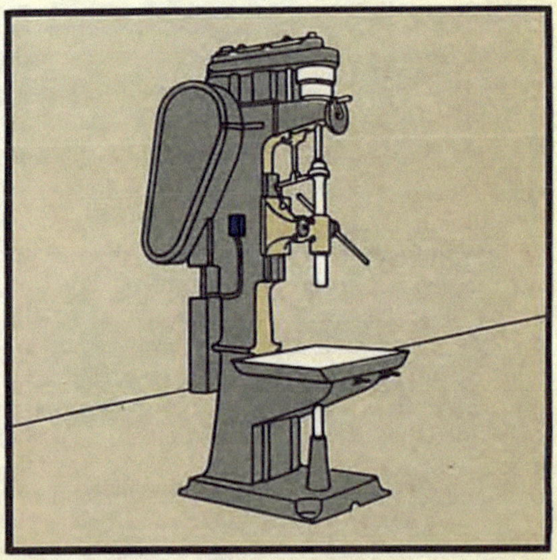

DRILL PRESS

SECTION 3–18, COMPRESSED GAS CYLINDER IDENTIFICATION

3–18–1, COLORS TO BE USED

The following colors which conform with MIL–STD–101, Color Code for Compressed Gas Cylinders and Pipelines, are to be used and lie within the tolerance limits of the code:

No. 14 Brilliant Yellow	No. 20 Medium Gray
	No. 29 Bright Blue
No. 18 International Orange	No. 33 Seal Brown
	No. 24 Spar
No. 13 Fire Red	White
No. 16 Bright Green	Black

3–18–2, TITLES

Exact identification of any material contained in a compressed gas cylinder is mandatory and shall be made only by means of the printed title. The title shall appear in two locations diametrically opposite and parallel to the longitudinal axis of the cylinder.

On cylinders 4 in. in diameter and larger the title shall be in approximately 2-in. high letters. On cylinders less than 4 in. in diameter they may be reduced in size.

Cylinders having a background color of yellow, orange, or spar shall have the title painted black. Cylinders having background colors of red, brown, black, blue, gray or green shall have the title painted white.

3–18–3, CYLINDER COLORS

The appearance on the body, top, or as a band on compressed gas cylinders of any of the following six colors shall provide a warning of danger from the hazard involved in handling the type of material contained in the cylinder.

Class	*Color*	*Class of materials*
1	Yellow	Flammable materials
2	Brown	Toxic and poisonous materials
3	Blue	Anesthetics and harmful materials
4	Green	Oxidizing materials
5	Gray	Physically dangerous materials
6	Red	Fire protection materials

The colors of the main portion of the body of cylinders are selected to group the materials conveniently and to simplify the code.

Extra cylinder color bands are to be used on the cylinder body. They serve as color warnings when yellow, brown, blue, green, or gray, and they provide color combinations to separate and distinguish cylinders for convenience in handling, storing, and shipping.

All cylinders shall be painted in accordance with the following table. Cylinders most frequently encountered in the Coast Guard are illustrated herewith.

The bottom and the lower portion of the cylinder body opposite the valve end may be used for commercial identification on cylinders not government owned. In this area, which shall not exceed one-sixth of the overall length of the cylinder, the use of a solid color other than the body color will not be permitted.

3–18–4, DECALCOMANIAS

Two decalcomanias may be applied on the shoulder of each cylinder and diametrically opposite and at right angles to the titles. They indicate the name of the gas, precautions for handling, and use. The background color of the decalcomania may or may not correspond to the cylinder color. Gas cylinder decalcomanias are listed in Class 51 of the Catalog of Navy Material.

Shatterproof cylinders shall be stenciled with the phrase NON-SHAT longitudinally and 90° from titles. Letters shall be black or white and approximately 1 in. in size.

TABLE OF COMPRESSED GAS CYLINDER COLORS

	Location on Cylinder			
	Top A	*Band B*	*Band C*	*Body*
MEDICAL ANESTHETIC GASES:				
Cyclopropane	Orange	Yellow	Blue	Blue
Ethylene	Yellow	Blue	Blue	Blue
Nitrous oxide	Blue	Blue	Blue	Blue
FUEL GASES:				
Acetylene	Yellow	Yellow	Yellow	Yellow
Hydrogen	Yellow	Black	Yellow	Yellow
Manufactured gases	Brown	Yellow	Yellow	Yellow
Petroleum (nonliquefied)	Yellow	White	Yellow	Yellow
Petroleum (liquefied)	Yellow	Orange	Yellow	Yellow
INDUSTRIAL GASES:				
Butadiene	Yellow	White	Spar	Spar
Ethylene oxide	Yellow	Blue	Spar	Spar
Ethyl chloride	Spar	Blue	Yellow	Spar
Propylene	Yellow	Gray	Spar	Spar
Vinyl chloride	Yellow	Orange	Spar	Spar
Vinyl methyl ether	Yellow	Black	Spar	Spar
Aerosol insecticide	Spar	Spar	Spar	Spar
Carboxide	Spar	Blue	Spar	Spar
TOXICS AND POISONOUS MATERIALS:				
Carbon monoxide	Yellow	Brown	Brown	Brown
Hydrogen sulfide	Brown	Yellow	Brown	Brown
Methyl bromide	Brown	Black	Brown	Brown
Boron trifluoride	Gray	Brown	Brown	Brown
Chlorine	Brown	Brown	Brown	Brown
Hydrogen chloride	Brown	White	Brown	Brown
Phosgene	Brown	Orange	Brown	Brown
Sulfur dioxide	Brown	Gray	Brown	Brown

	Location on Cylinder			
	Top A	*Band B*	*Band C*	*Body*
REFRIGERANTS:				
Ammonia	Brown	Yellow	Orange	Orange
Freons	Orange	Orange	Orange	Orange
Methyl chloride	Yellow	Brown	Orange	Orange
OXIDIZING GASES:				
Oxygen	Green	Green	Green	Green
Oxygen, aviator's	Green	White	Green	Green
Air, oil pumped	Black	Green	Green	Black
Air, water pumped	Black	Green	Black	Black
Helium, oxygen	White	Green	Black	Black
Oxygen, carbon dioxide	Gray	Green	Black	Black
INERT GASES:				
Argon, oil pumped	Gray	White	White	Gray
Argon, water pumped	Gray	White	Gray	Gray
Carbon dioxide	Gray	Gray	Gray	Gray

TABLE OF COMPRESSED GAS CYLINDER COLORS

INERT GASES:				
Helium, oil pumped	Gray	Orange	Gray	Gray
Helium, oil free	Gray	Orange	Orange	Gray
Nitrogen, oil pumped	Gray	Black	Gray	Gray
Nitrogen, water pumped	Gray	Black	Black	Gray
FIRE FIGHTING GASES:				
Carbon dioxide	Red	Red	Red	Red
Methyl bromide	Red	Brown	Red	Red

INSTRUCTIONS FOR PAINTING COMPRESSED GAS CYLINDERS

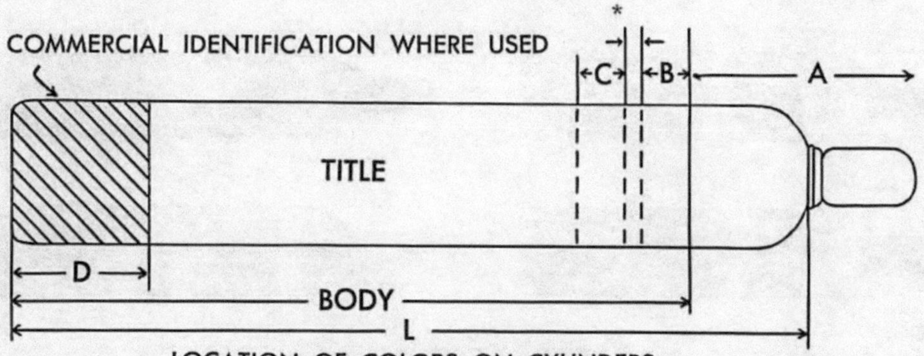

LOCATION OF COLORS ON CYLINDERS

LOCATION DIMENSIONS

OVERALL LENGTH	CAP AND SHOULDER COLOR	CYLINDER COLOR BAND(S)	COMMERCIAL IDENTIFICATION
L	A	B&C	D
OVER 30"	¼ OF (L+CAP)	3"	⅙ OF L
30" AND UNDER	¼ OF (L+CAP)	2"	⅙ OF L

*1" SPACE TO BE OMITTED IF BANDS B & C ARE OF DIFFERENT COLORS

ACETYLENE
PETROLEUM
LIQUEFIED
HYDROGEN
FREON
AMMONIA
METHYL CHLORIDE
OXYGEN
CARBON DIOXIDE
METHYL BROMIDE
OXYGEN
OXYGEN
AVIATOR'S
AIR
WATER PUMPED
AIR
OIL PUMPED
CARBON DIOXIDE
COMMERCIAL
CARBON DIOXIDE
FIRE FIGHTING
HELIUM
OIL PUMPED
NITROGEN
OIL PUMPED
NITROUS OXIDE
AEROSOL INSECTICIDE

SECTION 3–19, PIPING IDENTIFICATION

3–19–1, PIPING SYSTEM COLORS

Except as required below, all piping, including valve bodies, shall be painted to match adjacent bulkheads, walls, or overheads. Packing glands, valve stems, threads and similar working surfaces shall not be painted.

Gas and gasoline piping shall be painted No. 14 Brilliant Yellow throughout the piping system. Valves (except the threaded parts, valve stems and other working surfaces) shall be similarly painted.

Oxygen piping including valves (except the threaded parts, valve stems and other working surfaces) shall be No. 16 Bright Green.

Firemain valves, magazine sprinkler valves and fire hose connections (except threaded parts, valve stems and other working parts) shall be painted No. 13 Fire Red. The remainder of firemain and sprinkler main piping system shall be painted to match the surrounding areas. Aluminum sprinkler piping in magazines shall not be painted. Steel valve handles shall be painted No. 30 Equipment Gray unless otherwise specified.

Identification of piping by means of color bands is not authorized for Coast Guard use.

3–19–2, PIPING SYSTEM MARKINGS

Exact identification of piping to indicate contents or type of service is mandatory and shall be made only by means of stencilled titles. Titles shall be applied with black lettering ¾ in. high, except on pipe under 1½ in. O.D. on which the lettering shall be reduced proportionately to present a good appearance. Piping and tubing integral with machinery which are of such small sizes as to make lettering impractical need not be included in this requirement.

An arrow shall appear adjacent to each piping identification to indicate the normal direction(s) of flow of material in the system. Dimensions of the arrows shall be as shown herewith.

It is recommended that where the view is unobstructed, titles be stencilled on the two lower quarters of the pipe or covering. Lettering in this position is unlikely to be obscured by dirt or mechanical damage. In any case titles shall be clearly visible from operating positions, especially those adjacent to control valves. Each pipe line shall be identified at least once in each compartment.

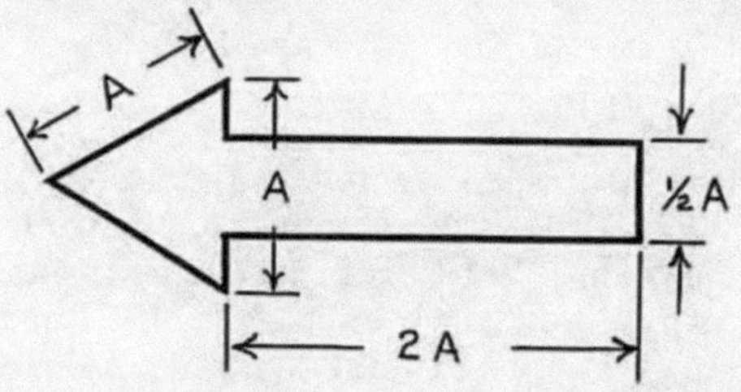

A = approximately ¾ of outside diameter of pipe or covering. (1½ in. maximum.)

PIPING IDENTIFICATION

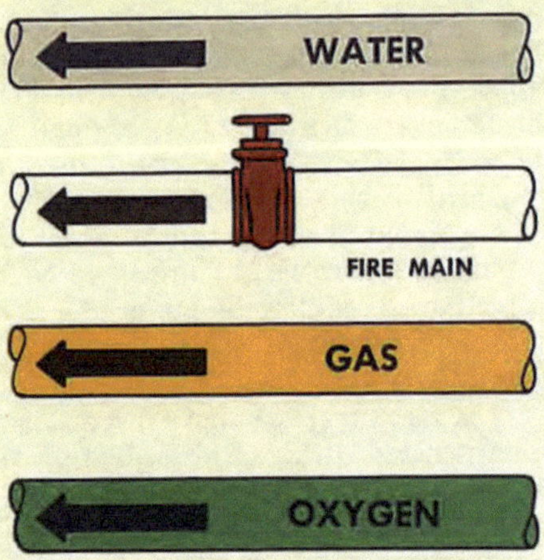

MISCELLANEOUS PIPING — PLAIN IDENTIFICATION ONLY. COLOR TO MATCH WALL OR OVERHEAD:

FIRE LINES SHOULD BE RED, OR WHITE WITH RED APPLIED TO VALVES AND CONNECTIONS.

GASOLINE AND FLAMMABLE GAS LINES SHOULD BE NO. 14. BRILLIANT YELLOW

OXYGEN LINES SHOULD BE NO. 16 BRIGHT GREEN.

SECTION 3–20, SIGNS

3–20–1, HIGHWAY SIGNS

Highway signs are to follow the standards established by the Manual on Uniform Traffic Control Devices for Streets and Highways, published by the Public Roads Administration, Federal Works Agency, Washington 25, D. C. In general, STOP signs shall be octagonal in shape and lettered with black on No. 14 Brilliant Yellow. Railroad crossing signs are circular in shape and are also black on yellow. Warnings signs referring to hills, bumps, narrow roads, schools, and arrows indicating road turns, crossroads, etc. shall be diamond shaped and lettered with black on No. 14 Brilliant Yellow.

Regulatory and informational signs referring to speed limits, traffic direction, also guide signs and route markers, shall have black lettering on a white background. Parking prohibition signs shall be rectangular and have No. 13 Fire Red lettering on white; limited parking signs shall be the same shape and have No. 16 Bright Green lettering on white.

No. 14 Brilliant Yellow shall be used for curb marking where parking is prohibited. Yellow stripes are used as regulatory lines over which it is unsafe or illegal to travel. White road striping is used for traffic control, lane marking and stop lines. Road hazards shall be indicated with black and white stripes.

Insofar as practicable, highway guide signs shall be used for shore units that are located less than twenty miles off main highways. These signs shall be 6 in. x 36 in. in size. Lettering shall be of a height suitable for the legend used and shall conform to the Standard Alphabet shown in Section 3–22, Letters and Numerals.

3–20–2, STANDARD STOCK SAFETY AND WARNING SIGNS

A number of standard safety and warning signs are stocked at Coast Guard Supply

Centers. These signs are manufactured to the specifications of the American Standards Association and are approved for Coast Guard use. The specifications for these signs are given below:

(A) DANGER SIGNS

Danger signs shall have a white background. The word DANGER shall appear in white letters on a red oval inside a black rectangular panel. Sign wording shall be black letters on the white background. Danger signs shall be used only where an immediate hazard exists.

(B) CAUTION SIGNS

Caution signs shall have a yellow background. Word CAUTION shall appear in yellow letters on black rectangular panel. Sign wording shall be in black letters on the yellow background. Such signs shall be used only to warn against potential hazards or to caution against unsafe practices.

(C) DIRECTIONAL SIGNS

Directional signs shall have a white background. Arrow shall be in white on black panel. Wording in arrow or below panel shall be in black. Used to indicate way to locations: exits, fire escapes, stairways, first aid rooms, etc.

(D) INFORMATIONAL SIGNS

Informational signs shall be white lettering on No. 29 Bright Blue background. Other detail lettering may be black on white. These signs convey information not necessarily of a safety nature. Operating instructions and signs which tend to avoid confusion and misunderstanding fall in this category.

(E) SAFETY INSTRUCTION SIGNS

Safety Instruction signs shall have a white or green background. If such words as THINK or SAFETY FIRST are used across the top, they should be in white letters on green rectangular panel. Other sign wording shall be in black letters on white background. These signs shall be used for general instructions and suggestions relative to safety measures.

A full description of these standard signs with the appropriate stock numbers and prices may be found in Class 42–S of the Coast Guard Supplement to the Catalog of Navy Material. Check the current catalog Bulletin for recent changes to sign stocks.

Locally made signs may follow the American Standards Association specifications given above or the below simplified specifications which are a modification of the ASA specifications.

3–20–3, MISCELLANEOUS SIGNS

No. 13 Fire Red, with white lettering, shall be used for fire regulation signs, regulations regarding cigarettes and matches, warning signs as to explosives and flammable liquids.

No. 14 Brilliant Yellow, with black lettering, shall be used for danger and warning signs, such as KEEP AWAY, MEN WORKING, also for floor load and clearance signs and for all such cautioning instructions.

White on No. 16 Bright Green shall be used for safety instruction signs, regulations concerning goggles, first aid equipment, the location of first aid stations.

Simple directional signs used for traffic direction, office and department designations, names, titles, shall be black on white.

White on No. 29 Bright Blue (or No. 29 Bright Blue on white) may be used for signs not covered above. This would refer in general to operating instruction signs, announcement and general information signs and to signs referring to miscellaneous current events.

All lettering on locally made signs shall conform to the Coast Guard Standard Alphabet shown in paragraph 3–22, Letters and Numerals.

3–20–4, CONTAMINATION SIGNS

Contaminated areas resulting from martial contamination (mine fields, gassed areas, biological and radioactive areas) are to be marked by a *triangular* sign (right angle isosceles triangle) mounted with the apex (right angle) downward. The sign shall meet the following specifications:

(a) Dimensions (see illustration).

(b) Colors:

(1) Surface of sign facing toward contam-

Regulatory and information signs shall be black on white

Stop signs shall be octagonal, with black on yellow

Warning signs shall be diamond-shaped, with black on yellow

Above illustrations show typical stock safety and warning signs

Above illustrations show miscellaneous signs described in text

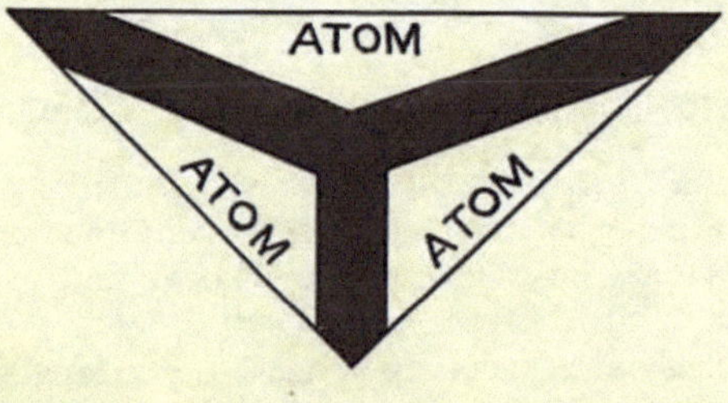

Contamination sign for radioactive area

ination or danger (inward side) shall be white. (2) Surface of sign facing away from contamination or danger (outward side) shall have a color scheme dependent upon the type of contamination as indicated below.

Mine fields, No. 13 Fire Red.

Gas, No. 14 Brilliant Yellow.

Biological, No. 14 Brilliant Yellow trisecting lines on No. 13 Fire Red background

Radioactive, Black trisecting lines on white background (illustrated).

(c) Lettering:

(1) The words MINES, GAS, GERMS or ATOM, as appropriate, are to be lettered in black along the three sides of the triangle on the outward side of the signs as shown in the illustration. (2) Details of the contamination or danger and the date shall appear on the white inward side of the sign.

Contaminated areas resulting from non-martial contaminates (refuse, garbage, etc.) shall be *rectangular* in shape, may be of any convenient dimensions, may have any background color other than used for martial contamination signs, and may be given any appropriate lettering.

3–20–5, SIGNS AND MARKERS FOR INACTIVE UNITS

Distinguishing Coast Guard signs and markers designed to catch the eye shall be removed from units which have been disestablished and whose names have been removed from OPFAC (CG–244). Standard warning notices should be posted on such property. These notices are available in vitreous enamel signs at Jersey City and Alameda Supply Centers as follows:

Stock No.	*Size – Inches*	*Wording*
CG–42–S–6615	9 x 20	Station Closed.
CG–42–S–6630	10 x 8	Warning. All persons are warned not to trespass on this property or to injure or disturb any property of the U. S. Coast Guard. All violations will be prosecuted.

SECTION 3–21, SHOP HELMETS

3–21–1, HELMET COLORS

At all Coast Guard units where shop helmets (hard hats) are worn, the color symbolism described below shall be applied. This color system, which is a modification of the Navy system, assures a rapid means of identifying personnel as to rating and shop in which employed.

Four types of color schemes shall be followed, as illustrated. That is, (A) the entire helmet shall be one solid color; (B) the crown and the brim shall have different colors; (C) the crown shall be divided with a different color on each side, but with a solid-colored brim; (D) the crown and brim shall be divided with a different color on each side. The complete color scheme is given in the accompanying table.

3–21–2, HELMET MARKING AT THE COAST GUARD YARD

In addition to the color schemes which identify the shops, the Yard shall apply the following classification markings.

(A) OFFICERS AND WARRANT OFFICERS

The helmet shall be all white. The appropriate commissioned or warrant officers cap device in gold or bronze decalcomania shall be placed on the front of the helmet. Below this insignia the officer's rank (abbreviated) and name followed by the department to which assigned (abbreviated) shall appear in ½ in. black letters.

(B) MASTERS AND FOREMEN

The helmet shall be all white. The masters' or foremen's insignia (gold or bronze decalcomania for masters—silver for foremen) shall be placed on the front of the helmet. Below this insignia the name, rating or shop shall appear in ½ in. black letters.

(C) QUARTERMEN AND OTHER IVa SUPERVISORS

Helmets shall be painted in accordance with the color scheme designated for the shop

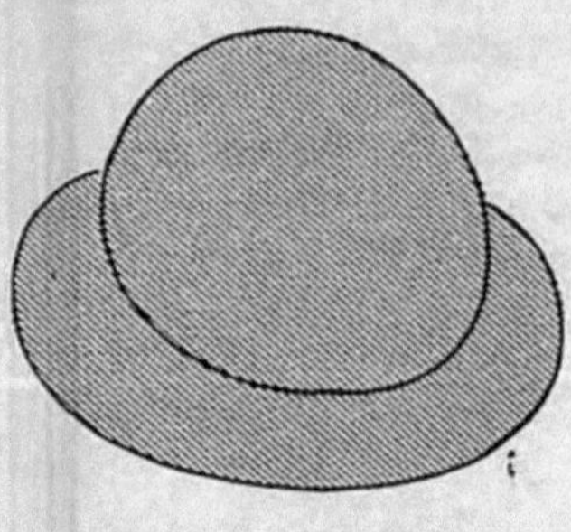

Color Scheme A

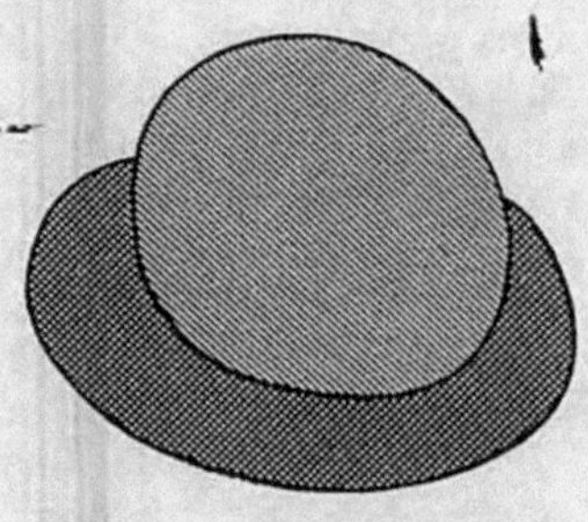

Color Scheme B

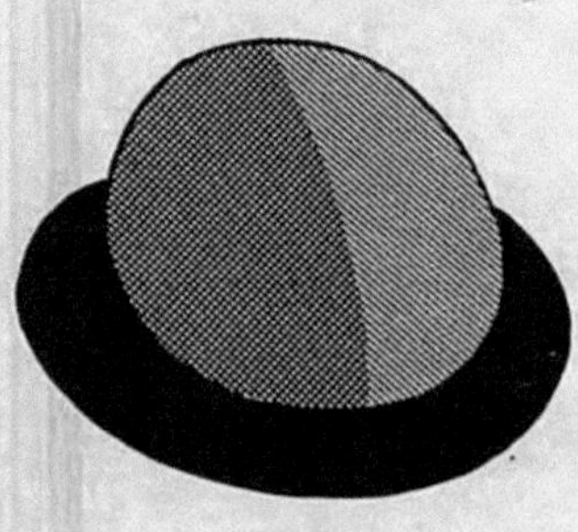

Color Scheme C

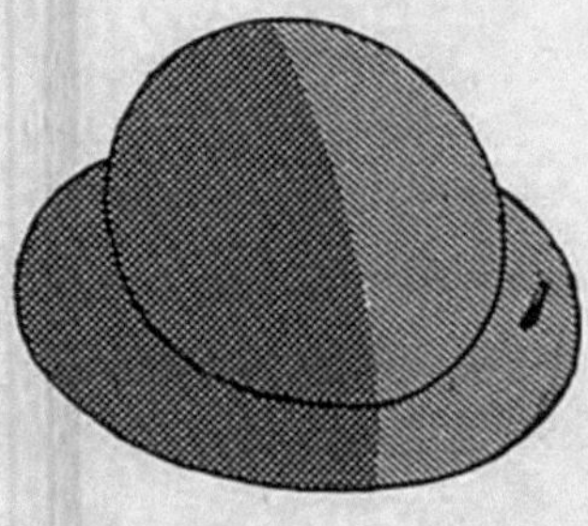

Color Scheme D

with an additional 2-in. stripe painted fore and aft from edge of brim to edge of brim. Chief Quartermen or Quartermen insignia (decalcomania) shall be placed on the front of the hat centered in the white stripe. Below this insignia the supervisor's name and rating shall be painted in ½-in. black letters.

(D) IVb EMPLOYEES AND IVb SUPERVISORS

Helmets shall be painted in accordance with the color scheme designated for the shop. The name of the activity, abbreviated as indicated below and in the case of supervisors, their title, such as Chief Planner. Lettering shall be in a contrasting color and ½ in. in height.

C.O. OFF.	Commanding Officer's Office
PROD.	Production Department
PLAN.	Planning Department, including Electronics and Ordnance offices
I.R.D.	Industrial Relations Department (Except Safety Division)
P.W.	Public Works Department
FIN.	Comptroller's Department
SUP.	Supply Department
ADMIN.	Administrative Department (except as hereinafter noted)
MED.	Medical and Dental Division, Administrative Department
F.D.	Fire Protection

(E) SHOP EMPLOYEES

Helmets shall be painted in accordance with the color scheme designated for the shop. In addition shop numbers, 2 in. in height, shall be painted on each side of the helmet in a contrasting color.

(F) APPRENTICES

In addition to the colors and markings prescribed for shop employees, the letter A in a contrasting color 2 in. in height shall appear on the front of the helmet.

COAST GUARD SHOP HELMET COLOR SCHEMES

Shop No. and Name	*Workman Position, Title*	*Color Scheme*	*Helmet Color: Crown*	*Helmet Color: Brim*
		Production Department		
11, Structural	Shipfitter; Boilermaker; Gas Burner; Calker & Chipper; Blacksmith	D	No. 6 Ivory No. 10 Light Blue	No. 6 Ivory No. 10 Light Blue
17, Sheet Metal	Sheet Metal Worker	D	No. 19 Light Gray No. 18 Inter. Orange	No. 19 Light Gray No. 18 Inter. Orange
26, Welding	Welders (Gas and Electric)	C	No. 6 Ivory No. 10 Light Blue	No. 16 Bright Green
31, Machine (Inside)	Machinist (inside); Machine Operator	A	No. 1 Light Green	No. 1 Light Green
35, Instrument	Instrument Maker	D	No. 1 Light Green No. 16 Bright Green	No. 1 Light Green No. 16 Bright Green
36, Ordnance	Machinist (Outside); Electrician	B	No. 1 Light Green	No. 18 Inter. Orange

COAST GUARD SHOP HELMET COLOR SCHEMES

Shop No. and Name	*Workman Position, Title*	*Color Scheme*	*Helmet Color Crown*	*Brim*
		Production Department		
38. Machine (Outside)	Machinist (Outside)	B	No. 1 Light Green	No. 16 Bright Green
39, Internal Combustion Eng.	Diesel Equipment Mechanic	B	No. 1 Light Green	No. 29 Bright Blue
51, Electric	Electrician	A	No. 18 Inter. Orange	No. 18 Inter. Orange
53, Pipe	Pipefitter	A	No. 8 Terra Cotta	No. 8 Terra Cotta
54, Electronics	Electronics Mechanic; Radio Mechanic	B	No. 18 Inter. Orange	No. 16 Bright Green
61, Shipwright	Shipwright	A	No. 29 Bright Blue	No. 29 Bright Blue
63, Joiner	Joiner; Carpenter	B	No. 29 Bright Blue	No. 16 Bright Green
68, Boat	Boatbuilder	B	No. 29 Bright Blue	No. 14 Brilliant Yellow
71, Paint	Painter; Sandblaster	D	No. 19 Light Gray No. 29 Bright Blue	No. 19 Light Gray No. 29 Bright Blue
72, Rigging Loft	Rigger	A	No. 14 Brilliant Yellow	No. 14 Brilliant Yellow
	Rigger Laborer	B	No. 14 Brilliant Yellow	No. 16 Bright Green
74, Sail Loft	Sailmaker; Upholsterer	A	No. 19 Light Gray	No. 19 Light Gray
85, Foundry	Foundryman; Molder	C	No. 1 Light Green	No. 18 Inter. Orange
94, Pattern	Patternmaker	B	No. 29 Bright Blue	No. 1 Light Green
95, Drydock	Shipwright; Electrician	B	No. 29 Bright Blue	No. 18 Inter. Orange
Production Office	Progressman	D	No. 16 Bright Green No. 19 Light Gray	No. 16 Bright Green No. 19 Light Gray
		Public Works Department		
Transportation (03)	Operator, automotive equipment (small portable crane); Chauffeur; Truck Driver; Engineman, hoisting & portable (Crane Operator); Auto Mechanic	B	No. 22 High-Light Buff	No. 18 Inter. Orange

COAST GUARD SHOP HELMET COLOR SCHEMES

Shop No. and Name	*Workman Position, Title*	*Color Scheme*	*Helmet Color Crown*	*Brim*
			Public Works Department	
Maintenance Shops (98) (b)	Maintenance Mechanic	A	No. 22 High-Light Buff	No. 22 High-Light Buff
Outside Maintenance and Janitors (98)	Laborer; Janitor; Gardener	B	No. 22 High-Light Buff	No. 29 Bright Blue
			Miscellaneous Administrative Groups	
Supply Department (01)	Property & Supply Clerk; Storekeeper; Warehouse Laborer; Operator, high lift truck	A	No. 4 Soft Yellow	No. 4 Soft Yellow
Planning Department (04)	Planner & Estimator; Engineer (design); Engineering Draftsman	D	No. 19 Light Gray No. 14 Brilliant Yellow	No. 19 Light Gray No. 14 Brilliant Yellow
Comptroller's Office	Property & Supply Clerk	B	No. 4 Soft Yellow	No. 16 Bright Green
Administrative Department (except Medical); Industrial Relations (except Safety), and all other offices		D	No. 16 Bright Green No. 19 Light Gray	No. 16 Bright Green No. 19 Light Gray
Safety	Safety Engineer	A	No. 16 Bright Green with white crosses	No. 16 Bright Green
Medical		A	White, with No. 16 Bright Green cross on each side	White
Fire Department		A	No. 13 Fire Red	No. 13 Fire Red

SECTION 3–22, LETTERS AND NUMERALS

3–22–1, STANDARD LETTERING SYSTEM

A standard lettering system as illustrated herewith has been adopted for Coast Guard use. This system was selected after considerable study. The testing of various styles of characters indicated that there is relatively little gain in legibility between the basic block lettering systems and so called high legibility systems. The standard alphabet and figures are known as rounded block letters. This system is legible, attractive and easily executed. The Standard Lettering System shall be used for the lettering of signs, ships' name and numbers, vehicles, buoys and elsewhere, unless another system is prescribed. Buoy numerals will be changed over to the standard system when stocks or present style numeral and letter stencils have been exhausted.

The characters are formed by drawing a grid 8 units in height, laying down the straight lines on the grid, then forming the curves by striking arcs to connect the straight lines. The large radius arcs have a radius of 2 units, while the small radius arcs have a radius of 3/4 unit. Stroke width of all characters is 1 1/4 units.

The spacing between letters in a word shall conform to the table included herewith. To determine the correct spacing between two letters find the first letter under the column titled First Letter, proceed horizontally to the right until the second letter is found. The column heading for the second letter indicates the spacing in terms of grid units.

The spacing between characters used in buoy identification, hull numbers, light vessel names and in similar cases where better than average legibility is desired shall be equal to 2 1/2 units. The spacing between words shall be 5 units.

3–22–2, DRAFT NUMERALS

Draft numerals shall conform to the system illustrated separately. Numerals shall always be 6 in. in height and all horizontal lines shall lay on even inches of numeral height. Draft numbers shall be placed on the hull so that the bottom of the draft mark rests at the draft it represents. When draft marks are placed on non-vertical surfaces the characters shall be projected onto the surface from a vertical plane.

3–22–3, AERIAL IDENTIFICATION AND AIRCRAFT LETTERING

Aerial identification of all types whether on vessels, communication trucks, ground markers, airfields or elsewhere shall follow the system specified in Civil Aeronautics Administration Bulletin No. 12, Air Marking. This bulletin may be procured from the Inquiry Branch, Civil Aeronautics Administration, Washington 25, D. C. This system of identification has been adopted as a uniform standard by the International Civil Aviation Organization and is also used by the Air Force and Navy.

Aircraft markings shall conform to those used by the Navy and are described in the current edition of Bureau of Aeronautics Specification 2R–2–Insignia, Markings and Color of Naval Aircraft.

STANDARD ALPHABET

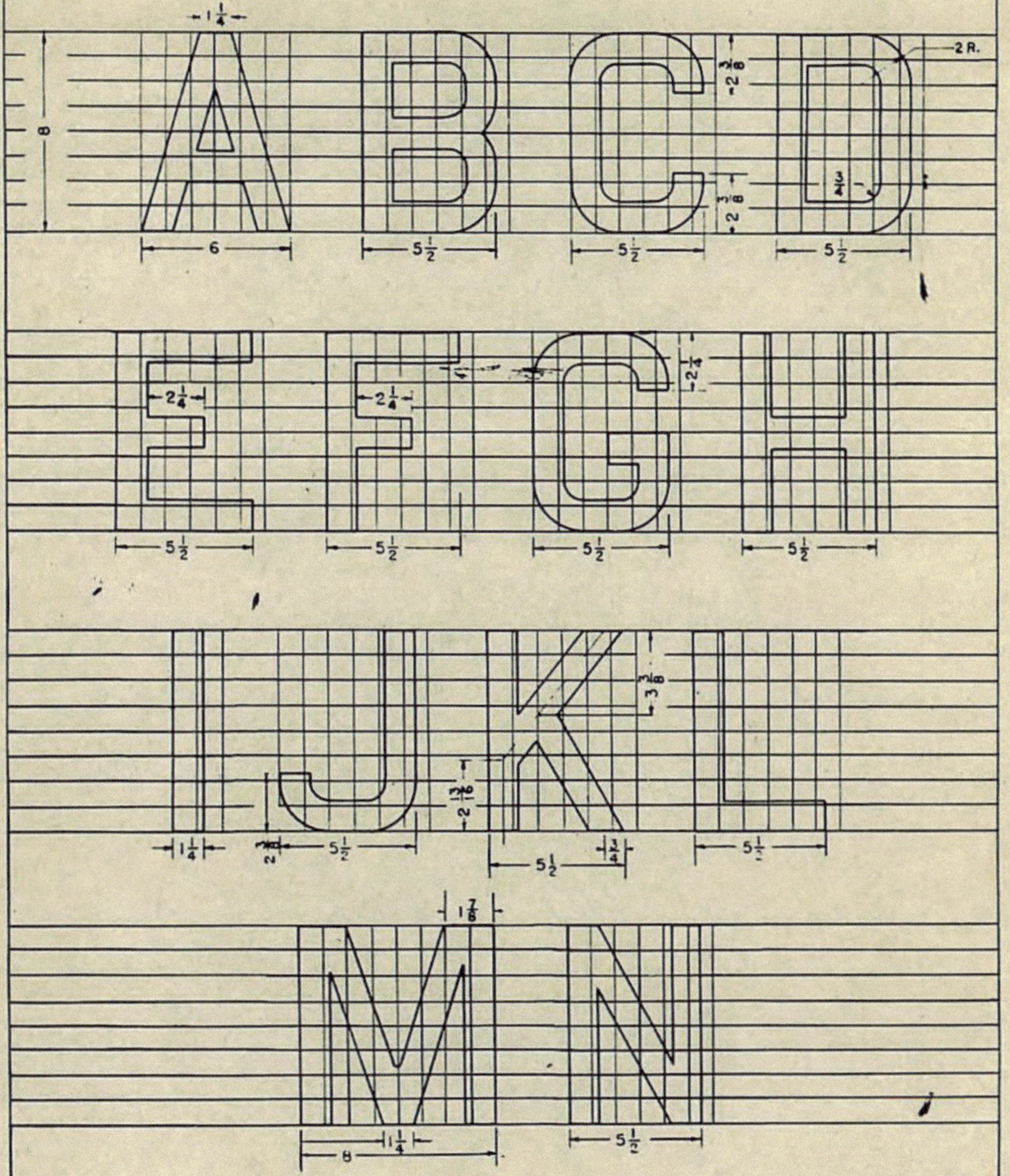

STANDARD ALPHABET

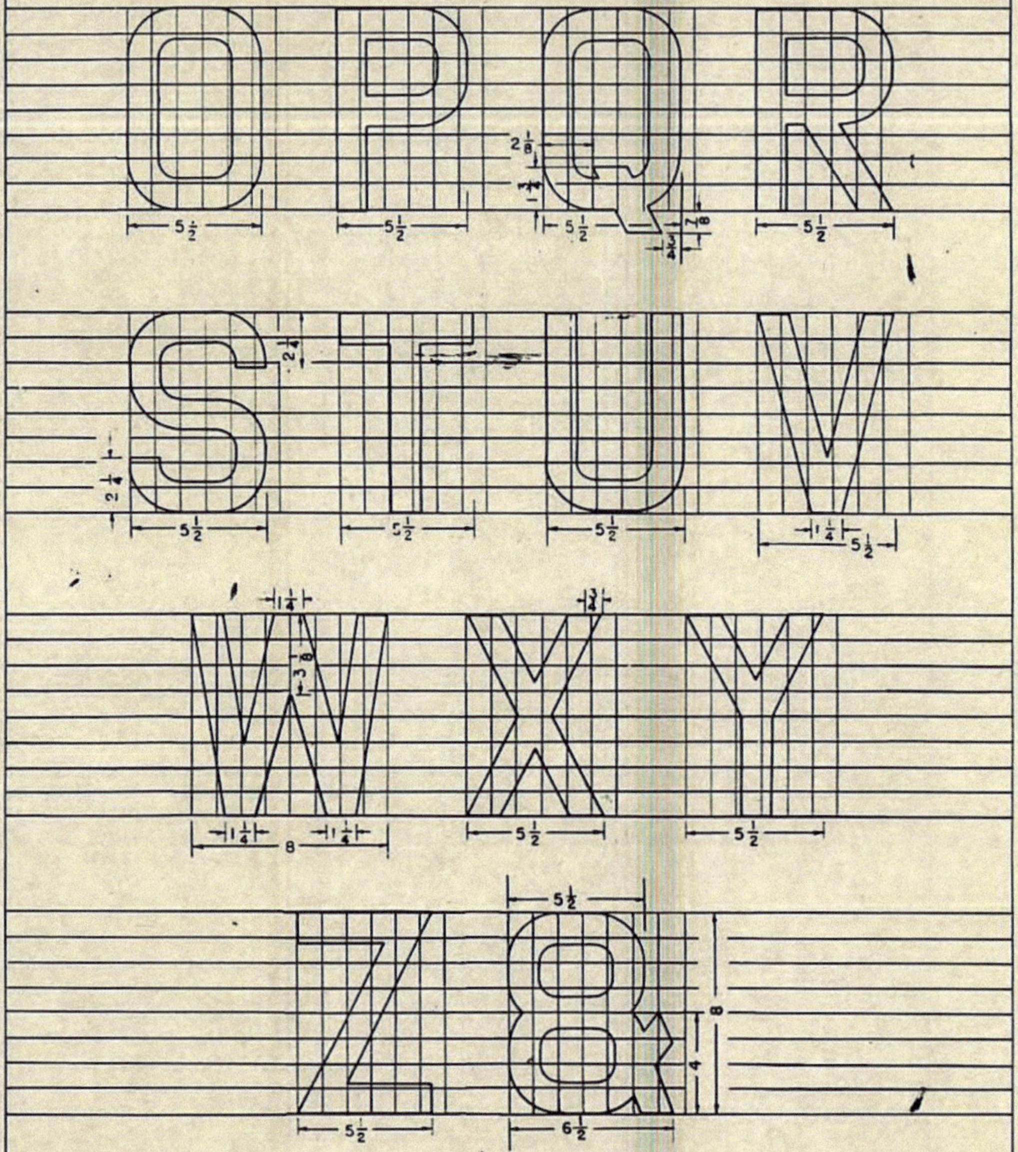

VISUAL CALL FIGURES

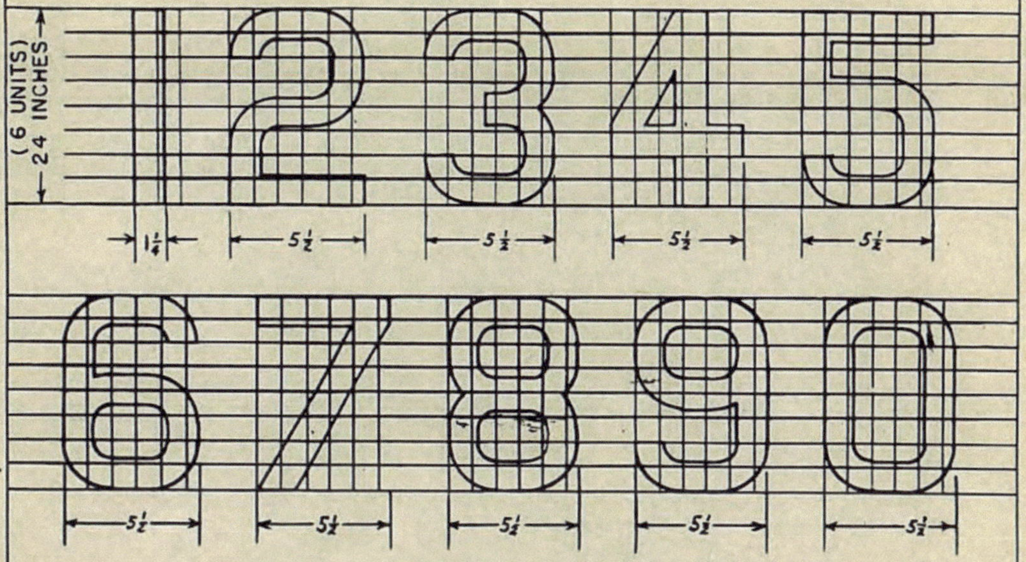

DRAFT FIGURES

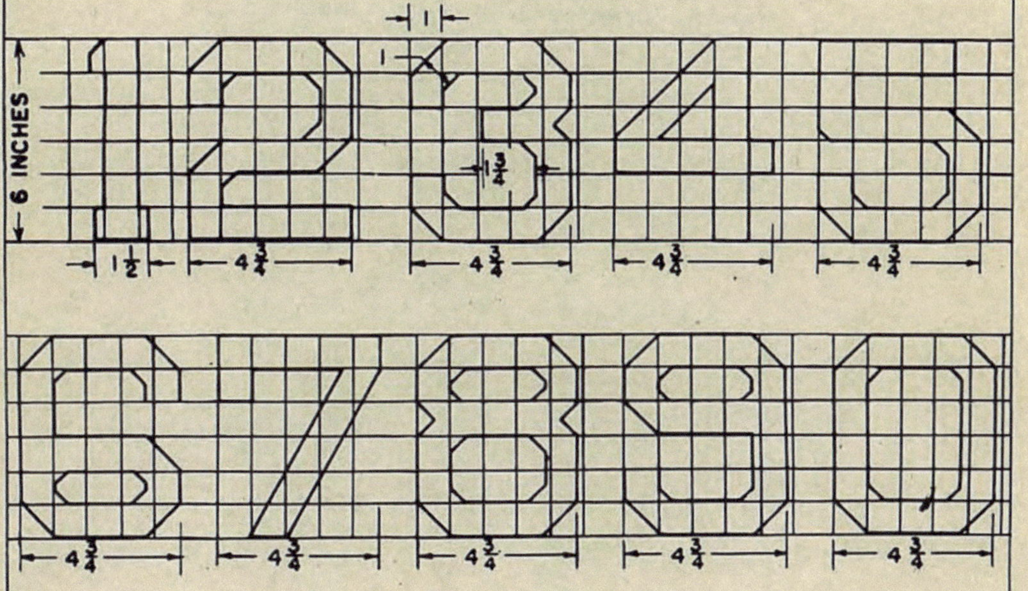

SPACING KEY

FIRST LETTER	NUMBER OF SPACES 2½	2¼	2	1¾	1½	1¼	1	¾	½	0
A					BDEFHI KLMNPR	U	CGOQSZ	AX	JTW	VY
BCDGOQRS			BDEFHI KLMNPR	CGOQSU	Z	X	AJTWY	V		
E				BDEFHI KLMNPR	CGOQSU	Z	X	AJTWY	V	
F						BDEFHI KLMNPR	U	CGOQST VWXYZ		AJ
HIMN	BDEFHI KLMNPR	U	CGOQSZ	X	AJTWY	V				
JU		BDEFHI KLMNPR	U	CGOQSZ	X	AJTWY	V			
K				BDEFHI KLMNPR	U	CGOQSZ	X	AJTWY	V	
L						BDEFHI KLMNPR	U	ACGOQS XZ	J	TVWY
P					BDEFHI KLMNPR	U	CGOQSZ	TVWXY	AJ	
TW					BDEFHI KLMNPR	U	CGOQSZ	TVWXY	A	J
V						BDEFHI KLMNPR	U	CGOQST VWXYZ		AJ
X				BDEFHI KLMNPR	U	CGOQSZ	X	AJTVW Y		
Y					BDEFHI KLMNPR	U	CGOQSZ	TVWXY		AJ
Z			BDEFHI KLMNPR	U	CGOQSZ	X	ATVWY	J		

NOTES

1. CORNER RADII: 2 UNITS TO OUTSIDE, ¾ UNITS TO INSIDE
2. 45° CORNERS OF DRAFT FIGURES CUT BACK ONE INCH
3. STROKE OF ALL LETTERS 1¼ UNITS WIDE
4. FIVE UNITS SHALL BE LEFT BETWEEN WORDS
5. VISUAL CALL FIGURES WILL HAVE TWO UNIT SPACING

SECTION 3–23, MISCELLANEOUS COLOR STANDARDS

The following references and publications, all relating to the subject of color, are included here. Copies may be secured as noted. These standards should be studied, checked, and applied where necessary.

3–23–1, AIDS TO NAVIGATION REGULATIONS

Coast Guard publication CG–208 contains descriptions, illustrations and color standards of various aids to navigation, such as buoys, intercoastal waterway markings, lighting requirements for bridges, etc. This publication may be obtained from the Commandant (CHS).

3–23–2, AIRCRAFT FLUID LINES, IDENTIFICATION OF

Current editions of AND 10375 and AN 8022, Army-Navy Aeronautical Design Standards, give color bands to be applied to fluid lines to aid identification, assembly and repairs. These design standards may be obtained from the Aircraft Specifications Branch, Technical Data Division, Bureau of Aeronautics, Navy Department, Washington 25, D. C.

3–23–3, AMMUNITION, MARKING OF

Current edition of Ordnance Pamphlet No. 4 (Restricted), Ammunition, Instructions for the Naval Service, gives general instructions for the handling of ammunition. References are made to colors for the marking of all types of projectiles, explosives, and boxes containing ammunition. Ordnance pamphlets are obtainable from the Commandant (OGR).

3–23–4, AIDS TO MARINE NAVIGATION OF THE UNITED STATES

Special booklet issued by Coast Guard, CG–193, describes buoyage maintained in the navigable waters of the United States and its possessions. It explains briefly the significance of the various colors of lighthouses and buoys, their markings and their shapes. This publication is obtainable from the Commandant (CHS).

3–23–5, LIGHT LIST, COAST GUARD

Light Lists, published annually by the U. S. Coast Guard and obtainable from the Commandant (CHS) contain color requirements for light towers, structures, vessels and buoys maintained as aids to navigation.

3–23–6, RADIATION HAZARDS

Various publications are available on the application of color symbols to the hazards of radioactive materials. A purple symbol on a yellow base is to be used as a warning where radioactive material is stored. Shipping containers for radioactive materials take a variety of color symbols depending upon the classification of the material. For further information on this subject write to Commandant (CHS).

3–23–7, CANISTERS

RBA Canisters: Approved standard canisters for Rescue Breathing Apparatus are painted gray and have instructions stencilled on the side with yellow paint. A yellow stripe approximately ½ in. by 3½ in. and additional instructions are also stencilled on the concave side of the approved canister.

Miscellaneous Canisters: A national standard is recognized for the marking of canisters used with gas masks. Colors for canisters indicate protection against the following:

White	Low concentrations of acid gases such as chlorine, formic acid, hydrogen sulfide, nitrogen peroxide, sulfur dioxide, etc.
Black	Low concentrations of organic vapors such as aniline, benzine, toluene, gasoline, etc.
Green	Ammonia gas.

Blue	Carbon Monoxide.
Yellow	Combinations of organic and acid fumes.
Brown	Acid and organic vapors and ammonia.
Black and White stripes	Dust, smoke, and mists in combination with any of the foregoing gases.
Red	Mixtures of the foregoing gases in air.

The colors used in the Manual may be employed, using No. 8 Terra Cotta where brown is required. For more detail and a description of materials to be used in canisters, reference may be made to Respirators, Gas Masks, Hose Masks and Breathing Apparatus, Safe Practices Pamphlet No. 64, published by the National Safety Council, 425 No. Michigan Avenue; Chicago 11, Ill.

3–23–8, ELECTRIC CABLES, SHIPBOARD, MARKING OF

Standards have been set up for the coding of electrical cables used on ships. A description of the symbolism and application of the colors is given in the current issue of MIL–C–915; Cables, Electric (Shipboard use). This publication may be obtained from U.S. Navy Purchasing Office, Main Navy Building, Code SPF–IA, 17th and Constitution Avenue, Washington 25, D. C.

3–23–9, ELECTRICAL INSULATION

Military Standard MIL–STD–104, 27 June 1951, describes light and dark limits for electrical insulation colors. This publication is obtainable from U.S. Navy Purchasing Office, Main Navy Building, Code SPF–IA, 17th and Constitution Avenue, Washington 25, D. C.

3–23–10, METAL BARS, INGOTS, WELDING RODS, ETC.

Color markings must be in accordance with the current edition of the Catalog of Navy Material, General Stores Section, Class 46, obtainable from Bureau of Supplies and Accounts, Navy Department, Washington 25, D. C.

3–23–11, PIPE, TUBES, TUBING (STOCK)

Color markings must be in accordance with current edition of the Catalog of Navy Material, General Stores Section, Class 44, obtainable from Bureau of Supplies and Accounts, Navy Department, Washington 25, D. C.

3–23–12, COLORS: (FOR) READY-MIXED PAINTS, FEDERAL SPECIFICATION

Publication TT–C–595, 12 January 1950, serves as a basis for practically all of the colors described and specified in this Manual. It is approved by the Commissioner, Federal Supply Service, General Services Administration, for the use of all Federal Agencies. This publication, which contains color standards for matching, may be used for reference by any Coast Guard supply sources. Copies may be purchased from the Superintendent of Documents, U. S. Government Printing Office, Washington 25, D. C. Price $4.50.

SECTION 3-24, COAST GUARD COLOR SPECIFICATIONS

The color standards shown on Charts I and II are for shade of color only, not for paint finish or paint quality. The tabulation below includes: Coast Guard numbers and designations; Federal specification numbers as given in publication TT-C-595, 12 January 1950 (see Article 3-23-12); and average light reflectances. It will be noted that no Federal specifications are given for No. 10 Light Blue and No. 32 Blue Gray, both of which have been standardized since the publication of the above reference. No color chip is presented for No. 34 Emblem Blue. The printed color shown on the cover of this Manual may be used as a standard. To avoid fading, keep the chips from excessive exposure to light.

Coast Guard Number and Name	*Federal Specification* TT-C-595	*Light Reflectance*
1. LIGHT GREEN	3555	55%
2. MEDIUM GREEN	3540	26%
4. SOFT YELLOW	2325	63%
6. IVORY	3360	63%
~~8. TERRA COTTA~~	1730	22%
9. PEARL GRAY	3635	50%
~~10. LIGHT BLUE~~	——	54%
11. SUN TAN	1335	58%
12. SPRUCE GREEN	1450	14%
13. FIRE RED	1110	8%
14. BRILLIANT YELLOW	1310	53%
16. BRIGHT GREEN	1475	21%
18. INTERNATIONAL ORANGE	1205	16%
19. LIGHT GRAY	2635	29%
20. MEDIUM GRAY	1615	16%
22. HIGH-LIGHT BUFF	1750	56%
24. SPAR	1735	35%
25. BEACH SAND	2330	65%
26. TILE RED	1020	5%
29. BRIGHT BLUE	1525	10%
30. EQUIPMENT GRAY	1640	41%
31. DECK GREEN	2435	16%
32. BLUE GRAY	——	9%
33. SEAL BROWN	1710	4%
34. EMBLEM BLUE	——	3%

13. FIRE RED Federal 1110	18. INTERNATIONAL ORANGE Federal 1205	14. BRILLIANT YELLOW Federal 1310
16. BRIGHT GREEN Federal 1475	29. BRIGHT BLUE Federal 1525	22. HIGH-LIGHT BUFF Federal 1750
30. EQUIPMENT GRAY Federal 1640	19. LIGHT GRAY Federal 2635	20. MEDIUM GRAY Federal 1615
31. DECK GREEN Federal 2435	32. BLUE GRAY	33. SEAL BROWN Federal 1710

COAST GUARD COLOR STANDARDS

These samples are for shade only, not paint finish or paint quality

CHART II

CHAPTER 4, MATERIALS

TABLE OF CONTENTS

SECTION 4–1, INTRODUCTION

Previous chapters give the names of the materials, colors of materials and numbers of coats of each material to be used on the various surfaces encountered in the Coast Guard. Chapter 4 gives the information necessary to procure the correct materials for the job.

Materials are grouped into certain broad categories for quick reference: i.e., Primers, Fillers, Sealers, and Preservatives; Paints and Enamels; Varnishes and Lacquers; Thinners; Compounds and Cements; Special Materials. Individual items are located by consulting the proper category in the Table of Contents. Where there is no material specified for a particular use, materials specified for similar uses will normally be found satisfactory.

Each item is described by a brief statement of its principal uses in the Coast Guard, general information on applying the paint, reference to all the articles in the Manual wherein the use of the item is prescribed, and the information necessary to order the paint. It should be noted that the name of the paint as used in the Manual, which is generally descriptive of its use in the Coast Guard, does not always correspond to the name of the paint as listed under Ordering Information. The reason for this is that the name given under Ordering Information is the name appearing in the Catalog of Navy Material or Coast Guard Supplement thereto. These names are descriptive of the use of the material in the Navy which is frequently different from the use of the material in the Coast Guard.

The information given under General Application Instructions should be read in conjunction with the appropriate articles listed under Specific Application Instructions in order to arrive at the number of coats of each material which comprise the coating system for the particular surface under consideration. Attention must be given to the thinners recommended in order to prevent adding a material which would be incompatible with the paint. In addition, the drying times listed are minimum times based on a temperature of 70° F.; lower temperatures will require longer time to obtain thoroughly dried paint films.

It should be noted that some of the stock numbers are preceded by the letters CG whereas others are not. The prefix CG indicates that the material is available through Coast Guard supply channels (and in some cases the Federal Supply Service channels). Where a stock number does not carry the prefix CG the material is available through both Coast Guard and Navy supply channels (and in some instances Federal Supply Service).

When ordering materials sufficient time should be allowed for the normal interval between submission of the requisition and receipt of materials. Additional time will be necessary for materials which are small volume items, since supply depots will not normally carry these materials on the shelf but will procure the material from their Navy supply support activity or Coast Guard Supply Centers only upon receipt of a requisition for the material. Since Stock Status Records assume that each requisition represents an amount of material to be used within a certain period and will result in a recurring demand, it is essential that when materials are requisitioned in quantities or for purposes that do not represent a recurring demand, the requisition clearly indicated the difference between those items which will be recurring and those which will not.

Unexpected delays in procurement should be investigated to determine the cause. Inadequate ordering information can be supplemented and delays avoided by adequate follow-up measures. In no case should overstocking be resorted to by units as a means of counteracting procurement delays. Only in very unusual circumstances is it permissible to procure more material than will be consumed by the unit in four months, since this practice creates further shortages in the supply system which subsequently results in over-stocking by Supply Depots and Centers due to false demand figures. Materials being temporarily out of stock at Supply Activities should not lead to excess stocks at units.

The inclusion of an individual paint material in the Paint and Color Manual shall not be construed as a necessity for stocking the item at Supply Depots or individual units. Provisions for stocking the most widely used items have been made. Stocks of additional items will be established on a basis of actual requests on the part of Supply Depots and units.

Paints should be used within 1 year of the date of manufacture stenciled on the container. Normally, paints should not be more than 6 months old when received by the unit. Inspectors' reports have shown paints to be in stock at certain units which were as much as 5 years old. Any paint in stock at a unit which is over a year old suggests that the item was over-stocked. Paints 2 years old or more indicate a disregard of painting instructions and supply procedures.

Paints are fairly stable in storage with respect to the degree of fire hazard which they present. However, upon prolonged storage in the order of 2 or more years, there is a possibility that the degree of fire hazard presented by the material may increase through chemical reaction between paint and container, chemical reaction within the paint itself resulting in pressure build-up in the container or through rusting of containers resulting in leaks. Another very real hazard in over-stocking is the fact that excessive quantities of combustible material are being stored.

All paints deteriorate in storage. The rate of deterioration is dependent on several factors. Therefore, no definite shelf life can be determined for a particular paint. However, every effort should be made to apply paint within 1 year of manufacture. Many paint failures have been found to be directly or indirectly the result of prolonged storage of the paint.

SECTION 4–2, PRIMERS, FILLERS, SEALERS AND PRESERVATIVES

4–2–1, PRETREATMENT-WASH PRIMER

USE: As a pretreatment on bare metal to produce a corrosion-resisting surface and provide mechanical adhesion between the metal and subsequent coats of paint. It is especially valuable in obtaining good paint bond to aluminum, brass, copper, and cadmium-plated and galvanized surfaces.

GENERAL APPLICATION INSTRUCTIONS

Mixing: Pretreatment-Wash Primer requires the mixing of 2 components before use, in the ratio of 4 volumes of resin component to 1 volume of acid component. The resin component should be first mixed thoroughly and the acid component slowly poured in while stirring. After addition of acid component, mix thoroughly. Mix only enough for use within an 8-hour period after mixing. While no visible change in the mixture may be apparent 8 hours after mixing, poor adhesion will result if the Pretreatment-Wash Primer is used over 8 hours after mixing. When mixed the pretreatment should have viscosity approximately the same as thin shellac.

Application: Pretreatment-Wash Primer may be applied by brush, spray, dipping or swabbing. It may be applied at any temperature above 5° F. Special care is needed to insure that painting equipment is thoroughly clean, since alcohols used in this material will dislodge hardened particles of conventional paints. Brushes or spray equipment should be cleaned before and after use with denatured ethyl alcohol. A difference in pressure between spray gun and tank should be 20-30 lbs. Because of rapid solvent evaporation, spray gun must be kept within 8-10 in. of the work. The coating must be applied so as to give a thoroughly wet appearance to the freshly sprayed surface. Since only 1 coat is to be applied, it should be applied liberally to insure thorough coverage and adequate film thickness. Solid hiding is not to be expected. With brush application apply liberally using brush to rapidly spread material. Do not attempt to brush out this material nor to eliminate overlaps or sags. Once the pretreatment has dried the next material should be applied as soon as possible to avoid contamination of the clean surface by salt spray, oils, and other contaminants. However, no adverse effect will result from allowing up to 30 days before application of subsequent coats, provided the surface is thoroughly cleaned to remove all contaminants.

Drying: Drying time before recoating is about 15 minutes at 60° F. Lower temperatures may require a longer drying time.

Thinning: If mixed in the proper proportions, thinner will not normally be needed. Thin if necessary and clean brushes and spray equipment with denatured ethyl alcohol, isopropyl alcohol, or butyl alcohol.

PRECAUTIONS: Pretreatment-Wash Primer is a flammable mixture presenting a fire and explosive hazard of the same order as shellac or other material containing ethyl alcohol. Insure positive and ample ventilation when mixing. Application on interior surfaces such as tanks and voids requires supervision to insure adequate ventilation and to exclude possible flames and sparks. Flash point 50° F.

SPECIFIC APPLICATION INSTRUCTIONS

Article 2–2–2(A), Exterior Ungalvanized Steel (Shore Establishments)

Article 2–2–2(B), Exterior Galvanized Steel and Corrosion-Resisting Metals other than Aluminum (Shore Establishments)

Article 2–2–2(C), Exterior Aluminum (Shore Establishments)

Article 2–2–2(D), Interior Ungalvanized Steel (Shore Establishments)

Article 2–2–2(E), Interior Galvanized Steel and Corrosion-Resisting Metals (Shore Establishments)

Article 2–2–2(F), Gutters, Downspouts and Flashing

Article 2–2–2(G), Furniture and Equipment (Shore Establishments)
Article 2–2–2(H), Machinery (Shore Establishments)
Article 2–2–2(N), Interior Piping (Shore Establishments)
Article 2–2–8, Interior of Refrigeration Boxes (Shore Establishments)
Article 2–2–11, Surfaces Requiring Acid Resistance (Shore Establishments)
Article 2–3–1(A), Protective Coatings During Construction, Steel Vessels
Article 2–3–2(A), Bottom Painting, Steel Vessels in Salt Water
Article 2–3–2(B), Bottom Painting, Steel Vessels in Fresh Water
Article 2–3–2(C), Bottom Painting, Steel 40-ft. Utility Boats
Article 2–3–3(A), Boot-top Painting, Steel Vessels in Salt Water
Article 2–3–3(B), Boot-top Painting, Steel Vessels in Fresh Water
Article 2–3–3(C), Boot-top Painting, Steel 40-ft. Utility Boats
Article 2–3–4(A), Exterior Ungalvanized Steel (Vessels)
Article 2–3–4(B), Exterior Galvanized Steel and Corrosion-Resisting Metals other than Aluminum (Vessels)
Article 2–3–4(C), Exterior Aluminum (Vessels)
Article 2–3–5(A), Interior Metal Surfaces Not Subject to Moisture (Vessels)
Article 2–3–5(B), Interior Metal Surfaces Subject to Moisture (Vessels)
Article 2–3–6, Decks and Floor Plates (Vessels)
Article 2–3–10, Voids and Cofferdams (Vessels)
Article 2–3–12, Bilges
Article 2–3–14, Spaces Subject to Heavy Sweating (Vessels)
Article 2–3–16, Piping (Vessels)
Article 2–3–17, Machinery (Vessels)
Article 2–3–18, Furniture and Equipment (Vessels)
Article 2–3–19, Electric Cable (Vessels)
Article 2–3–20, Radio and Radar Antennas (Vessels)
Article 2–3–21, Anchor Chain
Article 2–3–23, Propeller Shafting
Article 2–3–30, Sonar Domes and Transducers
Article 2–3–32, Surfaces Requiring Acid Resistance (Vessels)
Article 2–4–1, Exterior of Metal Buoys
Article 2–5–2, Repainting of Vehicles

ORDERING INFORMATION

Catalog Name: Pretreatment-Wash Primer
Specification: MIL–P–15328, Formula 117

Size Container	*Stock Number*
1 gal.	CG–52–P–20649–135
5 gals.	52–P–20649–150

4–2–2, QUICK DRYING RED LEAD PRIMER

USE: The standard steel primer for general Coast Guard use. Composed of red lead and zinc chromate. A high quality, heavy duty primer suitable for interior or exterior use under severe exposure conditions. When used over Pretreatment-Wash Primer this material is equal to Slow Drying Red Lead in all respects without the disadvantage of long drying periods.

GENERAL APPLICATION INSTRUCTIONS

Apply 1 liberal coat of Pretreatment-Wash Primer. Follow with appropriate number of coats of Quick Drying Red Lead Primer.

Drying: Sets in 2 hours. Dries hard for recoating in 6 to 8 hours.

Thinning: This primer is of brushing consistency as received. For spraying, thin with Paint Thinner.

SPECIFIC APPLICATION INSTRUCTIONS

Article 2–2–2(A), Exterior Ungalvanized Steel (Shore Establishments)
Article 2–2–2(B), Exterior Galvanized Steel and Corrosion-Resisting Metals other than Aluminum (Shore Establishments)
Article 2–2–2(D), Interior Ungalvanized Steel (Shore Establishments)
Article 2–2–2(F), Gutters, Downspouts and Flashing
Article 2–2–2(G), Furniture and Equipment (Shore Establishments)

Article 2–2–2(H), Machinery (Shore Establishments)
Article 2–2–2(N), Interior Piping (Shore Establishments)
Article 2–2–5(B), Metal Roofs
Article 2–2–8, Interior of Refrigeration Boxes (Shore Establishments)
Article 2–3–1(A), Protective Coatings During Construction, Steel Vessels
Article 2–3–3, Boot-top Painting
Article 2–3–4, Exterior of Vessels Above The Boot-top—General
Article 2–3–5, Interior of Vessels—General
Article 2–3–10, Voids and Cofferdams
Article 2–3–12, Bilges
Article 2–3–13, Inaccessible Spaces
Article 2–3–14, Spaces Subject to Heavy Sweating (Vessels)
Article 2–3–16, Piping (Vessels)
Article 2–3–17, Machinery (Vessels)
Article 2–3–18, Furniture and Equipment (Vessels)
Article 2–3–23(B), Propeller Shafting, Inboard
Article 2–5–2, Repainting of Vehicles

ORDERING INFORMATION

Catalog Name: Maintenance Exterior Primer

Specification: Formula 116

Color	*Size Container*	*Stock Number*
Red	1 gal.	52–P–20488–100
	5 gals.	52–P–20488–110

4–2–3, VINYL RED LEAD PRIMER

USE: This paint is intended for use as an anticorrosive coating only over a surface treated with Pretreatment-Wash Primer. It is not to be used over conventional paints nor as a substitute for red lead in conventional paint systems. When used over Pretreatment-Wash Primer, it is suitable for surfaces above and below water. Most common types of paint may be successfully applied over this paint. It is not a finish coat. It is designed for application of a finish coat of Vinyl-Alkyd Paint or Vinyl Antifouling Paint.

GENERAL APPLICATION INSTRUCTIONS

This paint may be applied at any temperature above 5° F. The surface must be dry. It may be applied over old well-adhering vinyl paint but not over conventional paints. This paint should never be applied to bare metal. Apply 1 liberal coat of Pretreatment-Wash Primer and follow with sufficient coats (usually 2 coats) of Vinyl Red Lead to give a dry film thickness of 4.0 mils. Use orange color for initial and final coats and brown color for alternate coats to eliminate holidays.

Spraying: This paint is designed for spray application. Because of rapid solvent evaporation, the spray gun must be kept within 8 to 10 in. of the work. The objective is to get a paint film that looks wet. Air pressure should be at least 15 psi above the liquid pressure. Each coat should consist of a vertical and a horizontal pass of the spray gun over the area. Special care is needed to insure that spray equipment is thoroughly clean before using this paint, since solvents used will dislodge hardened particles of conventional paints, causing the spray gun to clog frequently. Spray equipment should be cleaned with Vinyl Paint Thinner. Before use, charge equipment with Vinyl Paint Thinner and operate gun until all paint particles are cleared from the lines. Clean equipment immediately after use with Vinyl Paint Thinner.

Brushing: This paint is designed for spray application; however, it may be brushed on if applied properly. Take a full brush and spread it on without attempting to brush it out as with conventional paints. Due to the quick drying properties of this paint, do not attempt to cross work more than once. Stir paint frequently to prevent the pigment from settling.

Drying: Dries for recoating in 30 minutes at temperatures above 60° F. Lower temperatures may require slightly longer drying time.

Thinning: This paint is of spraying consistency as received. Since solvent evaporates rapidly, thinner may be required after the can has been open for some time. In very cold weather, store paint in a warm place 24 hours before using. Thinner may be required to reduce viscosity in cold weather. The proper thinner is Vinyl Paint Thinner. Do not attempt to use other thinners in this paint.

PRECAUTIONS: In mixing or using this paint in an enclosed area, particular attention should be given to the air temperature in the vicinity of application. If the air temperature is below the flash point of the paint (40° F.), the material can be applied with normal ventilation. When the air temperature is above that of the flash point, sufficient ventilation must be provided to insure that the vapor concentration does not exceed the lower limits of explosibility. In addition, usual paint precautions to keep open flames and sparks away from the painting area and the area where paints are mixed must be strictly enforced. Like most paint materials in common use, adequate ventilation is necessary to avoid prolonged breathing of the concentrated paint fumes.

SPECIFIC APPLICATION INSTRUCTIONS

Article 2–2–11, Surfaces Requiring Acid Resistance (Shore Establishments)
Article 2–3–2(C), Bottom Painting, Steel 40-ft. Utility Boats
Article 2–3–32, Surfaces Requiring Acid Resistance (Vessels)
Article 2–4–1, Exterior of Metal Buoys

ORDERING INFORMATION

Catalog Name: Vinyl Red Lead Primer
Specification: CGS–52P–4

Color	*Size Container*	*Stock Number*
Orange	1 gal.	CG–52–P–175–500
Orange	5 gals.	CG–52–P–175–510
Brown	1 gal.	CG–52–P–175–400
Brown	5 gals.	CG–52–P–175–410

4–2–4, SLOW DRYING RED LEAD PRIMER

USE: Commonly referred to as "Old Fashioned Red Lead." Used as an alternate for the Pretreatment-Wash Primer and Quick Drying Red Lead system where time will permit thorough drying of each coat. Since it is seldom practical to allow adequate drying time for this material, Quick Drying Red Lead, which gives equal performance, is recommended.

GENERAL APPLICATION INSTRUCTIONS

Apply directly to clean, dry, bare metal. In damp or cold weather apply 1 liberal coat of Pretreatment-Wash Primer and wait for suitable weather to apply the Red Lead.

Drying: Requires from 36 to 72 hours drying time for each coat. Allow adequate drying time before recoating.

Thinning: This primer is of brushing consistency as received. For spraying, thin with Paint Thinner.

SPECIFIC APPLICATION INSTRUCTIONS

Article 2–4–2, Interior of Metal Buoys

ORDERING INFORMATION

Catalog Name: Heavy-Bodied Red Lead Primer

Specification: Federal TT–P–86, Type I

Color	*Size Container*	*Stock Number*
Red	1 gal.	52–P–5944
Red	5 gals.	52–P–5945

4–2–5, YELLOW ZINC CHROMATE, HIGH ALKYD TYPE

USE: As a primer for aluminum. May be used on steel where a yellow colored primer eliminates need for a color coat, for example, acetylene cylinders.

GENERAL APPLICATION INSTRUCTIONS

Apply 1 liberal coat of Pretreatment-Wash Primer. Follow with appropriate number of coats of Yellow Zinc Chromate.

Drying: Sets in 2 hours. Dries hard for recoating in 6 to 8 hours.

Thinning: Suitable for brush or spray application as received. If necessary, thin with Paint Thinner.

SPECIFIC APPLICATION INSTRUCTIONS

Article 2–2–2(C), Exterior Aluminum (Shore Establishments)
Article 2–2–2(F), Gutters, Downspouts, and Flashing
Article 2–3–4(C), Exterior Aluminum (Vessels)
Article 2–3–19, Electric Cable (Vessels)
Article 2–3–20, Radio and Radar Antennas (Vessels)

ORDERING INFORMATION

Catalog Name: Yellow Zinc Chromate, High Alkyd Type
Specification: JAN–P–735, Formula 84/47

Color	*Size Container*	*Stock Number*
Yellow	1 gal.	52–P–20630–2
	5 gals.	52–P–20635–2

4–2–6, ZINC CHROMATE PRIMER FOR SONAR DOMES

USE: As an anticorrosive coating on sound-transparent windows of sonar domes.

GENERAL APPLICATION INSTRUCTIONS

Apply 1 liberal coat of Pretreatment-Wash Primer. Apply by spray 1 wet coat of Zinc Chromate Primer to deposit a semi-transparent greenish-yellow coat. A hiding coat of full yellow color should be avoided. A wet coat is necessary to secure maximum adhesion and corrosion-inhibiting properties. Follow with 8 spray coats of Antifouling Sonar Dome Paint.

Drying: Dries hard for recoating in 4 hours.

Thinning: For spraying, thin with 2 parts Synthetic Enamel Thinner to 1 part Zinc Chromate Primer.

CAUTION: Zinc Chromate Primer, High Alkyd Type is *not* suitable for this purpose.

SPECIFIC APPLICATION INSTRUCTIONS

Article 2–3–30, Sonar Domes and Transducers

ORDERING INFORMATION

Catalog Name: Zinc Chromate Primer
Specification: MIL–P–6889

Size Container	*Stock Number*
1 gal.	R52–P–20390–4–100

4–2–7, ANTICORROSIVE SHIPBOTTOM PRIMER

USE: As an anticorrosive primer on the underwater hulls of steel vessels in conjunction with Cold Plastic Antifouling Paint.

GENERAL APPLICATION INSTRUCTIONS

Apply 1 liberal coat of Pretreatment-Wash Primer. Apply appropriate number of coats of Anticorrosive Shipbottom Primer. On vessels in salt water, follow with appropriate number of coats of Cold Plastic Antifouling Paint. When applying this primer, use alternate coats of Formula 14N and 14ND. The only difference in these two formulations is one of color. The use of alternate color coats will eliminate holidays and insure adequate coverage.

Drying: Sets in 10 minutes. Dries hard for recoating in 1 hour.

Thinning: This primer is one of brushing and spraying consistency as received. If thinning is necessary, thin with Coal Tar Naphtha.

SPECIFIC APPLICATION INSTRUCTIONS

Article 2–3–2(A), Bottom Painting, Steel Vessels in Salt Water
Article 2–3–2(B), Bottom Painting, Steel Vessels in Fresh Water
Article 2–3–3(A), Boot-top Painting, Steel Vessels in Salt Water

ORDERING INFORMATION

Catalog Name:
Anticorrosive Shipbottom Paint
Specification:
Formulas 14N and 14ND

Formula No.	*Size Container*	*Stock No.*
14N	5 gals.	52–P–6510
14ND	5 gals.	52–P–6540

4–2–8, COPPER NAPHTHENATE WOOD PRESERVATIVE

USE: As a preservative for wood to prevent decay.

GENERAL APPLICATION INSTRUCTIONS

Apply by brush, spray, swab, dipping, soaking or flooding. Allow the wood to absorb as much preservative as possible. Paint, putty, seam compounds, or varnish may be applied over preservative without bleeding.

Drying: Allow preservative to dry 72 hours before applying other material.

Thinning: Thinning not normally needed. Thin if necessary with Paint Thinner.

CAUTION: Flammable. Irritating to the skin. Before using agitate thoroughly. Store and use outdoors or in adequately ventilated rooms.

SPECIFIC APPLICATION INSTRUCTIONS

Article 2–3–1(B), Protective Coatings During Construction, Wooden Vessels
Article 2–3–2(D), Bottom Painting, Wooden Boats
Article 2–3–3(D), Boot-top Painting, Wooden Boats
Article 2–3–4(D), Exterior Wood (Vessels)
Article 2–3–5(C), Interior Wood (Vessels)
Article 2–3–26, Wood Masts and Spars
Article 2–4–4, Wooden Spar Buoys

ORDERING INFORMATION

Catalog Name: Water-Repellent Compound, Wood Preservative, Copper Naphthenate, Type A

Specification: MIL–W–906

Color	*Size Container*	*Stock Number*
Green	1 gal.	CG–52–W–1084–940

4–2–9, PASTE WOOD FILLER

USE: To fill the pores of open-grained woods, including floors, to permit the application of varnish or finish with minimum absorption.

GENERAL APPLICATION INSTRUCTIONS

Thin to brushing consistency and apply to bare wood by brushing across the grain. On stained wood, apply filler after stain is completely dry. Allow the filler to "flatten out," as characterized by a surface drying appearance, after which it is wiped or rubbed into the pores with a pad made of burlap. Clean the excess from the surface with clean rags wiping across the grain.

Drying: After wiping clean, allow the filler to dry for 24 hours before applying finish coats.

Thinning: Thin the paste to brushing consistency with Paint Thinner.

SPECIFIC APPLICATION INSTRUCTIONS

Article 2–2–1(I), Interior Wood Floors
Article 2–2–1(J), Furniture Finishing

ORDERING INFORMATION

Catalog Name: Paste Wood Filler
Specification: Federal TT–F–336

Color	*Size Container*	*Stock Number*
Transparent	1 gal.	52–F–515

4–2–10, FLOOR SEALER

USE: For sealing wood or cork floors to provide a satisfactory foundation for finishing materials such as floor wax.

GENERAL APPLICATION INSTRUCTIONS

Apply 1 thin coat by brush or lamb's wool mop. This sealer seals the pores of the flooring by absorption and leaves no apparent surface film. Allow it to dry and buff it in. "Natural" color.

Drying: Sets to touch in 1 to 3 hours. Dries hard in 7 hours.

Thinning: Thinning is not normally needed. Clean brushes and thin if necessary with Paint Thinner.

SPECIFIC APPLICATION INSTRUCTIONS

Article 2–2–1(I), Interior Wood Floors

ORDERING INFORMATION

Catalog Name: Floor Sealer
Specification: TT–S–176, Class I

Size Container	*Stock Number*
1 gal.	52–S–130

Note: Available from Federal Supply Service under Stock No. 52–S–129–50.

4–2–11, EXTERIOR WOOD PRIMER

USE: As a priming coat on previously unpainted exterior or interior wood. This primer should never be used as a top coat nor be exposed to the weather for more than 2 weeks.

GENERAL APPLICATION INSTRUCTIONS

Unpainted Wood: Spot prime knots and pitch pockets in resinous wood with Ready-Mixed Aluminum Paint. Apply 1 coat of Exterior Wood Primer to clean, dry wood. Follow with number of coats and type of finish material prescribed for particular application.

Previously Painted Wood: Normally priming will not be needed. If old coatings have deteriorated to the extent that enough bare wood is exposed to warrant priming, apply 1 coat of Exterior Wood Primer. Follow with number of coats and type of finish material prescribed for particular application.

Drying: Sets to touch in 8 hours. Dries hard for recoating in 48 hours. Allow adequate drying time before applying top coats.

Thinning: This primer is of brushing consistency as received. If thinning is necessary, thin with Paint Thinner.

SPECIFIC APPLICATION INSTRUCTIONS

Article 2–2–1(A), Exterior Wood, General
Article 2–2–5(A), Wood Roofs

ORDERING INFORMATION

Catalog Name: Exterior Wood White Primer
Specification: Federal TT–P–25

Color	*Size Container*	*Stock Number*
White	1 gal.	CG–52–P–5346–80

4–2–12, INTERIOR WALL PRIMER-SEALER

USE: For sizing interior wall surfaces such as plaster, wallboard, brick, unglazed tile, and concrete to prevent suction spotting of finish coats.

GENERAL APPLICATION INSTRUCTIONS

This is an oil type pigmented wall sealer. New plaster contains much moisture and should not be primed with this paint until the plaster has dried at least 1 month, 6 months if possible. When a plaster surface contains severe suction spots and 1 coat of this primer does not give a uniform appearance, a second coat should be applied, or at least the suction spots touched up with a second coat of the primer before applying the finish coats. Follow with number of coats and type of finish material prescribed for particular surface being painted.

Drying: Sets to touch in 1½ to 4 hours. Dries hard for recoating in 24 hours.

Thinning: This primer is of brushing consistency as received. If thinning is necessary, thin with Paint Thinner.

SPECIFIC APPLICATION INSTRUCTIONS

Article 2–2–3(C), Interior Walls Above Grade
Article 2–2–4, Coating System for Dry Wall Construction

ORDERING INFORMATION

Catalog Name: Interior Wall Primer-Sealer
Specification: Federal TT–P–56

Color	*Size Container*	*Stock Number*
White	1 gal.	52-P-5375

Note: Available from Federal Supply Service under Stock No. 52–P–5377.

4–2–13, INTERIOR WHITE ENAMEL UNDERCOAT

USE: As an undercoat for interior gloss and semi-gloss enamels.

GENERAL APPLICATION INSTRUCTIONS

New, dry plaster, composition wallboard,

and masonry should first be sealed with Interior Wall Primer-Sealer. Apply 1 coat of Undercoat as received. Follow with the number of coats and type of finish material specified for the particular surface.

New Or Previously Unpainted Wood: Apply 1 coat of Undercoat as received. Follow with the number of coats and type of finish material specified for the particular surface being painted.

Metal Surfaces: Prime metal surfaces with the appropriate number of coats of Quick Drying Red Lead Primer (prime aluminum with Yellow Zinc Chromate). Follow with 1 coat of Undercoat and the number of coats and type of finish material specified for the particular surface being painted.

Drying: Sets to touch in ½ to 2 hours. Dries hard for recoating in 7 hours. Allow adequate drying time before applying finish material.

Thinning: This paint is of brushing consistency as received. If thinning is necessary, thin with Paint Thinner.

SPECIFIC APPLICATION INSTRUCTIONS

Article 2-2-1(H), Interior Wood Trim (Shore Establishments)
Article 2-2-2(D), Interior Ungalvanized Steel (Shore Establishments)
Article 2-2-2(E), Interior Galvanized Steel and Corrosion-Resisting Metals (Shore Establishments)
Article 2-2-3(C), Interior Walls Above Grade
Article 2-2-4, Coating System For Dry Wall Construction
Article 2-2-8, Interior of Refrigeration Boxes
Article 2-3-5(C), Interior Wood (Vessels)

ORDERING INFORMATION

Catalog Name: Interior White Enamel Undercoat
Specification: Federal TT-E-543

Color	*Size Container*	*Stock Number*
White	1 gal.	52-E-9600-5

4-2-14, THIN-FILM RUST PREVENTIVE

USE: Grade I: On ferrous and other corrodible metals where exposed to weathering. Resistant to salt spray.

Grade II: On ferrous and other corrodible metals where undercover storage is provided; on interior and on exterior surfaces of all machines and instruments, and on all interiors of gasoline and diesel engines (except water jackets).

GENERAL APPLICATION INSTRUCTIONS

Apply 1 coat by brush or spray gun which will give a film thickness of approximately 2.0 mils. Will not corrode steel, copper, brass or aluminum. Contains no abrasive. Transparent but with sufficient color to be discernible during the protective life of the film. This material may be removed by wiping with rags saturated with kerosene or Paint Thinner. Where Thin-Film Rust Preventive has been used in the lubrication systems of propulsion turbines and gears it shall be removed in accordance with the instructions set forth in BuShips Bulletin of Information No. 39, dated 1 July 1950.

Thinning: Not needed. Clean brushes and spray equipment with Paint Thinner.

Drying: May remain tacky for several months.

CAUTION: When spraying these compounds, the operator and others exposed to the mist shall wear respirators. Adequate means shall be used to entrap the mist near the operation. When brushing or dipping, adequate ventilation shall be maintained to reduce the concentration of the solvent vapor.

SPECIFIC APPLICATION INSTRUCTIONS

Article 2-2-2(M), Interior of Steel Tanks
Article 2-3-9, Lube Oil Tanks and Reservoirs
Article 2-3-13, Inaccessible Spaces
Article 2-3-15, DUKW Rubbing Rails
Article 2-3-25, Standing Rigging

ORDERING INFORMATION

Catalog Name: Thin-Film Rust Preventive
Specification: MIL-C-972

Size Container	*Stock Number*
Grade I: For Surfaces Exposed to Weathering	
1 gal.	52-C-3257
Grade II: For Surfaces where Undercover Storage is Provided	
1 gal.	52-C-3257-30
5 gals.	52-C-3257-38

4-2-15, CHAIN, GEAR AND WIRE-ROPE LUBRICATING GREASE

USE: As a protective coating against corrosion on wire ropes and cables and as a lubricant for exposed gears.

GENERAL APPLICATION INSTRUCTIONS

This grease is usually heated before application but may be applied cold. Apply to clean, dry surfaces.

SPECIFIC APPLICATION INSTRUCTIONS

Article 2-2-13, Surfaces Not To Be Painted (Shore Establishments)
Article 2-3-24, Running Rigging
Article 2-3-33, Surfaces Not To Be Painted (Vessels)

ORDERING INFORMATION

Catalog Name: Chain, Gear, And Wire-Rope Lubricating Grease
Specification: Federal VV-L-751, Type II

Color	*Size Container*	*Stock Number*
Grade A, For Cold Weather	25 lbs.	W14-L-161-25
Grade B, For Warm Weather, Open Air, and Underwater	25 lbs.	W14-L-165
Grade C, For Hot Weather	25 lbs.	W14-L-171-25

SECTION 4–3, PAINTS AND ENAMELS

4–3–1, EXTERIOR OIL PAINT, GENERAL PURPOSE

LUSTER: Semi-Gloss.

USE: As a finish on exterior wood or metal walls, fences, lattices, posts, etc., where a self-washing "house paint" is desired. A general purpose exterior paint to be used for all purposes where no special purpose paint is required. In areas where sulfide fumes are encountered, use Exterior Oil Paint, Fumeproof. For masonry use Masonry Oil Paint.

GENERAL APPLICATION INSTRUCTIONS

Unpainted Wood Surfaces: Prime with 1 coat of Exterior Wood Primer as received. Follow with 2 coats of paint as received.

Unpainted Metal Surfaces: Apply 1 liberal coat of Pretreatment-Wash Primer. Prime surfaces with appropriate number of coats of Quick Drying Red Lead Primer (prime aluminum with Yellow Zinc Chromate Primer). Follow with 2 coats of paint as received.

Previously Painted Surfaces: Priming is not needed. Apply 2 coats of the paint as received.

Drying: Sets to touch in 6 hours and dries hard for recoating in 18 hours. Allow adequate drying time between coats.

Thinning: This paint is of brushing consistency as received. For spraying, thin with Paint Thinner.

SPECIFIC APPLICATION INSTRUCTIONS

Article 2–2–1(A), Exterior Wood, General
Article 2–2–1(F), Previously Stained Wood Shingles and Siding
Article 2–2–2(B), Exterior Galvanized Steel and Corrosion-Resisting Metals Other Than Aluminum (Shore Establishments)

4–3–2, EXTERIOR OIL PAINT, FUMEPROOF

LUSTER: Semi-gloss.

USE: As a finish coat over a suitable primer on exterior wood, metal and concrete where sulfide fumes are encountered such as buildings in certain industrial areas or near sewer outlets. For a general purpose exterior paint use Exterior Oil Paint, General Purpose.

GENERAL APPLICATION INSTRUCTIONS

Unpainted Wood Surfaces: Prime with 1 coat of Exterior Wood Primer as received. Follow with 2 coats of the paint as received.

Unpainted Metal Surfaces: Apply 1 liberal coat of Pretreatment-Wash Primer. Prime surfaces with appropriate number of coats

---

ORDERING INFORMATION

Catalog Name: Exterior Oil Paint, General Purpose
Specification: Federal TT–P–102

Color	*Federal Color No.*	*Size Container*	*Stock Number*
White*		1 gal.	52–P–4505
		5 gals.	52–P–4507
No. 11 Sun Tan	1335	1 gal.	52–P–4515–2500
No. 12 Spruce Green	1450	1 gal.	52–P–4515–1400
No. 9 Pearl Gray	3635	1 gal.	52–P–4515–700

*Note: Also available from Federal Supply Service under same Stock Numbers.

ORDERING INFORMATION

Catalog Name: Exterior Oil Paint, Semi-gloss, Special Fumeproof
Specification: Federal TT-P-103

Color	Federal Color No.	Size Container	Stock Number
White		1 gal.	52-P-4509
No. 9 Pearl Gray	3635	1 gal.	52-P-4508-170
No. 11 Sun Tan	1335	1 gal.	52-P-4508-700

of Quick Drying Red Lead Primer (prime aluminum with Yellow Zinc Chromate Primer). Follow with 2 coats of paint as received.

Previously Painted Surfaces: Priming is not needed. Apply 2 coats of the paint as received.

Drying: Sets to touch in 6 hours and dries hard for recoating in 18 hours. Allow adequate drying time between coats.

Thinning: This paint is of brushing consistency as received. For spraying, thin with Paint Thinner.

SPECIFIC APPLICATION INSTRUCTIONS

Article 2-2-1(A), Exterior Wood, General

4-3-3, EXTERIOR VINYL ALKYD PAINT

LUSTER: Semi-gloss.
USE: Primarily as an exterior topcoat on metal surfaces which have been treated with Pretreatment-Wash Primer and primed with Vinyl Red Lead Primer.

Preliminary reports indicate excellent performance of this material when applied to masonry such as concrete, brick, etc. This material is not specified for such uses but may be so employed. Apply directly to bare masonry. No primer or sealer is necessary. Reports on actual performance are desired.

GENERAL APPLICATION INSTRUCTIONS

This paint may be applied over firmly adhering, age-hardened conventional paints (with the exception of paints based on coal tar and asphalt) on wood and metal surfaces. It will not adhere if applied directly to wood, and may lift conventional paint films recently applied. Best results are achieved only over paints of the vinyl system; therefore it is desirable, on a surface previously painted with other than vinyl paint, to remove the old coatings to bare metal. Apply 1 liberal coat of Pretreatment-Wash Primer to bare metal. Prime the surface with sufficient coats (usually 2 coats) of Vinyl Red Lead to give a dry film thickness of 4.0 mils. Follow with sufficient coats (usually 2 coats) of Vinyl Alkyd Paint to give a dry film thickness of 3.0 mils and a total dry film thickness of 7.5 mils for the system.

Spraying: This paint is designed for spray application. Because of rapid solvent evaporation, the spray gun must be kept within 8 to 10 in. of the work. The objective is to get a paint film that looks wet. Air pressure should be at least 40 psi above the liquid pressure. Each coat should consist of a vertical and a horizontal pass of the spray gun over the area. Special care is needed to insure that spray equipment is thoroughly clean before using this paint, since solvents used will dislodge hardened particles of conventional paints, causing the spray gun to clog frequently. Spray equipment should be cleaned with Vinyl Paint Thinner before use. Before use, charge equipment with Vinyl Paint Thinner and operate gun until all paint particles are cleared from the lines. Clean equipment immediately after use with Vinyl Paint Thinner.

Brushing: This paint is designed for spray application; however, it may be brushed on if applied properly. Take a full brush and spread it on without attempting to brush it out as with conventional paints. Due to the quick drying properties of this paint, do not attempt to cross work more than once. Stir paint frequently to prevent the pigment from settling.

Drying: Dries for recoating in 45 minutes at temperatures above 60° F. Lower temperatures may require slightly longer drying time.

Thinning: This paint is of spraying consistency as received. Since solvent evaporates rapidly, thinner may be required after the can has been open for some time. In very cold weather, store paint in a warm place 24 hours before using. Thinner may be required to reduce viscosity in cold weather. The proper thinner is Vinyl Paint Thinner. Do not attempt to use other thinners in this paint.

PRECAUTIONS: In mixing or using this paint in an enclosed area, particular attention should be given to the air temperature in the vicinity of application. If the air temperature is below the flash point of the paint (60° F.), the material can be applied with normal ventilation. When the air temperature is above that of the flash point, sufficient ventilation must be provided to insure that the vapor concentration does not exceed the lower limits of explosibility. In addition, usual paint precautions to keep open flames and sparks away from the painting area and the area where paints are mixed must be strictly enforced. Like most paint materials in common use, adequate ventilation to avoid prolonged breathing of the concentrated paint fumes is necessary.

SPECIFIC APPLICATION INSTRUCTIONS

Article 2-2-11, Surfaces Requiring Acid Resistance (Shore Establishments)
Article 2-3-32, Surfaces Requiring Acid Resistance (Vessels)
Article 2-4-1, Exterior of Metal Buoys
Article 2-4-2, Interior of Metal Buoys

4-3-4, EXTERIOR GLOSS ENAMEL

LUSTER: Glossy.

USE: On exterior and interior wood or metal where toughness, weather resistance, high gloss and freedom from chalking are desired. The following examples illustrate some of the typical uses of this paint:

- Exterior of vessels above the boot-top
- Automobiles
- Buoys (limited use)
- Signs
- Gas Cylinders
- Striping and Marking Enamel

ORDERING INFORMATION

Catalog Name: Exterior Vinyl Alkyd Paint
Specification: CGS-52P-5

Color	*Federal Color No.*	*Size Container*	*Stock Number*
Black	2710	1 gal.	CG-52-P-5270-50
		5 gals.	CG-52-P-5270-55
White		1 qt.	CG-52-P-5270-250
		1 gal.	CG-52-P-5270-255
		5 gals.	CG-52-P-5270-260
No. 13 Fire Red	1110	1 gal.	CG-52-P-5270-175
		5 gals.	CG-52-P-5270-180
No. 14 Brilliant Yellow	1310	1 qt.	CG-52-P-5270-300
No. 18 International Orange	1205	1 qt.	CG-52-P-5270-100

GENERAL APPLICATION INSTRUCTIONS

Unpainted Wood Surfaces: Prime with 1 coat of Exterior Gloss Enamel thinned by the addition of 1 pt. of Boiled Linseed Oil to each gal. of paint. Follow with 2 coats of the enamel as received.

Unpainted Metal Surfaces: Apply 1 liberal coat of Pretreatment-Wash Primer. Prime surfaces with appropriate number of coats of Quick Drying Red Lead Primer (prime aluminum with Yellow Zinc Chromate Primer). Follow with 2 coats of Exterior Gloss Enamel as received.

Previously Painted Surfaces: Priming not needed. Apply 1 coat enamel as received.

Thinning: For spraying, thin with Paint Thinner. For thinning in excess of 1 part thinner to 4 parts paint, use Synthetic Enamel Thinner in lieu of Paint Thinner.

SPECIFIC APPLICATION INSTRUCTIONS

Article 2-2-1(C), Exterior Wood Signs
Article 2-2-1(E), Creosoted Wood
Article 2-2-2(A), Exterior Ungalvanized Steel (Shore Establishments)
Article 2-2-2(B), Exterior Galvanized Steel and Corrosion-Resisting Metals other than Aluminum (Shore Establishments)
Article 2-2-2(C), Exterior Aluminum (Shore Establishments)
Article 2-2-8, Interior of Refrigeration Boxes

ORDERING INFORMATION

Catalog Name: Exterior Gloss Enamel
Specification: Federal TT-E-489a

Color	*Federal Color No.*	*Size Container*	*Stock Number*
No. 14 Brilliant Yellow	1310	1 qt.	CG-52-E-6022-282*
		1 gal.	52-E-6022-275*
No. 13 Fire Red	1110	1 qt.	CG-52-E-6022-215*
		1 gal.	CG-52-E-7000*
White		1 gal.	52-E-6022-250
		5 gals.	52-E-6022-255
Black	1775	1 gal.	52-E-6022-10
		5 gals.	52-E-6022-25
No. 16 Bright Green	1475	1 qt.	CG-52-E-53699-7525
No. 29 Bright Blue	1525	1 qt.	CG-52-E-52974-7925
No. 22 High-Light Buff	1750	1 qt.	CG-52-E-6022-44
No. 18 International Orange	1205	1 qt.	CG-52-E-6022-155
		1 gal.	52-E-6022-165
No. 33 Seal Brown	1710	1 gal.	52-E-6022-40
No. 24 Spar	1735	1 gal.	CG-52-E-6022-42
No. 19 Light Gray	2635	1 gal.	52-E-6022-81
No. 20 Medium Gray	1615	1 gal.	52-E-53000-500

Note: All items, except those prefixed by CG, are also available from Federal Supply Service under the same Stock Numbers.
*These items obtainable only from Coast Guard Supply. No. 14 Brilliant Yellow and No. 13 Fire Red as stocked by the Navy do not meet Coast Guard color requirements.

Article 2–3–3(B), Boot-top Painting, Steel Vessels in Fresh Water
Article 2–3–3(C), Boot-top Painting, Steel 40 ft. Utility Boats
Article 2–3–3(D), Boot-top Painting, Wooden Boats
Article 2–3–4(A), Exterior Ungalvanized Steel (Vessels)
Article 2–3–4(B), Exterior Galvanized Steel and Corrosion-Resisting Metals other than Aluminum (Vessels)
Article 2–3–4(C), Exterior Aluminum (Vessels)
Article 2–3–4(D), Exterior Wood (Vessels)
Article 2–3–19(B), Electric Cable, Exterior (Vessels)
Article 2–3–20, Radio and Radar Antennas (Vessels)
Article 2–3–26, Wood Masts and Spars
Article 2–4–3, Metal Buoys with High Loss Probability
Article 2–4–4, Wooden Spar Buoys
Article 2–4–5, Color Renewal on Station (Buoys)
Article 2–5–2, Repainting of Vehicles

4–3–5, MASONRY OIL PAINT

LUSTER: Eggshell.

USE: As a finish coat on interior and exterior concrete and masonry walls. For concrete floors use Basement Floor Paint or Deck and Floor Paint as applicable. Preliminary results indicate that Exterior Vinyl Alkyd Paint may be superior to this product for certain applications.

GENERAL APPLICATION INSTRUCTIONS

Unpainted Masonry: The application of oil paint should be deferred until new masonry has had time to dry. This may require from 3 months to 1 year depending on the thickness and porosity of the walls and weather conditions. If immediate painting is necessary, use Cement Powder Paint. Apply oil paint to clean, dry masonry. At least 1 week of dry weather should precede the application of the first coat of oil paint. Since masonry surfaces tend to chill and collect condensed moisture, painting early in the morning and late afternoon should be avoided except in dry climates. Apply 2 coats of Masonry Oil Paint as received.

Previously Painted Masonry: Old coatings of oil paint or Cement Powder Paint in sound condition need not be removed. Peeling, flaking and scaling paints or whitewash should be completely removed. Apply 2 coats of Masonry Oil Paint as received.

Drying: Sets to touch in 6 hours. Dries hard for recoating in 24 hours.

Thinning: This paint is of brushing consistency as received. If thinning is necessary, thin with Paint Thinner.

SPECIFIC APPLICATION INSTRUCTIONS

Article 2–2–3(A), Exterior Walls Above Grade

ORDERING INFORMATION

Catalog Name: Masonry Oil Paint
Specification: Federal TT–P–24

Color	*Federal Color No.*	*Size Container*	*Stock Number*
White		1 gal.	52–P–1274–50
No. 11 Sun Tan	1335	1 gal.	52–P–1249–70
No. 9 Pearl Gray	3635	1 gal.	52–P–1245–100
No. 18 International Orange	1205	1 gal.	CG–52–P–4377–3250
Black		1 gal.	CG–52–P–4377–300

4–3–6, CEMENT POWDER PAINT

LUSTER: Flat.

USE: On interior and exterior porous masonry walls, such as porous concrete, masonry blocks, and rough plaster. If the masonry is porous enough to absorb water, this paint can be successfully applied.

GENERAL APPLICATION INSTRUCTIONS

Cement Powder Paint cannot be applied over other types of paint. Apply this paint to clean, bare, damp masonry. One hour prior to painting, thoroughly wet the walls with a garden hose adjusted to a fine spray. The walls should be wet but not dripping when the paint is applied. In hot weather, redampen slightly just prior to painting. Apply this paint with a short, stiff fiber bristle brush such as an ordinary scrub brush. When 2 coats of paint are prescribed, moisten the dried first coat prior to applying the second coat.

Drying: Allow 24 hours between coats. After painting, it is desirable to sprinkle the freshly painted surface 2 or 3 times a day with a fog spray, such as is used for dampening walls prior to painting, and it is recommended that this be done between coats and for 2 days after the final coat, starting as soon as the paint has set, usually 6 to 12 hours after application.

Thinning: While thick films are to be avoided, there is a tendency to use too much water in cement-water paint and to brush it out too thin. Coatings applied in this manner may look well at first but will generally lose their opacity and protective value much sooner than thicker films.

SPECIFIC APPLICATION INSTRUCTIONS

Article 2–2–3(A), Exterior Walls Above Grade
Article 2–2–3(B), Exterior Walls Below Grade
Article 2–2–3(C), Interior Walls Above Grade
Article 2–2–3(D), Interior Walls Below Grade

4–3–7, TILE RED ROOF PAINT

USE: As a finish on wood and metal roofs.

GENERAL APPLICATION INSTRUCTIONS

Unpainted Wood: Prime with 1 coat of Exterior White Wood Primer. Follow with 2 coats of paint as received.

Unpainted Metal: Prime with 1 liberal coat

ORDERING INFORMATION

Catalog Name: Cold-Water Cement Powder Paint
Specification: Federal TT–P–21, Type I

Color	*Size Container*	*Stock Number*
Without Siliceous Aggregate, Class A, For all surfaces except Open Texture Masonry		
White	50 lb. bag	52–P–8995
With White Siliceous Aggregate, Class B, For Open Texture Masonry such as Cinder Block		
White	50 lb. bag	52–P–8997

ORDERING INFORMATION

Catalog Name: Tile Red Roof Paint
Specification: Federal TT-P-31

Color	Federal Color No.	Size Container	Stock Number
No. 26 Tile Red	1020	1 gal.	52-P-3725

of Pretreatment-Wash Primer and 1 coat of Quick Drying Red Lead Primer. Follow with 2 coats of paint as received.

Previously Stained Wood: Prime and seal with 2 coats of Ready-Mixed Aluminum Paint. Follow with 2 coats of paint as received.

Previously Painted Surfaces: Priming is not needed. Apply 1 coat of paint as received.

Drying: Dries overnight to a soft film. Several days drying time needed between coats.

Thinning: This paint is of brushing consistency as received. For spraying, thin with Paint Thinner.

SPECIFIC APPLICATION INSTRUCTIONS

Article 2-2-5(A), Wood Roofs
Article 2-2-5(B), Metal Roofs

4-3-8, DECK AND FLOOR PAINT

USE: As a finish coat on interior or exterior metal and wood decks, floors, steps, bleachers, concrete floors above grade, and miscellaneous traffic areas. On concrete floors in contact with the ground, use Basement Floor Paint.

GENERAL APPLICATION INSTRUCTIONS

Unpainted Wood Surfaces: Prime with 1 coat of Deck and Floor Paint thinned with 1 qt. of thinning mixture (2/3 Boiled Linseed Oil and 1/3 Paint Thinner) per gal. of paint. On interior floors, follow with 1 coat of Deck and Floor Paint as received. On exterior floors follow the prime coat with 2 coats of Deck and Floor Paint as received.

Unpainted Metal Surfaces: Apply 1 liberal coat of Pretreatment-Wash Primer. Prime surfaces with appropriate number of coats of Quick Drying Red Lead Primer (prime aluminum with Yellow Zinc Chromate Primer). Follow with 2 coats of Deck and Floor Paint as received.

Unpainted Concrete Above Grade: After aging for 1 year, prepare the surface and apply 2 coats of paint as received.

ORDERING INFORMATION

Catalog Name: Deck and Floor Enamel
Specification: Federal TT-E-487

Color	Federal Color No.	Size Container	Stock Number
No. 26 Tile Red	1020	1 gal.	CG-52-E-4032-4650
No. 31 Deck Green	2435	1 gal.	CG-52-E-4032-4550
No. 19 Light Gray	2635	1 gal.	CG-52-E-4032-4520
No. 20 Medium Gray	1615	1 gal.	CG-52-E-4032-4525
No. 32 Blue Gray	——	1 gal.	CG-52-E-4032-4500

Previously Painted Surfaces: Priming not needed. Apply 2 coats paint as received.

Drying: Sets to touch in 2 hours. Dries hard for recoating in 18 hours.

Thinning: This paint is of brushing consistency as received. For spraying, thin with Paint Thinner.

SPECIFIC APPLICATION INSTRUCTIONS

Article 2–2–1(B), Exterior Floors, Decks, Steps, Bleachers
Article 2–2–1(I), Interior Wood Floors
Article 2–2–2(I), Metal Traffic Areas
Article 2–2–3(E), Concrete Floors Above Grade
Article 2–3–6, Decks and Floor Plates (Vessels)

4–3–9, BASEMENT FLOOR PAINT

USE: A rubber base paint for use on indoor concrete floors, especially basement and other concrete floors in direct contact with the ground which are subject to dampness. It is not intended for exterior use. This paint cannot be applied over other types of paint.

GENERAL APPLICATION INSTRUCTIONS

Unpainted Concrete: Apply to clean, bare, dry concrete floors. New concrete floors must be aged for 2 months prior to applying this paint. Floors with hard slick surfaces should be acid etched by flooding (1 gal. to 100 sq. ft.) with a mixture of 1 gal. of muriatic acid added to 4 gal. of water. After 15 minutes, hose off the acid using plenty of water. Allow the floor to dry thoroughly before painting. At least 2 coats should be applied and for extreme wear resistance 3 coats are recommended.

Previously Painted Floors: If a non-rubber base paint was previously used it should be removed with paint remover and the surface thoroughly scrubbed with soap and water. Treat as unpainted concrete.

CAUTION: This paint should be stored under good conditions (40°-90° F.).

Drying: Sets to touch in 1 hour. Dries hard for recoating in 8 hours. Allow 48 hours after last coat before subjecting the floor to traffic.

Thinning: This paint is of brushing consistency as received. For spraying, thin with Synthetic Enamel Thinner.

SPECIFIC APPLICATION INSTRUCTIONS

Article 2–2–3(F), Concrete Floors Below Grade

4–3–10, EQUIPMENT ENAMEL

LUSTER: Glossy.

USE: On furniture and equipment such as laundry machinery and galley equipment and miscellaneous equipment of all types. On ship's joiner doors, switch boxes, controllers, switchboards, gauge boards and connection boxes. On interior machinery surfaces which do not exceed 300° F.

GENERAL APPLICATION INSTRUCTIONS

Unpainted Wood Surfaces: Prime with 1 coat of Exterior Wood White Primer. Follow with 2 coats of enamel.

Unpainted Metal: Apply 1 liberal coat of

ORDERING INFORMATION

Catalog Name: Basement Floor Paint
Specification: Federal TT–P–91

Color	*Federal Color No.*	*Size Container*	*Stock Number*
No. 26 Tile Red	1020	1 gal.	CG–52–P–194–3950
No. 19 Light Gray	2635	1 gal.	CG–52–P–194–2900

ORDERING INFORMATION

Catalog Name: Light Gray Equipment Enamel, Medium Air Drying, Type 1.
Specification: MIL-E-15090.

Color	*Federal Color No.*	*Size Container*	*Stock Number*
No. 30 Equipment Gray	1640	1 qt.	CG-52-E-4034-25
		1 gal.	52-E-4034-55

Pretreatment-Wash Primer. Prime surfaces with appropriate number of coats of Quick Drying Red Lead Primer (prime aluminum with Yellow Zinc Chromate Primer). Follow with 2 coats of the enamel.

Drying: Sets to touch in 1 hour. Dries hard for recoating in 8 hours.

Thinning: This paint is of brushing consistency as received. If thinning is necessary, thin with Paint Thinner.

SPECIFIC APPLICATION INSTRUCTIONS

Article 2-2-2(G), Furniture and Equipment (Shore Establishments)
Article 2-2-2(H), Machinery (Shore Establishments)
Article 2-3-17(A), Interior Machinery (Vessels)
Article 2-3-18(A), Furniture and Equipment, Interior (Vessels)
Article 2-3-23(B), Propeller Shafting, Inboard (Vessels)

4-3-11, INTERIOR GLOSS ENAMEL

LUSTER: Glossy.

USE: For general purpose use on the interiors of vessels. On interior walls, ceilings and woodwork of galleys, sculleries, heads, sick bays and other sanitary spaces in shore establishments where the surfaces are frequently scrubbed. Also for general use on wood and metal doors, door sash, window sash, baseboard, dadoes, etc., which must be frequently scrubbed. For a general purpose wall and ceiling paint in other than sanitary spaces, use Interior Synthetic Rubber Emulsion Paint.

Interior Fire Retardant Paint was formerly specified for use in many of the places where Interior Gloss Enamel is now specified. The present fire retardant paints have increased maintenance problems without commensurate reduction of fire hazards. When new formulations of fire retardant interior paints now in development become available, this section will be amended to direct their use where applicable.

GENERAL APPLICATION INSTRUCTIONS

Unpainted Plaster, Masonry and Porous Wallboard Surfaces: Seal surface with 1 coat of Interior Wall Primer-Sealer. Prime with 1 coat of Interior White Enamel Undercoat. Follow with 1 coat of Interior Gloss Enamel as received.

Unpainted Wood and Non-Porous Wallboard Surfaces: Prime with 1 coat of Interior White Enamel Undercoat as received. Follow with 1 coat of Interior Gloss Enamel as received.

Unpainted Metal Surfaces: Apply 1 liberal coat of Pretreatment-Wash Primer. Prime surfaces with appropriate number of coats of Quick Drying Red Lead Primer (prime aluminum with Yellow Zinc Chromate Primer). Follow with 1 coat of White Enamel Undercoat and 1 coat of Interior Gloss Enamel as received.

Previously Painted Surfaces: Priming is not needed if the surface is in good condition and not too glossy. If the surface has a high gloss it may be necessary to either roughen the surface with sandpaper or to apply a coat of the Enamel Undercoat as received. Apply 1 coat of Interior Gloss Enamel.

Thinning: This paint is of brushing and

spraying consistency as received. If thinning is found necessary, thin with Synthetic Enamel Thinner and compensate for reduction in paint film thickness by applying a heavier than normal coat.

Drying: Sets to touch in 2 hours and dries hard for recoating in 7 hours. Allow adequate drying time between coats.

SPECIFIC APPLICATION INSTRUCTIONS

Article 2–2–1(H), Interior Wood Trim
Article 2–2–2(D), Interior Ungalvanized Steel (Shore Establishments)
Article 2–2–2(E), Interior Galvanized Steel and Corrosion-Resisting Metals (Shore Establishments)
Article 2–2–3(C), Interior Walls Above Grade
Article 2–2–4, Coating System For Dry Wall Construction
Article 2–3–5(A), Interior Metal Surfaces Not Subject to Moisture (Vessels)
Article 2–3–5(B), Interior Metal Surfaces Subject to Moisture (Vessels)
Article 2–3–5(C), Interior Wood (Vessels)
Article 2–3–5(D), Insulation (Vessels)
Article 2–3–10, Voids and Cofferdams
Article 2–3–14, Spaces Subject to Heavy Sweating (Vessels)
Article 2–3–16(A), Interior Piping (Vessels)
Article 2–3–19(A), Electric Cable, Interior (Vessels)

4–3–12, INTERIOR SYNTHETIC RUBBER EMULSION PAINT

LUSTER: Semi-gloss.
USE: On interior walls, ceilings, and woodwork as a general purpose paint on properly primed plaster, wallboard, wood, masonry and metal surfaces. Good washability after 30 days.

GENERAL APPLICATION INSTRUCTIONS

Unpainted Plaster, Masonry, Wood and Wallboard Surfaces: Priming normally not needed unless surface is extremely porous in which case apply a coat of Interior Wall Primer-Sealer followed by the Interior Synthetic Rubber Emulsion Paint.

Unpainted Metal Surfaces: Apply 1 liberal coat of Pretreatment-Wash Primer. Prime surface with appropriate number of coats of Quick Drying Red Lead Primer (prime Aluminum with Yellow Zinc Chromate). Follow with 1 coat of paint as received.

Previously Painted Surfaces: Priming not

ORDERING INFORMATION

Catalog Name: Interior Gloss Enamel
Specification: Federal TT–E–506

	Color	*Federal Color No.*	*Size Container*	*Stock Number*
White			1 gal.	52–E–6019
No. 1	Light Green	3555	1 gal.	52–E–5935–100
No. 2	Medium Green	3540	1 gal.	52–E–5943–100
No. 4	Soft Yellow	2325	1 gal.	52–E–6021–40
No. 6	Ivory	3360	1 gal.	52–E–5950–200
No. 25	Beach Sand*	2330	1 gal.	52–E–5994–800
No. 8	Terra Cotta	1730	1 gal.	52–E–6004–500
No. 9	Pearl Gray	3635	1 gal.	52–E–5918–300
No. 10	Light Blue	—	1 gal.	52–E–5892
No. 19	Light Gray	2635	1 gal.	52–E–5918–100

*When ordering this color from Navy sources, specify No. 7 Peach, Federal Color No. 2330, Catalog No. 52–E–5994–800.

---

needed. Apply 1 coat of the paint as received.

Thinning: This paint should be satisfactory for either brush or spray application as received. If thinning becomes necessary, thin with water. Clean brushes and spray equipment with water or soap and water.

Drying: Sets to touch in 20 minutes and dries hard for recoating in 24 hours. Allow adequate drying time between coats. A period of 30 days should be allowed for a curing process with this paint. During this period the surfaces should not be washed. After the process is completed the washability approaches that of an oil paint.

SPECIFIC APPLICATION INSTRUCTIONS

Article 2–2–1(G), Interior Wood Walls and Wood Wainscoting (Shore Establishments)

Article 2–2–2(D), Interior Ungalvanized Steel (Shore Establishments)

Article 2–2–2(E), Interior Galvanized Steel and Corrosion-Resisting Metals (Shore Establishments)

Article 2–2–3(C), Interior Walls Above Grade

Article 2–2–3(D), Interior Walls Below Grade

Article 2–2–4, Coating Systems for Dry Wall Construction

4–3–13, STRIPING PAINTS

Exterior Gloss Enamel, Section 4–3–4, is to be used as a striping paint where necessary. Exterior Gloss Enamel is suitable for all purposes for which a striping paint is used and is more durable than the striping paints formerly stocked.

Some of the colors of Exterior Gloss Enamel are stocked in 1 qt. containers while in other colors 1 gal. containers are the smallest stocked.

In the event that enough units request smaller container sizes, steps will be taken to stock the necessary colors in 1 pt. or ½ pt. cans.

4–3–14, RED BOOT-TOPPING PAINT

USE: As boot-topping paint on steel vessels with black hulls.

GENERAL APPLICATION INSTRUCTIONS

Apply 1 liberal coat of Pretreatment-Wash Primer. Apply 3 coats of Quick Drying Red Lead Primer. Follow with 3 coats of Red Boot-topping Paint.

ORDERING INFORMATION

Catalog Name: Interior Synthetic Rubber Emulsion Paint

Specification: GSA No. 828 Glidden Co. "Spred Satin" or equal

	Color	*Federal Color No.*	*Size Container*	*Stock Number*
White			1 gal.	52–P–3620–350
No. 1	Light Green	3555	1 gal.	52–P–3620
No. 2	Medium Green	3540	1 gal.	52–P–3620–50
No. 4	Soft Yellow	2325	1 gal.	52–P–3620–100
No. 6	Ivory	3360	1 gal.	52–P–3620–150
No. 25	Beach Sand*	2330	1 gal.	52–P–3620–200
No. 9	Pearl Gray	3635	1 gal.	52–P–3620–250
No. 10	Light Blue	—	1 gal.	52–P–3620–300

*When ordering this color from Navy sources, specify No. 7 Peach, Federal Color No. 2330, Catalog No. G52–P–3620–200.

Drying: Sets to touch in 30 minutes. Dries hard in 4 hours.

Thinning: This paint is of brushing consistency as received. For spraying, thin with Paint Thinner.

SPECIFIC APPLICATION INSTRUCTIONS

Article 2-3-3(A), Boot-top Painting, Steel Vessels in Salt Water

ORDERING INFORMATION

Catalog Name: Red Boot-topping Paint
Specification: Maritime Commission 52-MC-153

Color	*Size Container*	*Stock Number*
Red	1 gal.	CG-52-P-1450-10

4-3-15, COLD PLASTIC ANTIFOULING PAINT, FORMULA 146/50

USE: As boot-topping paint on steel hulls in salt water.

GENERAL APPLICATION INSTRUCTIONS

Mixing: In all antifouling paints the pigment is heavy and has a tendency to settle. Thorough mixing is necessary before use and constant agitation during use is imperative. This paint has a tendency to thicken upon standing and it may appear that thinning is necessary for spraying. However, such dilution is not generally required. The paint should be mixed until it reaches spraying consistency. This will require at least 15 minutes mixing time with a propeller type mechanical mixer. In cold weather store paint in a warm room for 24 hours before use.

Application: Spraying is the preferred method although brushing is permissible. Apply 3 coats over properly primed surface.

Drying: Allow 2 hours drying time between coats and 4 hours after last coat before undocking.

Thinning: This paint is of spraying consistency as received. If necessary thin with Coal Tar Naphtha.

SPECIFIC APPLICATION INSTRUCTIONS

Article 2-3-3(A), Boot-top Painting, Steel Vessels in Salt Water

ORDERING INFORMATION

Catalog Name: Cold Plastic Antifouling Paint
Specification: Navy Formula No. 146/50

Color	*Size Container*	*Stock Number*
Black	1 gal.	CG-52-P-555-985
	5 gals.	52-P-556

4-3-16, COLD PLASTIC ANTIFOULING PAINT, FORMULA 105

USE: As an antifouling bottom paint on steel hulls in salt water.

GENERAL APPLICATION INSTRUCTIONS

Mixing: In all antifouling paints the pigment is heavy and has a tendency to settle. Thorough mixing is necessary before use and constant agitation during use is imperative. This paint has a tendency to thicken upon standing and it may appear that thinning is necessary for spraying. However, such dilution is not generally required. The paint should be mixed until it reaches spraying consistency. This will require at least 15 minutes mixing time with a propeller type mechanical mixer. In cold weather store paint in a warm room for 24 hours before use.

Application: Spraying is the preferred method although brushing is permissible. Apply 3 coats over properly primed surface.

Drying: Allow 2 hours drying time between coats and 4 hours after last coat before undocking.

Thinning: This paint is of spraying consistency as received. If necessary thin with Coal Tar Naphtha.

SPECIFIC APPLICATION INSTRUCTIONS

Article 2-3-2(A), Bottom Painting, Steel Vessels in Salt Water

ORDERING INFORMATION

Catalog Name: Cold Plastic Antifouling Paint
Specification: Navy Formula No. 105

Color	*Size Container*	*Stock Number*
Red	5 gals.	52–P–6578

4–3–17, VINYL ANTIFOULING PAINT

USE: As an antifouling paint for the bottom of wood and steel vessels and the underwater areas of metal and wood buoys.

GENERAL APPLICATION INSTRUCTIONS

This paint may be applied at any temperature above 5° F. The surface must be dry. It may be applied over well-adhering vinyl paint but not over old conventional paint. This paint should never be applied to bare unprimed metal as galvanic corrosion may result. Upon immersion in salt water the reddish brown color usually undergoes a gradual change to various shades of blue and green. This paint should not be used on the interior of water tanks or on roofs due to the high copper content of the paint.

Metal Surfaces: Apply 1 liberal coat (0.5 mil dry film thickness) of Pretreatment-Wash Primer. Prime surface with sufficient coats (usually 2 coats) of Vinyl Red Lead to give a dry film thickness of 4.0 mils. Follow with sufficient coats (usually 2 coats) of Vinyl Antifouling Paint to give a dry film thickness of 3.5 mils and a total dry film thickness of 8.0 mils for the system.

Wood Surfaces: Apply Vinyl Antifouling Paint to bare, clean, dry wood. If old coatings are other than vinyl paints they must be completely removed. Vinyl Antifouling may be applied over old, firmly adhering vinyl paint. Apply 1 coat thinned by the addition of 1 pt. of thinner per gal. of paint. Follow with 2 coats of the paint as received.

Spraying: This paint is designed for spray application. Because of rapid solvent evaporation, the spray gun must be kept within 8 to 10 in. of the work. The objective is to get a paint film that looks wet. Air pressure should be at least 40 psi above the liquid pressure. Each coat should consist of a vertical and a horizontal pass of the spray gun over each area. Special care is needed to insure that spray equipment is thoroughly clean before using this paint, since solvents used will dislodge hardened particles of conventional paints, causing the spray gun to clog frequently. Spray equipment should be cleaned with Vinyl Paint Thinner. Before use, charge equipment with Vinyl Paint Thinner and operate gun until all paint particles are cleared from the lines. Clean equipment immediately after use with Vinyl Paint Thinner.

Brushing: This paint is designed for spray application. However, it may be brushed on if applied properly. Take a full brush and spread it on without attempting to brush it out as with conventional paints. Due to the quick drying properties of this paint, do not attempt to cross work more than once. Stir paint frequently to prevent the cuperous oxide pigment from settling.

Drying: Dries for recoating in 30 minutes at temperatures above 60° F. Lower temperatures may require slightly longer drying time.

Thinning: This paint is of spraying consistency as received. Since solvent evaporates rapidly, thinner may be required after the can has been open for some time. In very cold weather, store paint in a warm place 24 hours before using. Thinner may be required to reduce viscosity in cold weather. The proper thinner is Vinyl Paint Thinner. Do not attempt to use other thinners in this paint.

PRECAUTIONS: In mixing or using this paint in an enclosed area, particular attention should be given to the air temperature in the vicinity of application. If the air temperature is below the flash point of the paint (40° F.), the material can be applied with normal ventilation. When the air temperature is above that of the flash point, sufficient ventilation must be provided to insure that the vapor concentration does not exceed the lower limits of explosibility. In addition, usual paint precautions to keep open flames and sparks away from the painting area and the area where paints are mixed must be strictly enforced. Like most paint materials in common use, adequate ventilation to avoid prolonged breathing of the concentrated paint fumes is necessary.

SPECIFIC APPLICATION INSTRUCTIONS

Article 2-3-2(C), Bottom Painting, Steel 40-ft. Utility Boats
Article 2-3-2(D), Bottom Painting, Wooden Boats
Article 2-4-1(B), Exterior of Metal Buoys, Below Waterline
Article 2-4-4(B), Wooden Spar Buoys, Below the Waterline

ORDERING INFORMATION

Catalog Name: Vinyl Antifouling Paint
Specification: CGS-52P-3

Color	*Size Container*	*Stock Number*
Reddish Brown	1 qt.	CG-52-P-179-300
	1 gal.	CG-52-P-179-500
	5 gals.	CG-52-P-179-510

4-3-18, SONAR DOME ANTIFOULING PAINT

USE: As an antifouling paint for use on sound-transparent windows of sonar domes.

GENERAL APPLICATION INSTRUCTIONS

Apply 1 liberal coat of Pretreatment-Wash Primer. Apply by spray 1 liberal coat of Zinc Chromate Primer, Specification MIL-P-6889 thinned by the addition of 2 parts of Synthetic-Enamel Thinner to 1 part of the Zinc Chromate. Follow with 8 spray coats of the Antifouling paint thinned by the addition of 3 parts of Synthetic-Enamel Thinner to 4 parts of Antifouling Paint.

Drying: Allow 1½ hours drying time between coats and 24 hours between last coat and undocking.

Thinning: Thin with Synthetic Enamel Thinner.

SPECIFIC APPLICATION INSTRUCTIONS

Article 2-3-30(C), Painting of Corrosion-Resistant, Sound-Transparent Windows

ORDERING INFORMATION

Catalog Name: Antifouling Paint for Sonar Purposes
Specification: MIL-P-5051

Size Container	*Stock Number*
1 gal.	52-P-178

4-3-19, RUBBER ANTIFOULING PAINT

USE: As an antifouling paint for use on rubber sound projector domes and other rubber components subject to immersion in fouling waters.

GENERAL APPLICATION INSTRUCTIONS

Apply 4 coats by spray using an air pressure at the gun of 35 psi and a fluid pressure at the gun of 10 to 15 psi. Wet coats should be applied but precautions should be taken to avoid runs and sags. An application is desired which will avoid trapping air bubbles and eliminate film porosity as much as possible.

Drying: Allow 30 minutes drying time between coats. The final coat should dry at least 24 hours before being placed in water. A drying time of 96 hours to 1 week is highly desirable if time will permit.

Thinning: For spraying thin with 1 part Paint Thinner to 5 parts Rubber Antifouling Paint.

SPECIFIC APPLICATION INSTRUCTIONS

Article 2-3-30(D), Painting of Sound-Transparent Rubber Surfaces of Sonar Domes and the Diaphragms of Sound Transducers.

ORDERING INFORMATION

Catalog Name: Rubber Antifouling Paint
Specification: MIL-P-15152, Formula 15 (NRL)

Size Container	*Stock Number*
1 gal.	52-P-6325

4–3–20, READY-MIXED ALUMINUM PAINT

USE: For sealing creosoted or stained surfaces and knots and pitch pockets in wood prior to the application of color coats. Not a heat-resistant paint. For surfaces above 300° F. use Heat-Resisting Aluminum Paint.

GENERAL APPLICATION INSTRUCTIONS

Creosoted or Stained Wood: Allow newly creosoted wood to age 6 months. Remove all dirt and oil from the surface. Apply 2 coats of paint as received.

Knots and Pitch Pockets: Scrape excess resin and smooth the surface. Apply 1 coat of paint as received to resinous areas.

Drying: Sets to touch in 2 hours and dries hard for recoating in 16 hours.

Thinning: This paint is of brushing consistency as received. For spraying, thin as necessary with Paint Thinner.

CAUTION: The stability or shelf-life of this paint is less than that of most paints. Do not overstock.

SPECIFIC APPLICATION INSTRUCTIONS

Article 2–2–1(A), Exterior Wood, General
Article 2–2–1(B), Exterior Floors, Decks, Steps, Bleachers
Article 2–2–1(E), Creosoted Wood
Article 2–2–1(F), Previously Stained Wood Shingles and Siding
Article 2–2–1(H), Interior Wood Trim
Article 2–2–1(I), Interior Wood Floors
Article 2–2–5(A), Wood Roofs

ORDERING INFORMATION

Catalog Name: Ready-Mixed Aluminum Paint
Specification: CGS–52P–2

Size Container	*Stock Number*
1 qt.	CG–52–P–162
1 gal.	CG–52–P–164

4–3–21, HEAT-RESISTING ALUMINUM PAINT

USE: As a primer and finish coat on heated surfaces such as boiler drums and casings, steam lines, turbines and other steam machinery which are subjected to temperatures in excess of 300° F. This is not a ready-mixed paint. For sealing creosoted or resinous wood use Ready-Mixed Aluminum Paint.

GENERAL APPLICATION INSTRUCTIONS

This paint is packaged in double compartment containers with the aluminum pigment paste in one compartment and the mixing vehicle in the other. Two pounds of the paste should be added to each gal. of the vehicle. Do not mix more than is needed for each day's use. Continued stirring is necessary to keep the pigment from settling. Apply directly while the surface is cold.

Drying: Allow 16 hours between coats and 16 hours after last coat before subjecting the surface to high temperatures.

Thinning: Normally, this paint will not need thinning. If thinning is necessary, thin with Paint Thinner.

SPECIFIC APPLICATION

Article 2–2–2(K), Hot Surfaces (Shore Establishments)
Article 2–3–16(A), Interior Piping (Vessels)
Article 2–3–17(A), Interior Machinery (Vessels)
Article 2–3–28, Smoke Pipes (Vessels)

ORDERING INFORMATION

Catalog Name: Heat-Resisting Paint
Specification: Navy N:52–P–19

Size Container	*Stock Number*
1 gal.	52–P–3029–950

4–3–22, BITUMINOUS EMULSION COATING

USE: As a protective coating for chain lockers.

GENERAL APPLICATION INSTRUCTIONS

The surface must be cleaned as for all painting. The surface should be moistened just prior to application.

Mixing: A mechanically operated propeller type mixer is necessary. This emulsion has a tendency to settle and thicken upon standing and it may appear that thinning is necessary for spraying. However, such dilution is not generally required. The mixture should be agitated until it reaches a creamy consistency. This will require at least 15 minutes of mixing time. No water should be added prior to mixing.

Application: Spraying is the preferred method of application, although brushing is permissible. Both spraying and brushing should be carried out in exactly the same manner as for Cold Plastic Antifouling Paint except that the emulsion is not to be warmed as is permissible with the Cold Plastic. When brush applications are made, the brush must be soaked in water for 24 hours prior to use to prevent setting of the material in the bristles. Two coats should be applied. When brushing, the material should be "buttered on" in lieu of being brushed out.

Drying: This material will look dry in 1 to 2 hours, but a 24-hour drying time should be allowed where practicable.

Thinning: This material thins out after thorough mixing. Do not thin unless absolutely necessary. Proper thinner is water.

CAUTION: Since this material is a water emulsion, it must be protected from freezing while in storage. It must not be applied in spaces in which the temperature is below freezing.

SPECIFIC APPLICATION INSTRUCTIONS

Article 2–3–11, Chain Locker

ORDERING INFORMATION

Catalog Name: Bituminous Emulsion Coating

Specification: MIL–C–15203

Size Container	*Stock Number*
5 gals.	52–C–2100

4–3–23, TRAFFIC PAINT

USE: For traffic lines and safety zone markings on concrete, brick or asphalt pavement. Do not procure more than 3 months' supply because of settling tendencies of the paint.

GENERAL APPLICATION INSTRUCTIONS

Apply directly to dry, clean pavement.

Thinning: This paint is of brushing consistency as received. For application by paint striping machines, thin as necessary with Paint Thinner.

Drying: Sets to touch in 5 to 30 minutes. Dries hard in 1 hour at temperatures between 70° and 80° F.

SPECIFIC APPLICATION INSTRUCTIONS

Article 2–2–9, Road and Runway Markings

ORDERING INFORMATION

Catalog Name: Traffic Paint
Specification: Federal TT–P–115

Color	*Size Container*	*Stock Number*
White	1 gal.	52–P–8295
Yellow	1 gal.	52–P–8315
Black	1 gal.	52–P–8293–5

4–3–24, ZINC DUST PAINT

USE: For use on the interior of metal drinking water and feed water tanks.

GENERAL APPLICATION INSTRUCTIONS

This paint is issued as a two-package unit. Fifty-five and one-half lbs. of dry zinc dust and 5 gals. of the base paint comprise one unit. Mix as follows:

To 1 gal. of the liquid portion, the zinc dust shall be added in increments of 2½ to 5 lbs., and the mixture stirred with a paddle until no lumps or dry particles of zinc dust remain. The remaining 4 gals. of liquid shall be added in increments of ½ gal. and stirring continued after each addition until the mixture is homogeneous. Zinc dust shall not be mixed with the liquid portion until just before application. Only enough paint shall be mixed for immediate use.

Mixed paint shall not be stored. (After the material is mixed, a gas pressure may build up within closed containers.)

Zinc Dust Paint shall be applied in conformance with the following instructions:

The tank shall be thoroughly cleaned of all oil paint coatings. (It is especially important to remove entirely the zinc chromate after-pickling paint in drinking water tanks.) Removal of the paint coatings should be done by sandblasting if practicable. A power wirebrush may be used. Apply 1 coat of zinc dust paint. Unheated air shall then be circulated through the tank for 12 hours and a second coat applied. Unheated air shall again be circulated for 12 hours. Tanks shall be flushed twice with fresh water before being placed in service.

Drying: Sets to touch in 3 minutes and dries hard for recoating in 12 hours.

SPECIFIC APPLICATION INSTRUCTIONS

Article 2-2-2(L), Water Tanks (Shore Establishments)
Article 2-3-7, Water Tanks (Vessels)

ORDERING INFORMATION

Catalog Name: Water Resisting Zinc Dust Paint
Specification: MIL-P-15145

Size Container	*Stock Number*
Base Paint 5 gals. pail and Zinc Dust 55½ lbs. pail	52-P-9224-500

4-3-25, ANTISWEAT-COATING BINDER

USE: As a binder for Expanded Vermiculite Pigment to provide antisweat protection on piping and bulkheads subject to condensation.

GENERAL APPLICATION INSTRUCTIONS

Apply 1 liberal coat of Pretreatment-Wash Primer and 2 coats of Quick Drying Red Lead Primer. Follow with 1 liberal coat of Antisweat-Coating Binder. Allow the Binder to become tacky and apply 1 coat of Expanded Vermiculite by flocking or by hand sprinkling. Follow with 2 coats of Interior Gloss Enamel.

Drying: Sets to touch in ½ to 2 hours. Dries hard for recoating in 8 hours.

Thinning: Thin as necessary with Paint Thinner.

SPECIFIC APPLICATION INSTRUCTIONS

Article 2-3-14, Spaces Subject to Heavy Sweating

ORDERING INFORMATION

Catalog Name: Antisweat-Coating Binder
Specification: MIL-P-15144, Formula 34

Size Container	*Stock Number*
1 gal.	52-P-8350

4-3-26, EXPANDED VERMICULITE

USE: In conjunction with Antisweat-Coating Binder as a coating to prevent condensation on walls and pipes. Light golden color similar to color of ground cork.

GENERAL APPLICATION INSTRUCTIONS

Apply 1 liberal coat of Pretreatment-Wash Primer and 2 coats of Quick Drying Red Lead Primer. Follow with 1 liberal coat of Antisweat-Coating Binder. Allow the binder to become tacky and apply 1 coat of the Expanded Vermiculite by flocking or by hand sprinkling. Follow with 2 coats of Interior Gloss Enamel.

Drying: See Antisweat-Coating Binder.

SPECIFIC APPLICATION INSTRUCTIONS

Article 2-3-14, Spaces Subject to Heavy Sweating

ORDERING INFORMATION

Catalog Name: Expanded Vermiculite Pigment
Specification: MIL-V-15196

Size Container	*Stock Number*
50 lb. bag (Approx. 50 gals.)	52-P-19995

4-3-27, TINTING AND COLOR MATCHING CAMOUFLAGE MATERIALS

USE: For tinting white paint to match Federal Color Numbers 1615 and 2635 for application of camouflage on short notice.

GENERAL INFORMATION

Tinting materials will be carried by vessels as specified in Article 3–11–1(U) in lieu of camouflage paints. Amount of tinting material to be carried shall not exceed 1 gal. per 20 gals. of white paint carried on board except that a minimum of 1 gal. is authorized for all classes of vessels specified in Article 3–11–1(U). Five gals. each of Deck Gray Navy, Formula 20 and Haze Gray, Formula 5H are authorized for color matching purposes.

SPECIFIC APPLICATION INSTRUCTIONS

Article 2–3–4, Exterior of Vessels above the Boot-top—General
Article 3–11–1(U), Camouflage Instructions

ORDERING INFORMATION

Color Matching Material

Catalog Name: Dark Gray Exterior Deck Paint
Specification: JAN–P–699, Formula No. 20

Color	*Federal Color Number*	*Size Container*	*Stock Number*
Dark Gray	1615	5 gals.	52–P–1406–50
	Catalog Name: Haze Gray No. 27, Exterior Camouflage Paint Specification: MIL–P–15130, Navy Formula 5H		
Haze Gray	2635	5 gals.	52–P–961
	Tinting Material		
	Catalog Name: Lampblack Paste Pigment Specification: TT–P–381		
Black	—	½ pt. can	52–P–17190
	Catalog Name: Blue Black Tinting Material Specification: N:52–P–42, Navy Formula No. 5–TM		
Blue Black	—	1 qt. 1 gal.	G52–T–841 G52–T–842

SECTION 4–4, VARNISHES AND LACQUERS

4–4–1, SPAR VARNISH

USE: A high quality phenolic varnish suitable for general use. Meets almost all requirements of exterior and interior varnishes. Suitable for rubbed finishes on furniture. May be used on brightwork to prevent corrosion. Has excellent durability in marine atmosphere and salt spray.

GENERAL APPLICATION INSTRUCTIONS

Apply the appropriate number of coats to clean, dry, bare wood or metal. Sand lightly between coats.

Drying: Sets to touch in ½ to 2 hours. Dries hard for recoating in 8 hours.

Thinning: Thinning not normally required. If necessary thin with Paint Thinner.

SPECIFIC APPLICATION INSTRUCTIONS

Article 2–2–7, Window and Door Screens
Article 2–3–4(C), Exterior Aluminum (Vessels)
Article 2–3–4(D), Exterior Wood (Vessels)
Article 2–3–5(C), Interior Wood (Vessels)
Article 2–3–27, Wood Ladders, Gangplanks, Staging and Boatswain's Chairs

ORDERING INFORMATION

Catalog Name: Spar and Mixing Varnish
Specification: MIL–V–1174, Formula 80

Color	*Size Container*	*Stock Number*
Clear	1 qt.	CG–52–V–1679–950
	1 gal.	52–V–1680

4–4–2, ELECTRICAL INSULATING VARNISH

USE: A black, insulating, air drying varnish for painting electrical windings and other parts of motors, generators, transformers, and electrical parts in general. Oil and waterproof. Baking varnishes and high turpentine silicone varnishes are also available through Standard Stock to those shops which have baking ovens.

GENERAL APPLICATION INSTRUCTIONS

Apply to clean, bare metal or over clean, dry insulation or well-adhering old coatings of insulating varnish.

Drying: Dries hard for recoating in 8 hours.

Thinning: If necessary, thin with Paint Thinner.

SPECIFIC APPLICATION INSTRUCTIONS

Article 2–3–19, Electric Cable (Vessels)

ORDERING INFORMATION

Catalog Name: Electrical Insulating Varnish, Grade BA, Air Drying.
Specification: JAN–V–1137, Type N

Color	*Size Container*	*Stock Number*
Black	1 gal.	52–V–1100

4–4–3, ASPHALT VARNISH

USE: For painting anchor chains.

GENERAL APPLICATION INSTRUCTIONS

Apply 1 liberal coat of Pretreatment-Wash Primer. Follow with 1 coat of Asphalt Varnish.

Drying: Sets to touch in 3 hours. Dries hard in 24 hours.

Thinning: If necessary, thin with Paint Thinner.

SPECIFIC APPLICATION INSTRUCTIONS

Article 2–3–21, Anchor Chain

ORDERING INFORMATION

Catalog Name: Asphalt Varnish
Specification: Federal TT–V–51

Color	*Size Container*	*Stock Number*
Jet Black	1 gal.	52–V–310

4–4–4, SPRAYING LACQUER

USE: For applying color code to shop helmets. Designed for application by spray only. Will not lift undercoats.

GENERAL APPLICATION INSTRUCTIONS

Clean helmet and wipe with Lacquer Thinner to remove all traces of grease and oil. Apply 2 coats of the lacquer by spray. Do not attempt to brush the lacquer.

Drying: Dries to tack-free within 10 minutes. Dries hard in 24 hours.

Thinning: Thinning is not normally needed. If found to be necessary, use Lacquer Thinner.

SPECIFIC APPLICATION INSTRUCTIONS

Article 2–2–12, Shop Helmets

4–4–5, CLEAR BRUSHING LACQUER FOR BRIGHTWORK

USE: On brightwork to prevent corrosion and reduce labor expended in polishing. Not suitable for furniture finishing.

GENERAL APPLICATION INSTRUCTIONS

Polish brightwork. Wipe with a cloth dampened in Lacquer Thinner. Apply 1 coat of Clear Brushing Lacquer. Old coatings of Lacquer may be removed by wiping with a cloth wetted with Lacquer Thinner.

Drying: Dries tack-free in 30 minutes.

Thinning: Thinning is not normally needed. Clean brushes and thin if necessary with Lacquer Thinner.

SPECIFIC APPLICATION INSTRUCTIONS

Article 2–2–10, Brightwork (Shore Establishments)

ORDERING INFORMATION

Catalog Name: Spraying Lacquer, General Use, Pigmented, Type II
Specification: Federal TT–L–58

Color		*Federal Color No.*	*Size Container*	*Stock Number*
No. 1	Light Green	3555	1 qt. can	52–L–978–5
No. 4	Soft Yellow	2325	1 qt. can	52–L–1005–250
No. 6	Ivory	3360	1 qt. can	52–L–979–20
No. 25	Beach Sand*	2330	1 qt. can	52–L–985–500
No. 8	Terra Cotta	1730	1 qt. can	52–L–990–155
No. 9	Pearl Gray	3635	1 qt. can	52–L–976–175
No. 10	Light Blue	—	1 qt. can	52–L–957–50
No. 13	Fire Red	1110	1 gal. can	52–L–989
No. 14	Brilliant Yellow	1310	1 qt. can	52–L–998
No. 16	Bright Green	1475	1 qt. can	52–L–976–760
No. 18	International Orange	1205	1 qt. can	52–L–983
No. 19	Light Gray	2635	1 qt. can	52–L–976–80
No. 22	High-Light Buff	1750	1 qt. can	52–L–962–60
White		1755	1 qt. can	52–L–992
No. 101	Black	1775	1 gal. can	52–L–951

*When ordering this color from Navy sources, specify No. 7 Peach 1 qt., 52–L–985–500.

Article 2–2–13, Surfaces Not To Be Painted (Shore Establishments)
Article 2–3–31, Brightwork (Vessels)
Article 2–3–33, Surfaces Not To Be Painted (Vessels)

ORDERING INFORMATION

Catalog Name: Brushing Lacquer for Copper and Brass
Specification: JAN–L–1118

Size Container	*Stock Number*
1 qt.	52–L–420

4–4–6, SPRAY-TYPE PLASTIC COATING

USE: As a water-clear protective coating on brightwork. For waterproofing ignition systems, shipping tags, and shipping and package labels. For weatherproofing stencils on shipping containers.

GENERAL APPLICATION INSTRUCTIONS

This material is furnished in Aerosol-type automatic spray dispenser. A transparent, nonflammable liquid for spray application at room temperature. Resistant to water, alcohol, alkali, acids, oils, grease, and discoloration at high temperatures. Apply wet coats.

Drying: Dries hard for recoating in 1 to 2 minutes.

SPECIFIC APPLICATION INSTRUCTIONS

Article 2–3–31, Brightwork (Vessels)

ORDERING INFORMATION

Catalog Name: Spray-Type Plastic Coating
Specification: Krylon, Inc., "Krylon" or equal

Color	*Size Container*	*Stock Number*
Transparent	12 oz. can (Spray dispenser)	52–C–2215–250

SECTION 4–5, THINNERS

4–5–1, PAINT THINNER

USE: A general purpose paint thinner suitable for thinning all Coast Guard and Navy paints with the exception of those few which require thinners of higher solvent power. The correct thinner is specified for each paint listed in this Chapter. Paint Thinner may also be used for cleaning oil, grease, and wax from surfaces preparatory to painting.

GENERAL INSTRUCTIONS

This thinner is composed of volatile mineral spirits or petroleum spirits and replaces turpentine as the standard general purpose paint thinner for use in the Coast Guard. Clean brushes and spray equipment with the thinner specified for the paint in which the equipment was used. Where a thinner of higher solvent power is desired, use Synthetic-Enamel Thinner.

ORDERING INFORMATION

Catalog Name: Paint Thinner
Specification: Federal TT–T–291, Grade 1

Size Container	*Stock Number*
1 gal.	CG–52–T–725
5 gals.	52–T–725–5

Note: These items available from Federal Supply Service under Stock Numbers 52–T–725 and 52–T–725–5.

4–5–2, SYNTHETIC-ENAMEL THINNER

USE: As a thinner for synthetic resin base paints which require a thinner of higher solvent power than Paint Thinner. May be used as a substitute for Paint Thinner to avoid delaying a job in case Paint Thinner is not readily available.

GENERAL INSTRUCTIONS

This thinner is composed of aromatic hydrocarbons and is of intermediate solvent power. Use this thinner in Exterior Gloss Enamel when thinning in excess of 1 part thinner to 4 parts enamel. Paint Thinner is compatible with Exterior Gloss Enamel up to 1 part thinner in 4 parts enamel.

ORDERING INFORMATION

Catalog Name: Synthetic-Enamel Thinner
Specification: Federal TT–T–306

Size Container	*Stock Number*
1 gal.	52–T–445

Note: Available from Federal Supply Service under the same Stock Number.

4–5–3, VINYL PAINT THINNER

USE: For thinning vinyl type paints and for cleaning brushes and equipment used in vinyl paints.

GENERAL INSTRUCTIONS

A clear thinner consisting of equal parts by weight of toluene and methyl isobutyl ketone. High solvent power. Rapid drying. Not suitable for thinning Pretreatment-Wash Primer.

CAUTION: Flammable. Keep away from heat and open flames. Avoid prolonged breathing of vapors. Provide adequate ventilation.

ORDERING INFORMATION

Catalog Name: Vinyl Paint Thinner
Specification: CGS–52–T–1

Size Container	*Stock Number*
1 qt.	CG–52–T–822–200
1 gal.	CG–52–T–822–210
5 gals.	CG–52–T–822–220

4–5–4, LACQUER THINNER

USE: For thinning spraying or brushing lacquers, clear or pigmented. For removing old coatings of clear lacquer from brightwork.

GENERAL INFORMATION

Composed of butyl and ethyl acetate, butyl and ethyl alcohol and petroleum naphtha. Ketones such as methyl ethyl ketone may be present to a maximum of 10 per cent.

CAUTION: Flammable. Keep away from heat and open flames. Avoid prolonged breathing of vapors. Provide adequate ventilation.

ORDERING INFORMATION

Catalog Name: Lacquer Thinner
Specification: TT-T-266

Size Container	*Stock Number*
1 qt.	52-T-500
1 gal.	52-T-520

4-5-5, COAL TAR NAPHTHA

USE: As a thinner for Cold Plastic Antifouling Paints. For cleaning brushes and spray equipment.

GENERAL INSTRUCTIONS

Cold Plastic Antifouling Paints do not normally need to be thinned. Thorough mixing will usually bring the material to spraying consistency. In cold weather store paint in a warm room for 24 hours before use. This is a noncorrosive, high solvent power thinner. If thinning becomes necessary, use only the minimum amount necessary to restore spraying consistency.

ORDERING INFORMATION

Catalog Name: Naphtha
Specification: MIL-N-15178

Size Container	*Stock Number*
5 gals.	52-N-300

4-5-6, BOILED LINSEED OIL

USE: As a drying oil protective coating on wood such as stages, boatswain's chairs, gratings, and ladders where protection of the surface is desired without obscuring defects in the wood. Also used to thin priming coats.

GENERAL APPLICATION INSTRUCTIONS

Apply appropriate number of coats to dry, clean wood. For diluting paint, follow directions specified in applicable painting system.

Drying: Dries hard for recoating in 18 hours.

CAUTION: Dispose of all rags soaked with linseed oil as soon as possible after using. Do not allow oil or paint soaked rags to accumulate. Linseed oil dries by an oxidation-reduction chemical reaction which gives off heat resulting in spontaneous combustion if confined and not allowed to dissipate.

SPECIFIC APPLICATION INSTRUCTIONS

Article 2-2-1(B), Exterior Wood Floors, Decks, Steps, Bleachers
Article 2-2-1(D), Wood Docks, Wharves, Piers, Ramps
Article 2-2-1(I), Interior Wood Floors
Article 2-2-1(L), Wood Windows and Doors
Article 2-2-13, Surfaces Not to Be Painted (Shore Establishments)
Article 2-3-4(D), Exterior Wood (Vessels)
Article 2-3-5(C), Interior Wood (Vessels)
Article 2-3-27, Wood Ladders, Gangplanks, Staging and Boatswain's Chairs
Article 2-3-33, Surfaces Not to be Painted (Vessels)

ORDERING INFORMATION

Catalog Name: Kettle Boiled (Heat-Bodied) Linseed Oil

Specification: TT-O-364

Size Container	*Stock Number*
1 gal.	CG-52-O-520

SECTION 4–6, COMPOUNDS AND CEMENTS

4–6–1, WHITE PUTTY

USE: To fill seams, dents, and holes in wood.

GENERAL APPLICATION INSTRUCTIONS

Consists of linseed oil, whiting, and white lead. Work excess linseed oil into putty by kneading. Fill holes, seams, etc. *after* priming coat has dried. If putty is applied to unprimed wood, linseed oil in the putty will be absorbed by the wood, the putty will dry chalky, crack and fall out. Paint over putty as soon as a firm skin is formed on the surface.

Drying: Surface skin forms in 3 to 10 days depending on exposure and weather conditions. Do not paint over fresh putty as the paint will not dry.

Thinning: Boiled linseed oil may be added if necessary.

SPECIFIC APPLICATION INSTRUCTIONS

Article 1–5–4, Preparation of Wood Surfaces

ORDERING INFORMATION

Catalog Name: White Linseed Oil Putty
Specification: BuShips, Formula No. 69

Color	*Size Container*	*Stock Number*
White	1 lb. can	CG–52–P–20750
	5 lb. can	CG–52–P–20755
	12½ lb. can	52–P–20760

4–6–2, OIL-TYPE CALKING COMPOUND FOR WOODEN VESSELS

USE: As a durable compound for filling exterior hull seams of wooden vessels. For putty knife application.

GENERAL APPLICATION INSTRUCTIONS

Calk seams with oakum. Soak seams and oakum with Copper Naphthenate Wood Preservative. After the Preservative has dried, pay seams with Oil-Type Calking Compound. Finish seams with a hollow or concave surface to prevent forcing the compound out of the seams when the planking swells.

Thinning: Remains plastic and easily workable after a long period of time in storage. Modification of formulation by the addition of oils or thinners is not permitted.

Drying: May be painted over after 24 hours.

SPECIFIC APPLICATION INSTRUCTIONS

Article 1–5–4(E), Calking Wooden Vessels
Article 2–3–1(B), Protective Coatings During Construction, Wooden Vessels
Article 2–3–26, Wood Masts and Spars

ORDERING INFORMATION

Catalog Name: Oil-Type Calking Compound for Wooden Vessels
Specification: JAN–C–168

Size Container	*Stock Number*
1 gal. can	52–C–3086–25

4–6–3, OIL-TYPE CALKING COMPOUND FOR METAL SEAMS AND AIR PORTS

USE: Remains plastic. Corrosion-resistant and waterproof. Application at room temperature. Use on lightweight sheet metal and as a seal in lieu of white lead putties. Also use for sealing air ports and window frames on vessels.

GENERAL APPLICATION INSTRUCTIONS

Type I may be applied by putty knife or hand-operated calking gun. Type II is knife grade only. Where wood is in contact with aluminum, apply one coat of Spar Varnish to the wood and allow it to dry before joining the faying surfaces. Whether the joint or seam is intended to be watertight or not, make the joint tight by calking from both sides. Label plates, regardless of material, to be installed on aluminum structures are to be imbedded in Type I Compound.

SPECIFIC APPLICATION INSTRUCTIONS

Article 2–2–2(C), Exterior Aluminum (Shore Establishments)
Article 2–3–4(C), Exterior Aluminum (Vessels)

ORDERING INFORMATION

Catalog Name: Compound, Calking, for Metal Seams
Specification: MIL–C–16231

Size Container	*Stock Number*
Gun and Knife Application, Type I 1 gal. can	52–C–3089–70
Knife Application, Type II 12½ lb. can	52–C–3092–70

4–6–4, PLASTIC CALKING COMPOUND

USE: For sealing joints in wood or masonry buildings or joints between various combinations of structural materials such as wood and metal, masonry and wood, etc.

GENERAL APPLICATION INSTRUCTIONS

Apply to clean, dry surfaces. This material will not stain the surface and will give better service without the use of primers. Grade 1 is of a soft consistency that can be applied by hand gun. Grade 2 is for knife application and is about the consistency of glazing putty. Since Grade 1 shrinks considerably and sometimes wrinkles, Grade 2 is often used on the more conspicuous parts of buildings. Calking should not be attempted when the temperature is below 40° F. In cold weather (above 40° F.) the material may be softened by warming it in the container. Calking compounds usually undergo changes in consistency in the container and should not be kept in storage for a long period before use.

Drying: Remains plastic. A surface drying forms a skin which may be painted over 24 hours after application.

Thinning: No modification of the formulation is permitted. In cold weather (above 40° F.) the material may be softened by heating in the container.

SPECIFIC APPLICATION INSTRUCTIONS

Article 1–5–4(F), Calking Cracks on Buildings
Article 1–5–5(D), Sealing and Waterproofing the Surface, Masonry

ORDERING INFORMATION

Catalog Name: Plastic Calking Compound
Specification: Federal TT–C–598

Color	*Size Container*	*Stock Number*
	Grade 1, For Gun Application	
Gray	1 gal.	52–C–3092–3275
	Grade 2, For Knife Application	
Near White	1 gal.	52–C–3092–3675

4–6–5, PLASTIC WOOD

USE: For filling holes, cracks and other defects in wood. For building up or filling in parts of wood pattern or joiner work.

GENERAL APPLICATION INSTRUCTIONS

This material is light in color and may be stained while plastic to match dark colored woods. It cannot be stained after it has hardened. Apply with a putty knife and leave a slight excess on the surface to compensate for shrinkage. After setting, sand smooth. Adheres to wood or metal. The Solvent Fluid is suitable for softening the Plastic Wood after it has hardened. Plastic wood may be cut, sawed, bored, reamed, or filled when dry and will hold nails.

Drying: Sets quickly. Hardens for working in 4 hours.

Thinning: Soften when necessary with Plastic Wood Fluid Solvent.

SPECIFIC APPLICATION INSTRUCTIONS

Article 1–5–4(A), Preparation of Wood Surfaces, Bare Wood
Article 2–2–1(J), Furniture Finishing

ORDERING INFORMATION

Catalog Name: Plastic Wood-Substitute and Solvent
Specification: MIL-W-15880

	Size Container	*Stock Number*
Plastic Wood	1 lb. can	52-W-1100
Fluid Solvent	½ pt. can	52-S -1600

4-6-6, MARINE GLUE

USE: For waterproofing wooden deck seams on ships.

GENERAL APPLICATION INSTRUCTIONS

Preparing the Seams: Clean seams out thoroughly and remove all traces of pitch, putty, rosin or other seam compositions. Calk the seams carefully, leaving about ½ in. space above the calking for the Marine Glue. Calking irons should be dipped in water, coal oil, or other volatile fluid and not in linseed oil or grease as these will prevent the glue from adhering to the planking.

Preparing the Glue: Melt the glue slowly in cast iron or heavy sheet iron pots or electrically heated glue pots. It should be heated to approximately 300° F. or until it becomes entirely fluid. Do not heat the glue to its boiling point of 350° F. When properly melted, it should drip like oil without stringing out. It should be used as promptly as possible in this consistency.

Applying the Glue: Pour the glue into the clean dry seams from a paying shell or ladle with a spout issuing a fine stream. The spout should be held about 1 in. away from the seam to allow air to escape; it should not be drawn on the seams as this permits air to be enveloped which cannot escape and the resultant air pockets leave the seams hollow and unsound. Apply the glue in two operations; the first filling to consist of a small quantity payed into the bottom of the seam and allowed to set before applying the second one. Allow the glue to overflow the seams slightly, permitting the excess to remain for several days. Remove the excess by scraping diagonally across the seam. In hot weather when the glue is soft, application of water to the deck will facilitate scraping. Leave the glue flush with the deck.

Drying: Sets upon cooling.

Thinning: Altering the formulation of Marine Glue by the addition of solvents is not permitted.

SPECIFIC APPLICATION INSTRUCTIONS

Article 2-3-1(B), Protective Coatings During Construction, Wooden Vessels

ORDERING INFORMATION

Catalog Name: Marine Glue
Specification: MIL-G-413, Class 2

Color	*Size Container*	*Stock Number*
Jet Black	28 lb. box	52-G-1200
Jet Black	300 lb. bbl. or drum	52-G-1220

4-6-7, SMOOTHING CEMENT FOR IRON OR STEEL

USE: For filling in rough or pitted iron or steel surfaces prior to painting.

GENERAL APPLICATION INSTRUCTIONS

Apply to clean dry metal. Mix cement thoroughly to obtain a smooth, uniform consistency. Apply with a putty knife, pressing the cement firmly into pits. Smooth off the surface while the cement is plastic. The surface may be sanded smooth after the cement has hardened.

Drying: Sets to touch in ½ to 2 hours. Dries hard in 24 hours.

Thinning: May be thinned if necessary with Paint Thinner.

SPECIFIC APPLICATION INSTRUCTIONS

Article 2-2-2(H), Machinery (Shore Establishments)

Article 2-3-17, Machinery (Vessels)

ORDERING INFORMATION

Catalog Name: Smoothing Cement For Iron or Steel
Specification: MIL-C-15202, Formula No. 62

Size Container	*Stock Number*
1 gal.	52–C–1650

4–6–8, HYDRAULIC CEMENT FOR IRON OR STEEL

USE: For closing minor defects such as surface porosity in castings; not to be used for sealing cracks or hot tears.

GENERAL APPLICATION INSTRUCTIONS

Clean the crack or pit thoroughly by wire-brushing. Press the cement into the opening firmly. This compound will not chip or break out when the work is sawed or filed. It will stand quenching in 60° F. water from 1500° F. without separating from the metal to which it is applied.

Drying: Allow the cement to dry for 24 hours at normal air temperature before placing in service.

Thinning: Mix with water.

SPECIFIC APPLICATION INSTRUCTIONS

Article 2–3–22, Propellers

ORDERING INFORMATION

Catalog Name: Hydraulic Cement for Iron or Steel

Specification: MIL–C–1219, Class B

Size Container	*Stock Number*
1 lb. can	52–C–1900
5 lb. can	52–C–1920

SECTION 4–7, SPECIAL MATERIALS

4–7–1, WET-SANDBLASTING CHEMICALS

USE: As a rust inhibitor solution in wet sandblasting slurry and in the final wash.

GENERAL APPLICATION INSTRUCTIONS

Slurry: A 4 to 1 mixture of Diammonium Phosphate and Sodium Nitrite shall be used as follows: Add 2 lbs. of the mixture (fill a qt. can ⅘ full with Diammonium Phosphate and then to the top with Sodium Nitrite) to 15 gals. of water and 300 lbs. of sand in the sandblasting unit.

Final Wash: In the final operation of washing down the sand from the blasted area, 2 lbs. of the mixture are added to 40 gals. of water to make up the rust inhibiting wash.

CAUTION: Since this solution decomposes to form gases on standing, stock solutions of the inhibitor shall not be employed.

SPECIFIC APPLICATION INSTRUCTIONS

Article 1–5–1, Preparation of Metallic Surfaces (Except Aluminum and Galvanized Steel)

ORDERING INFORMATION

Catalog Name: Diammonium Phosphate
Formula: $(NH_4)_2HPO_4$

Form	*Size Container*	*Stock Number*
White Powder	100 lb. bag	51–D–143–8759

Catalog Name: Sodium Nitrite
Formula: $NaNO_2$

Form	*Size Container*	*Stock Number*
Yellowish Crystals	150 lb. bbl.	51–S–3457–150

4–7–2, LIQUID FLOOR WAX

USE: As a protective coating on all types of floors including asphalt and rubber.

GENERAL APPLICATION INSTRUCTIONS

Apply in thin coats with a cloth or mop. Can be diluted in any proportion with distilled water but is generally used without dilution. Dries to a transparent, lustrous, and practically colorless film. Dries to a hard film in 20 minutes. Polishing is not necessary but a high gloss may be obtained by machine buffing.

CAUTION: Protect from freezing.

SPECIFIC APPLICATION INSTRUCTIONS

Article 2–2–1(I), Interior Wood Floors
Article 2–2–13, Surfaces Not To Be Painted (Shore Establishments)
Article 2–3–33, Surfaces Not To Be Painted (Vessels)

ORDERING INFORMATION

Catalog Name: Water-Emulsion Floor Wax

Size Container	*Stock Number*
1 gal.	52–W–479–10

4–7–3, AUTOMOBILE POLISH

USE: For polishing all vehicles which have lacquer, baked enamel, or synthetic enamel finishes.

GENERAL APPLICATION INSTRUCTIONS

A liquid polish containing an abrasive in suspension. Volatile portion is essentially water. Shake often while using to prevent settling of the abrasive.

CAUTION: Protect from freezing.

SPECIFIC APPLICATION INSTRUCTIONS

Section 2–5, Coating Systems For Vehicles

ORDERING INFORMATION

Catalog Name: Automobile Polish
Specification: Federal P–P–546

Size Container	*Stock Number*
1 pt.	51–P–1046–5

4–7–4, PAINT AND VARNISH REMOVER

USE: To remove old coatings of paint or varnish from metal or wood.

GENERAL APPLICATION INSTRUCTIONS

Apply with a full brush without brushing out. Do not break the wax film which forms on the surface. The wax retards evaporation and allows the remover to lift the paint. After the paint or varnish blisters, remove it with a broad putty knife. Wash the work with Paint Thinner to remove any traces of Paint Remover before refinishing.

CAUTION: Volatiles may be toxic. Avoid prolonged breathing of vapors. Proper ventilation is imperative. Prevent evaporation as far as practicable during actual use of the material.

SPECIFIC APPLICATION INSTRUCTIONS

Article 1–5–2, Preparation of Aluminum Surfaces
Article 1–5–4(D), Use of Paint Remover
Article 2–2–1(J), Furniture Finishing
Article 2–2–3(F), Concrete Floors Below Grade

ORDERING INFORMATION

Catalog Name: Paint and Varnish Remover
Specification: Federal TT–R–251, Type II

Size Container	*Stock Number*
1 gal.	52–R–410

4–7–5, PAINT CLEANER

USE: A synthetic organic detergent for removing dirt and grease from painted surfaces with soft, hard, or sea water. Will not harm high gloss finish of automotive vehicles and construction equipment. Contains no abrasive or fatty acid soaps.

SPECIFIC APPLICATION INSTRUCTIONS

Article 1–5–4(B), Preparation of Wood Surfaces, Previously Painted
Article 1–5–6, Cleaning Painted Surfaces

ORDERING INFORMATION

Catalog Name: Painted-Surface Cleaner
Specification: Federal P–C–431

Form	*Size Container*	*Stock Number*
Type I, Powder	5 lb. can	51–C–1313–350
Type II, Flake	5 lb. can	51–C–1313–250

INDEX

This is a simplified index with a minimum of duplicate and cross references. Two numbers, 3-4, indicate complete sections bearing on any subject. Three numbers, 3-4-2, indicate articles within sections. Numbers having an appended letter, 3-4-2(A), indicate paragraphs within articles. Needless repetitions have been avoided. Where the specifications for any items such as fences, flag poles and the like appear in numerous places, and are all identical, only one or two references have been included for the sake of convenience.

A

B

C

INDEX

D

E

F

G

H

I

J K

L

INDEX

M

N

O

P

Q

R

S

INDEX

T

U

V

W

Made in the USA
Monee, IL
05 November 2026

67858287R00166